Gemstones: Mineralogy and Geochemistry

Gemstones: Mineralogy and Geochemistry

Edited by Serena Brown

Larsen & Keller
www.larsen-keller.com

Gemstones: Mineralogy and Geochemistry
Edited by Serena Brown
ISBN: 979-8-88836-074-3 (Hardback)

Larsen & Keller
Published by Larsen and Keller Education,
5 Penn Plaza,
19th Floor,
New York, NY 10001, USA

Cataloging-in-Publication Data

Gemstones : mineralogy and geochemistry / edited by Serena Brown.
p. cm.
Includes bibliographical references and index.
ISBN 979-8-88836-074-3
1. Precious stones. 2. Gemology. 3. Minerals. 4. Mineralogy. 5. Geochemistry. I. Brown, Serena.
QE392 .G46 2023
553.8--dc23

For more information regarding Larsen and Keller Education and its products, please visit the publisher's website www.larsen-keller.com

Table of Contents

Preface

I am honored to present to you this unique book which encompasses the most up-to-date data in the field. I was extremely pleased to get this opportunity of editing the work of experts from across the globe. I have also written papers in this field and researched the various aspects revolving around the progress of the discipline. I have tried to unify my knowledge along with that of stalwarts from every corner of the world, to produce a text which not only benefits the readers but also facilitates the growth of the field.

Gemstones refer to pieces of mineral crystals that have been cut and polished, and are used for creating jewelry and other adornments. They come in a wide variety of colors and shades. Mineralogy refers to the study of the chemistry, crystal structure and physical properties of minerals and mineralized artifacts such as gemstones. Geochemistry refers to the study of the chemical composition of the Earth and its rocks and minerals. As gemstones are formed and found beneath the Earth's surface and may contain traces of minerals, they are studied under the subjects of mineralogy and geochemistry. They can be categorized as crystalline or amorphous. Minerals with a definite and regular internal atomic structure are called crystalline gemstones. Amorphous gemstones have no naturally occurring form and no organized internal atomic structure. This book includes some of the vital pieces of works being conducted across the world, on various topics related to the mineralogy and geochemistry of gemstones. With its detailed analyses and data, it will prove immensely beneficial to professionals and students involved in this area of study at various levels.

Finally, I would like to thank all the contributing authors for their valuable time and contributions. This book would not have been possible without their efforts. I would also like to thank my friends and family for their constant support.

Editor

Preface

1

Mineralogy and Geochemistry of Nephrite Jade from Yinggelike Deposit, Altyn Tagh (Xinjiang, NW China)

Ying Jiang [1], Guanghai Shi [1,*], Liguo Xu [2] and Xinling Li [3]

1 State Key Laboratory of Geological Processes and Mineral Resources, China University of Geosciences, Beijing 100083, China; 3009150004@cugb.edu.cn
2 Geological Museum of China, Beijing 100034, China; ytxlgzidane@hotmail.com
3 Xinjiang Uygur Autonomous Region Product Quality Supervision and Inspection Institute, Xinjiang 830004, China; lxl2326164@163.com
* Correspondence: shigh@cugb.edu.cn

Abstract: The historic Yinggelike nephrite jade deposit in the Altyn Tagh Mountains (Xinjiang, NW China) is renowned for its gem-quality nephrite with its characteristic light-yellow to greenish-yellow hue. Despite the extraordinary gemological quality and commercial significance of the Yinggelike nephrite, little work has been done on this nephrite deposit, due to its geographic remoteness and inaccessibility. This contribution presents the first systematic mineralogical and geochemical studies on the Yinggelike nephrite deposit. Electron probe microanalysis, X-ray fluorescence (XRF) spectrometry, inductively coupled plasma mass spectrometry (ICP-MS) and isotope ratio mass spectrometry were used to measure the mineralogy, bulk-rock chemistry and stable (O and H) isotopes characteristics of samples from Yinggelike. Field investigation shows that the Yinggelike nephrite orebody occurs in the dolomitic marble near the intruding granitoids. Petrographic studies and EMPA data indicate that the nephrite is mainly composed of fine-grained tremolite, with accessory pargasite, diopside, epidote, allanite, prehnite, andesine, titanite, zircon, and calcite. Geochemical studies show that all nephrite samples have low bulk-rock Fe/(Fe + Mg) values (0.02–0.05), as well as low Cr (0.81–34.68 ppm), Co (1.10–2.91 ppm), and Ni (0.52–20.15 ppm) contents. Chondrite-normalized REE patterns of most samples exhibit strong to moderate negative Eu anomalies (0.04–0.67), moderate LREE enrichments, nearly flat HREE patterns, and low ΣREE contents (2.16–11.25 ppm). The nephrite samples have $\delta^{18}O$ and δD values of 5.3 to 7.4‰ and –74.9 to –86.7‰, respectively. The mineralogy, bulk-rock chemistry, and O–H isotope characteristics are consistent with the dolomite-related nephrite classification. Based on mineral paragenetic relationships, three possible mineral crystallization stages are recognized: (1) diopside formed by prograde metasomatism; (2) nephrite jade formed by retrograde metasomatism and replacement of Stage I anhydrous minerals; (3) hydrothermal alteration after the nephrite formation. Features of transition metal contents indicate that the color of the Yinggelike nephrite is likely to be controlled by the Fe^{2+}, Fe^{3+}, and Mn. Yellowish color is related to Mn and especially Fe^{3+}, while greenish color is related to Fe^{2+}. Our new mineralogical and geochemical results on the Yinggelike nephrite provide better constraints on the formation of other nephrite deposits in the Altyn Tagh Mountains, and can facilitate future nephrite prospecting and research in the region.

Keywords: nephrite jade; dolomite-related nephrite; Altyn Tagh Mountains; Xinjiang (NW China); tremolite; mineralogy; geochemistry

1. Introduction

Nephrite is essentially a near-monomineralic rock, normally comprises >95% microcrystalline amphibole of the tremolite-actinolite series, and shifts close to tremolite ($Ca_2(Mg,Fe)_5Si_8O_{22}(OH)_2$) in high-quality nephrite jade [1]. The essential mineralogical feature of nephrite is its microstructure, which is characterized by randomly oriented bundles of felted/twisted fabrics [2]. This key microstructure gives nephrite its extreme toughness and makes it a valuable carving material. Nephrite has been used since prehistoric times and has been treasured for its color and toughness in different cultures throughout the history, especially in Europe and Asia, e.g., in Italy [3], Switzerland [4], Russia [5,6], Bulgaria [7,8], Hungary [9], China [2,10,11], Mongolia [12], Pakistan [13], Taiwan (China), Philippines, Malaysia, Vietnam, Thailand, and Cambodia [14].

Nephrite deposits are widespread around the world, but gem-quality ones are uncommon [15]. Major nephrite deposits are located in China [16–24], Russia [25,26], Canada [27], Australia [1], New Zealand [28–30], USA [1], and South Korea [31]. Nephrite can be divided into serpentinite-related (S-type) and dolomite-related (D-type) types according to its origin [1]. S-type nephrite commonly forms along the contact between serpentinite/peridotite and more silicic rocks [1]. Famous and representative S-type nephrite deposits have been found in British Columbia in Canada [27], East Sayan in Russia [25,26], New Zealand [28–30], Jordanów and the adjacent areas in SW Poland [32,33], Taiwan and Manasi in China [34,35]. D-type nephrite forms along the contact between dolomitic marble and igneous intrusions [1], and is reported in many localities in China [17–22], together with Chuncheon (Korea) [31], Cowell (South Australia) [1], Vitim (Siberia, Russia) [25,26], Złoty Stok (SW Poland) [33,36], and some new discoveries were made in Europe [1]. Previous studies have suggested that S-type and D-type nephrites are of metasomatic and metamorphic origin, respectively [27]. However, recent studies have suggested that D-type nephrite is formed by contact metasomatic replacement of dolomitic marble by Si-rich aqueous fluids during felsic pluton emplacement [1,18,19,31].

In NW China, the giant Hetian nephrite belt extends about 1300 km from the West Kunlun Mountains to the Altyn Tagh Mountains, forming the largest nephrite belt in the world and hosting tens of important nephrite deposits. Magmatic intrusions in this belt were likely the main driving force behind the nephrite formation [16]. Compared to the many studies on the nephrite deposits in the West Kunlun [17–21], nephrite deposits in the Altyn Tagh have been much less investigated [37–40]. With more nephrite deposits being discovered in the recent years, the Altyn Tagh Mountains have become a major nephrite production region in China, with over half of the nephrites mined in the Qiemo and Ruoqiang Counties [40].

The Yinggelike nephrite deposit (Ruoqiang County) is famous for producing the highly valued yellow-hue nephrite, which is locally called the "Huangkou variety" (i.e., (greenish)-yellow variety). The deposit was first mined in the late Manchurian "Qing" Dynasty (early 1900s), paused for over 100 years due to geopolitical instability, and the production restarted in 2004 [41]. In recent years, the demand for and price of the high-quality yellow nephrite has increased markedly [42]. Despite the extraordinary gemological quality and commercial significance of the Yinggelike nephrite, it has not been well studied due to the geographic remoteness and inaccessibility of the deposit, including tough mountain traffic, difficult terrains, and high elevation. Previous studies only briefly introduced the geological features of the Yinggelike deposit [43,44], but no research has been carried out on the nephrite mineralogy and geochemistry of this deposit. Therefore, this study provides the first systematic geological investigation, mineralogical and geochemical analyses, with the aims of establishing the mineral crystallization stages and the gemological significance of the nephrite.

2. Geological Background

2.1. Hetian Nephrite Belt

The Hetian nephrite belt is located in the southern part of the Tarim Basin, along the Kunlun and Altyn Tagh Mountains (Figure 1a) [16]. This belt is divided into three geographic sections, i.e., (from west to east) Shache-Yecheng (76–78° E), Hetian-Yutian (78–82° E), and Qiemo-Ruoqiang (85–90° E) (Figure 1a) [16]. Recent studies show that this belt contains over 20 nephrite deposits with similar geological setting, mostly located at the contact between dolomitic marble and intermediate-felsic intrusion [19]. The nephrite deposits from the Shache-Yecheng, and Hetian-Yutian sections are distributed in the Kunlun Mountains, while those from Qiemo-Ruoqiang section are distributed in the Altyn Tagh Mountains (Figure 1a).

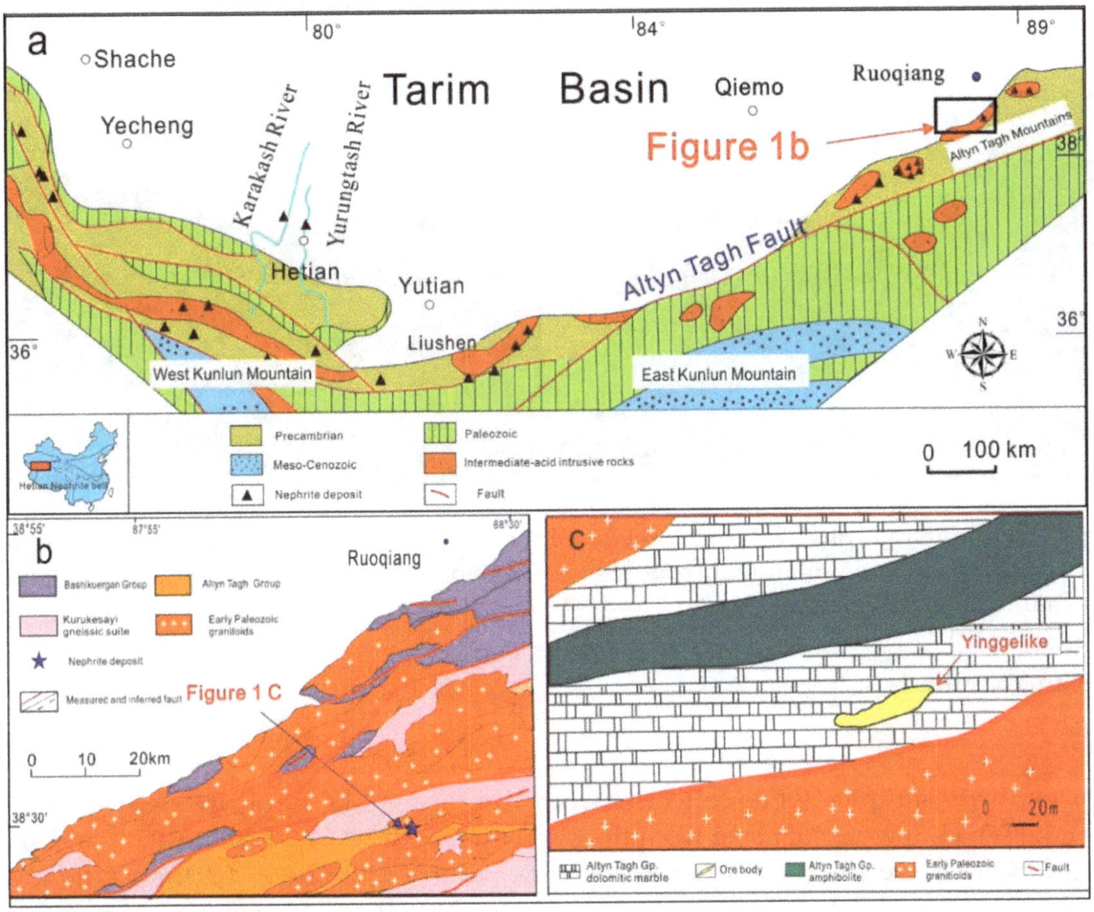

Figure 1. (**a**) Simplified geologic map of the Hetian nephrite belt, showing the three major sections of Shache-Yecheng (76–78° E), Hetian-Yutian (78–82° E), and Qiemo-Ruoqiang (85–90° E) (modified after [16,18]); (**b**) Geologic map of the study area, showing the location of Yinggelike nephrite deposit (modified after [43]); (**c**) Geologic map of the Yinggelike nephrite orebody (modified after [44]).

2.2. Qiemo-Ruoqiang Section (85–90° E)

The Qiemo-Ruoqiang nephrite belt is located in the Altun Tagh Mountains. Altyn Tagh is situated between the Tarim Craton in the north, and the Qaidam block, Qilian and East Kunlun orogens, and the Tibetan plateau in the south [45–49]. Tectonically, the Altyn Tagh is bounded to the north by the south-dipping North Altyn Tagh thrust fault, and to the south by the Altyn Tagh sinistral strike-slip fault [45–49].

The Yinggelike nephrite deposit (Ruoqiang County) is located geologically inside the Central Altyn massif [45]. Local stratigraphy comprises mainly the Bashikuergan Group, the Altyn Tagh Group, and the Kurukesayi gneissic suite (Figure 1b) [43]. The Bashikuergan Group contains weakly metamorphosed clastic sedimentary rocks, interpreted to be Mesoproterozoic. The Paleoproterozoic Altyn Tagh Group is mainly composed of amphibolite, dolomitic marble, schist, gneiss, quartzite and mafic-ultramafic rocks. The Neoproterozoic Kurukesayi gneissic suite is a set of metamorphosed granodiorite-monzogranite-K-feldspar granite. This gneissic suite was intruded and dismembered by the locally widespread early Paleozoic granitoids. Major faults in the study area are ENE-trending. This area has likely experienced multiple orogenic processes, accompanied by multiphase and intensive tectono-magmatic activities [43].

The Yinggelike open pit is situated around 70 km south of Ruoqiang County and is more than 3200 m above sea-level. The Paleoproterozoic dolomitic marble and amphibolite of Altyn Tagh Group, and early Paleozoic granitoids are exposed in the mine (Figures 1c and 2a,b). The nephrite is found in the folded grayish-white/-black banded dolomitic marble, and is restricted by faults developed along the contact zone between dolomitic marble and granitoids (Figure 2a,b).

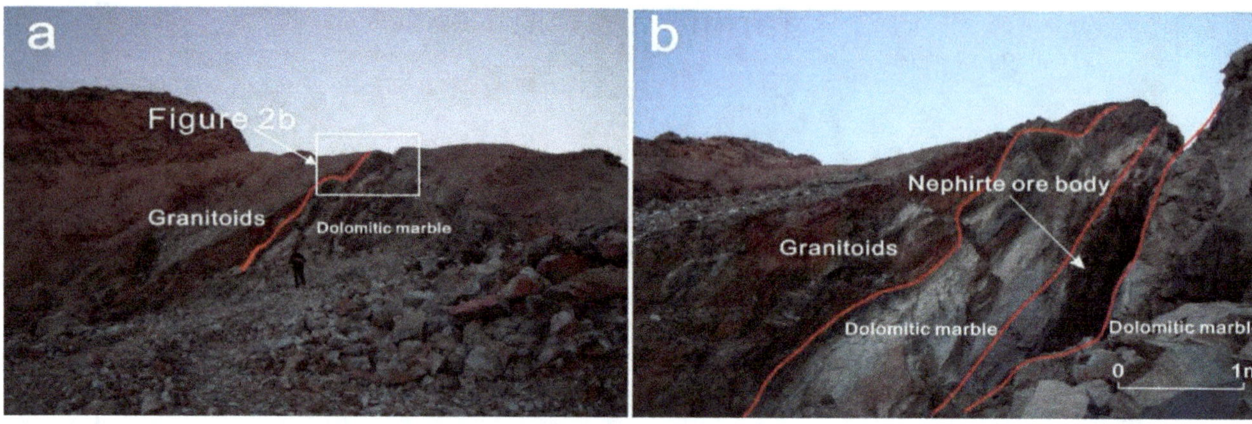

Figure 2. Field photographs of the Yinggelike nephrite mine. (**a**) Contact between granitoids and dolomitic marble; (**b**) zoom-in of Figure 2a, showing a nephrite ore vein occurring near the intrusive contact in the dolomitic marble.

3. Samples and Methods

3.1. Samples

Eleven nephrite samples were collected from the Yinggelike nephrite deposit, of which five (H-1, H-2, H-3, H-4, H-5) were collected from the open pit (N 38°29′26″, E 88°18′05″) and six (Q-1, Q-2, Q-3, Q-4, Q-5, Q-6) were purchased from the mine. Refractive index (RI), specific gravity (SG), and Vickers hardness of all the samples were determined. The RI and SG were examined by the distant vision method and hydrostatic weighing method, respectively, following the procedures outlined by [50,51]. The Vickers hardness was examined by the Microvickers hardness tester and converted to Mohs hardness by calculation [52]. During the hardness test, 1 kgf was used as the load value. The gemological properties are listed in Table 1.

Table 1. Gemological properties of the nephrite jade samples from the Yinggelike deposit in Altyn Tagh, NW China.

Samples	Color	Luster	Refractive Index	Specific Gravity	Vickers Hardness	Mohs Hardness
H-1	Light-yellow	Sub-vitreous	1.60	2.94	638	5.8
H-2	Greenish-yellow	Greasy	1.60	2.88	753	6.1
H-3	Yellow	Sub-vitreous	1.60	2.93	702	6.0
H-4	Light-yellow	Greasy	1.60	2.96	657	5.9
H-5	Light-yellow	Greasy	1.60	2.93	784	6.2
Q-1	Greenish-yellow	Sub-vitreous	1.61	2.90	713	6.0
Q-2	Greenish-yellow	Sub-vitreous	1.61	2.92	733	6.1
Q-3	Greenish-yellow	Sub-vitreous	1.62	2.93	738	6.1
Q-4	Light-yellow	Sub-vitreous	1.62	2.96	690	6.0
Q-5	Light-yellow	Sub-vitreous	1.62	2.90	698	6.0
Q-6	Light-yellow	Sub-vitreous	1.61	2.92	750	6.1

3.2. Methods

Chemical compositions of nephrite and backscattered electron (BSE) images were acquired at the Institute of Geology, Chinese Academy of Geological Sciences (CAGS) in Beijing, China, using a JXA-8100 Electron Microprobe Analyzer (EMPA). Analysis conditions included 15 kV acceleration voltage, 10 nA beam current, and 3 μm spot size. The analytical precision was ±2%. Matrix corrections were carried out using the ZAF correction program supplied by the EMPA manufacturer. The EMPA standards included: andradite for Si and Ca, rutile for Ti, corundum for Al, hematite for Fe, eskolaite for Cr, rhodonite for Mn, bunsenite for Ni, periclase for Mg, albite for Na, K-feldspar for K and barite for Ba.

Bulk-rock chemical compositions were acquired using an AXIODmAX X-ray fluorescence (XRF) spectrometer at the Hebei Institute of Regional Geology and Mineral Resources (HIRGMR) in Langfang, China. All samples were cleaned, crushed, and milled to <200 mesh with an agate mortar. Whole-rock powder samples (0.7 g) were mixed with 5.2 g $Li_2B_4O_7$, 0.4 g LiF, 0.3 g NH_4NO_3 (500 g/L) in a 25 mL porcelain crucible. The powder mixture was transferred to a platinum alloy crucible, and 1 mL LiBr solution was added before the sample was dried. The powder mixture was then melted in an automatic flame fusion machine at 1200 °C for 10 min and casted into glass discs for XRF analysis. Standard samples were prepared using the same procedure. The XRF operating conditions included an accelerating voltage of 30 kV (for Na-P, K, Ca), 40 kV (Ti) and 60 kV (Mn, Fe), beam current of 120 mA (for Na-P, K, Ca), 90 mA (Ti) and 60 mA (Mn, Fe) and collimators of 300 μm (for Al, Si) and 700 μm (all others). Analytical uncertainties were generally better than 1%. FeO was determined by $K_2Cr_2O_7$ titration after the samples were dissolved in HF and H_2SO_4.

Concentrations of trace elements (including rare earth elements (REEs)) were obtained using an X Series II inductively coupled plasma mass spectrometer (ICP-MS) at HIRGMR. The samples were ground in an agate mortar to 200 mesh. The analysis procedure followed the Silicate Rocks-Part 30as method recommended by the State Standard of the People's Republic of China (GB) [53]. The operating conditions included an analogue detector voltage of 1890 V, PC detector voltage of 2890 V, vacuum level of 2 to 2.4 mbar, scanning time of 60 s, and integration time per element of 0.5 s. The internal standard Rh was used to correct for any analytical drift and matrix effects. The standard deviations for trace elements in the blank samples varied from 0.0005 to 0.8759 and were generally less than 0.05.

Oxygen and hydrogen isotope analyses were acquired in the laboratory at the Beijing Research Institute of Uranium Geology. Nephrite samples were first crushed to <60 mesh to separate the minerals. The rare impurities were removed by hand-picking under the microscope to ensure high purity. The purified samples were then ground to 200 mesh in an agate mortar for oxygen and hydrogen isotope analyses. Oxygen was liberated from the samples by reaction with BrF_5 [54] and converted to CO_2 on a platinum-coated graphite rod. Oxygen isotope compositions were measured with a Delta V Advantage Isotope Ratio Mass Spectrometer. Analyses of GBW-04409 (quartz) during this study yielded an average $\delta^{18}O$ value of 11.11‰ (recommended value = 11.11‰; Chinese Standard Reference Material GBW-04409) [55]. Precision of the oxygen isotope analyses is better than 0.2‰. For hydrogen isotopic compositions, the samples were first degassed of labile volatiles by heating under vacuum to 120 °C for 3 h. Constitutional water was released by heating the samples to approximately 1200 °C in an induction furnace. The released water was converted to hydrogen by passage over heated zinc powder at 400 °C [56]. The hydrogen isotope compositions were analyzed using a MAT-253 Mass Spectrometer with ±2‰ precision. Analyses of Peking University standard water during this study gave an average δD value of −64‰ (recommended value = −64.8‰; Chinese Standard Reference Material PKU Standard Water) [55]. Stable isotope data for hydrogen and oxygen are reported as δD and $\delta^{18}O$ per mille (‰) relative to SMOW.

Samples H-1~H-5 were cut into small slabs, and polished thin sections were prepared for petrographic observations. Samples Q-1~Q-6 were cut to uniform-sized ($5 \times 2 \times 1$ cm^3) polished slabs to facilitate the color comparison (Figure 4).

4. Results

4.1. Gemological Properties

The conventional gemological properties of the 11 Yinggelike nephrite samples are listed in Table 1. The nephrite samples appear similar to typical nephrites except for the color. All the samples are light-yellow to greenish-yellow with fine texture (Figures 3 and 4), and are translucent with sub-vitreous to greasy luster. Polished samples have commonly sub-vitreous luster. Their reflective index (RI) ranges from 1.60 to 1.62, and specific gravity (SG) from 2.88 to 2.96. The Mohs hardness ranges from 5.8 to 6.2. The gemological properties are similar to typical nephrite jade [23,57].

Figure 3. Representative hand specimen photographs of nephrite collected from the Yinggelike open pit. (**a**) Representative greenish-yellow sample; (**b**) representative light-yellow sample.

Figure 4. Nephrite samples purchased from the Yinggelike mine. Samples Q-1, Q-2, and Q-3 are greenish-yellow, and Q-4, Q-5, and Q-6 are light-yellow.

4.2. Petrography

All the nephrite samples comprise predominantly tremolite (Tr; >98 vol.%), with minor pargasite (Prg), diopside (Di), epidote (Ep), allanite (Aln), andesine (Ads), prehnite (Prh), titanite (Ttn), zircon (Zrn), calcite (Cal), etc. Occurrences of these minerals are described below.

4.2.1. Tremolite

Based on the occurrence, paragenesis and chemical composition, two generations of tremolite can be classified: undeformed (Tr-I) and deformed (Tr-II). Tr-I is very rare and occurs as euhedral porphyroblasts of approximately 200 μm long and 60 μm wide (Figure 5a). Some Tr-I grains form rhombohedral pseudomorphs of carbonates (Figures 5b and 6a) and were partially replaced by Tr-II. In some cases, Tr-I grains also replaced rounded/oval diopside, and were themselves replaced by Tr-II (Figure 6b). The fine-grained, non-directional micro/cryptocrystalline Tr-II grains are the main form of tremolite in the Yinggelike nephrite (Figure 5). Aggregates of Tr-II commonly exhibit fine fibrous-felted texture (Figure 5c).

4.2.2. Pargasite

Pargasite is commonly found in the Yinggelike nephrite. Some pargasite grains appeared as euhedral crystals in the nephrite, with sizes of about 400 μm in length and 260 μm in width (Figure 5d), while some grains were partially replaced by tremolite (Figure 6c,d). Pargasite also replaced diopside locally (Figure 6c), indicating that it had formed before tremolite but after diopside. It is noted that this mineral is rare in other nephrite deposits in Hetian nephrite belt [18,19,37,38].

4.2.3. Diopside

Diopside grains are generally irregularly shaped (Figure 5e) and relatively common in the Yinggelike nephrite. They were commonly replaced by tremolite along their grain boundary or cracks (Figures 5e and 6b).

4.2.4. Epidote, Allanite, and Prehnite

Most epidote and allanite grains are idiomorphic columnar (Figure 5f–g) with local tremolite replacement (Figure 6e,f). Prehnite displays commonly metasomatic relict texture via tremolite replacement (Figure 6g).

4.2.5. Andesine, Titanite, Zircon, and Calcite

Andesine, titanite, and zircon are only distributed sporadically in some samples (Figure 6f,h,i). Andesine, which exhibits metasomatic relict texture, was replaced by tremolite. The titanite and zircon are relatively euhedral, and were formed before tremolite. Late-stage calcite veins intruded the Tr-II fractures locally (Figure 5h).

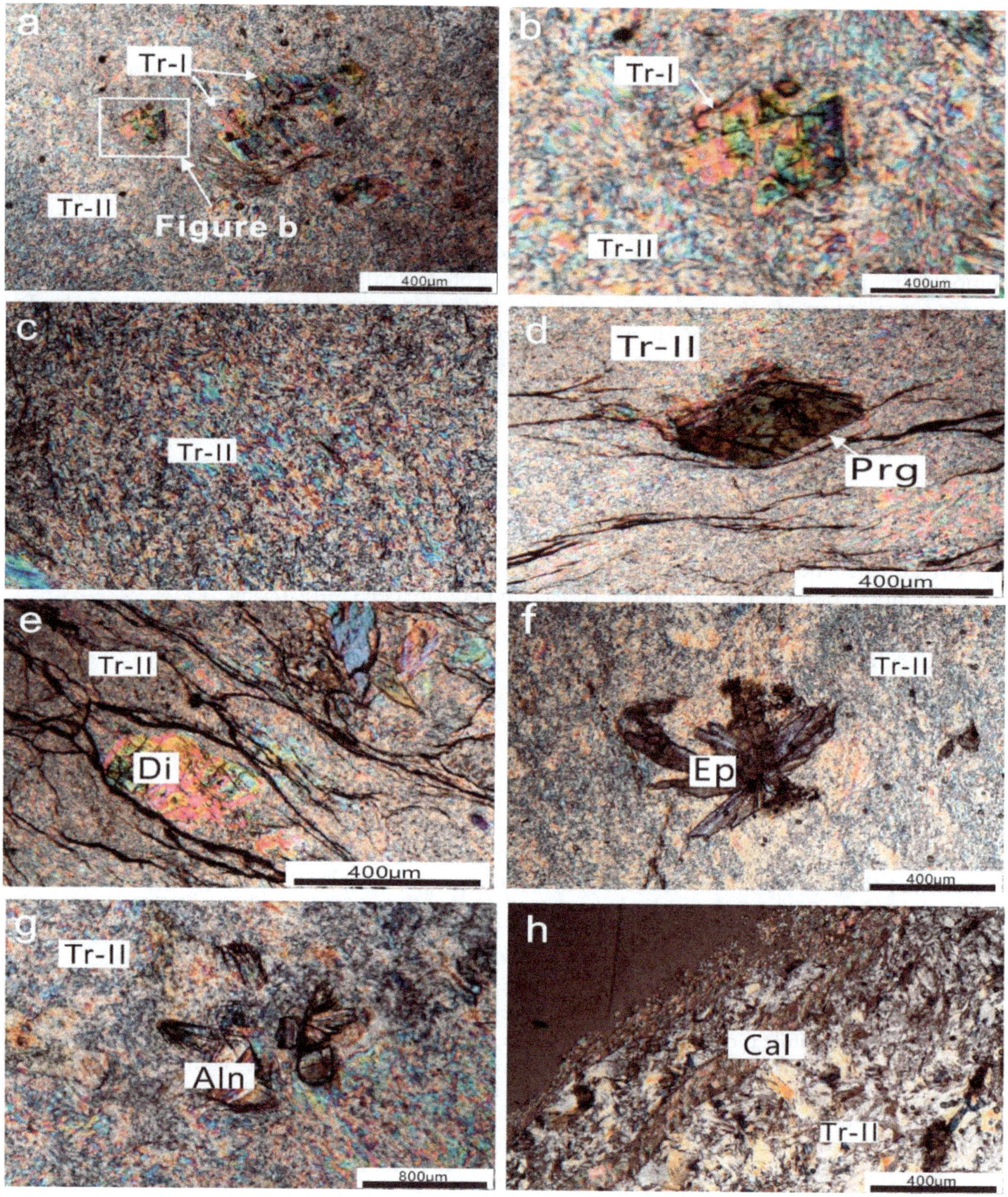

Figure 5. Photomicrographs of the Yinggelike nephrite (crossed-polarized light). The nephrite groundmass comprises mainly fine-fibrous Tr-II. (**a**) Coarse-grained Tr-I in nephrite (H-2); (**b**) Tr-I grains in carbonate pseudomorph (H-2), zoom-in of Figure 5a; (**c**) Fine-grained Tr-II (H-1); (**d**) Single pargasite crystal in nephrite (H-4); (**e**) Replacement of diopside by fine-fibrous Tr-II (H-3); (**f,g**) Idiomorphic epidote (H-3) and allanite (H-5) grains in nephrite; (**h**) Late-stage calcite vein in nephrite (H-3).

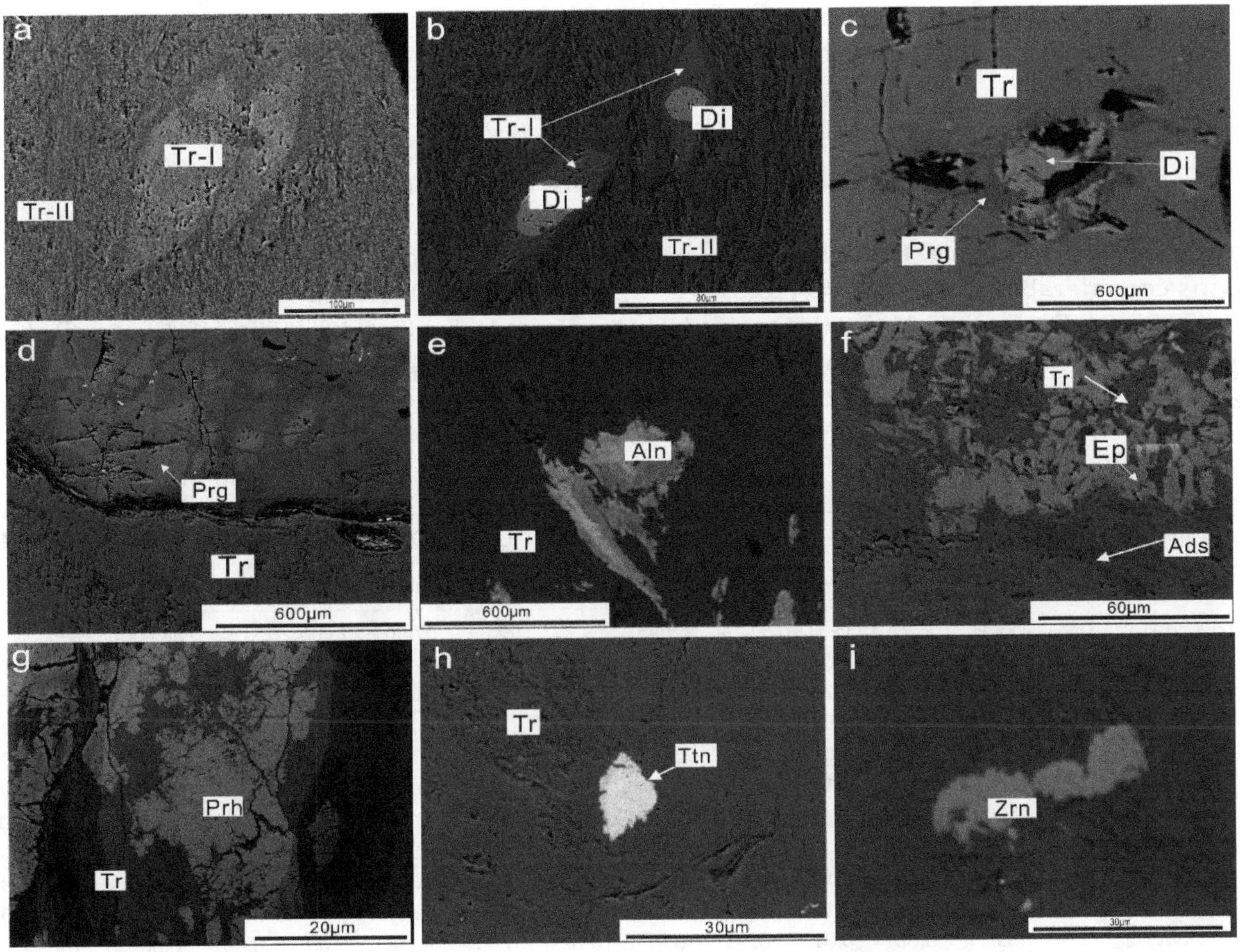

Figure 6. Back-scattered electron (BSE) images of the Yinggelike nephrite. (**a**) Tremolite grains in carbonate pseudomorph (H-2); (**b**) Rounded/oval-shaped diopside replaced by Tr-I, which was in turn replaced by Tr-II (H-5); (**c**) Diopside replaced by pargasite, which was in turn replaced by tremolite (H-5); (**d**) Replacement of pargasite by tremolite (H-5); (**e**–**g**) Metasomatic relicts of allanite (H-5), epidote, andesine and prehnite (H-3); (**h**) Idiomorphic titanite (H-1); (**i**) Irregularly shaped zircon crystal (H-5).

4.3. *Mineral Chemistry*

4.3.1. Amphibole

According to the amphibole classification diagram (Figure 7), all the Yinggelike amphiboles belong to the calcic group, and are classified as tremolite and pargasite [58]. Tremolite is mainly composed of (in wt.%) SiO_2 (56.96–59.74), CaO (12.54–13.89), MgO (21.05–24.62), total Fe as FeO (FeO_T = 0.27–3.48), with atoms per formula unit (a.p.f.u.) Si = 7.88–8.00, Ca on B site >1.50, (Na + K) on A site <0.5 (Table 2). Their Mg/(Mg + Fe^{2+}) ratios vary from 0.92 to 1.00. Tr-I grains have higher FeO_T content than that of Tr-II. Pargasite is composed of (in wt.%) SiO_2 (44.90–45.90), CaO (13.96–14.50), MgO (18.99–19.76), FeO_T (0.23–0.52). The pargasite grains are featured with Si = 6.29–6.39 a.p.f.u., Ca on B site >1.50, (Na + K) on A site >0.5, and Mg/(Mg + Fe^{2+}) = 0.99–1.00 (Table 2). FeO_T contents in Tr-II and pargasite are low. The calculation of Fe^{3+} in amphibole is based on the average estimation method of Schumacher [58].

4.3.2. Diopside

Diopside grains are mainly composed of (in wt.%) SiO_2 (54.20–55.03), CaO (25.93–26.94), MgO (16.50–18.47), FeO_T (0.05–0.40), with (a.p.f.u.) Si = 1.96–1.98, Mg = 0.89–0.99, and Ca = 1.00–1.04

(Table 3). It is almost homogeneous, and compositionally is close to the diopside end-member [59]. The FeO and Fe_2O_3 contents were calculated by charge balance method.

4.3.3. Epidote, Allanite, Andesine, Prehnite, Titanite, Zircon, and Calcite

The chemical compositions of other minor minerals are presented in Table A1. Epidote is mainly composed of (in wt.%) SiO_2 (37.14–37.75), CaO (25.71–25.72), Al_2O_3 (28.05–28.30), FeO_T (3.40–3.78). Both allanite and epidote belong to the epidote group. Different from epidote, the allanite contains considerable REE contents (10.89–12.18), as well as SiO_2 (35.64–36.25), CaO (18.14–20.11), Al_2O_3 (23.22–24.73), FeO_T (5.11–7.24). One andesine grain contains (in wt.%) SiO_2 (61.19), Al_2O_3 (24.32), CaO (7.27), and Na_2O (7.23). Ca/(Ca+Na) ratio is 0.36. The prehnite grains contain (in wt.%): SiO_2 (44.75–44.85), CaO (27.01–27.23), Al_2O_3 (22.86–23.53), and FeO_T (0.17–0.22).

Table 2. Chemical compositions of tremolite (Tr-I, Tr-II) in the Yinggelike nephrite (wt.%).

Samples	H-1	H-2	H-3	H-4	H-5	Q-1	Q-2	Q-3	Q-4	Q-5	Q-6	H-2	H-5
Minerals	Tr-II	Tr-II	Tr-II	Tr-II	Tr-II	Tr-II	Tr-II	Tr-II	Tr-II	Tr-II	Tr-II	Tr-I	Tr-I
SiO_2	58.34	58.77	58.31	59.74	59.35	59.21	59.09	57.90	58.06	58.62	58.40	56.96	57.07
TiO_2	0.00	0.00	0.01	0.01	0.00	0.02	0.00	0.00	0.02	0.04	0.03	0.00	0.00
Al_2O_3	0.76	0.72	0.78	0.60	0.63	0.63	0.72	0.69	0.72	1.62	0.76	1.08	2.02
FeO	0.97	0.98	0.94	0.43	0.37	0.35	0.27	0.45	1.96	1.13	1.06	3.48	2.66
Cr_2O_3	0.00	0.00	0.01	0.00	0.00	0.00	0.06	0.00	0.00	0.02	0.01	0.02	0.00
MnO	0.09	0.08	0.11	0.01	0.03	0.01	0.00	0.03	0.00	0.02	0.10	0.50	0.84
MgO	24.09	23.75	23.95	23.90	23.19	24.62	24.17	24.07	23.42	22.69	23.94	21.63	21.05
CaO	12.93	12.82	13.02	12.89	13.59	12.77	12.54	13.89	13.36	13.54	13.09	13.78	12.88
Na_2O	0.09	0.10	0.12	0.09	0.10	0.13	0.33	0.09	0.10	0.21	0.13	0.12	0.31
K_2O	0.14	0.08	0.12	0.08	0.05	0.11	0.20	0.07	0.08	0.06	0.15	0.08	0.14
NiO	0.00	0.03	0.01	0.00	0.01	0.02	0.00	0.00	0.00	0.00	0.00	0.03	0.00
Total	97.42	97.33	97.39	97.75	97.31	97.86	97.38	97.19	97.72	97.93	97.67	97.68	96.97
TSi	7.96	8.00	7.96	8.00	8.00	8.00	8.00	7.91	7.93	8.00	7.95	7.88	7.94
TAl	0.04	0.00	0.04	0.00	0.00	0.00	0.00	0.09	0.07	0.00	0.05	0.12	0.07
SumT	8.00	8.00	8.00	8.00	8.00	8.00	8.00	8.00	8.00	8.00	8.00	8.00	8.00
CAl	0.08	0.12	0.08	0.10	0.10	0.10	0.11	0.03	0.05	0.26	0.08	0.06	0.27
CFe^{3+}	0.00	0.00	0.00	0.05	0.04	0.00	0.03	0.00	0.00	0.00	0.00	0.00	0.00
CMg	4.90	4.82	4.87	4.77	4.66	4.90	4.85	4.91	4.77	4.61	4.86	4.46	4.36
CFe^{2+}	0.02	0.07	0.04	0.00	0.00	0.00	0.00	0.05	0.18	0.12	0.06	0.40	0.31
CMn	0.00	0.00	0.00	0.00	0.00	0.00	0.00	0.00	0.00	0.00	0.00	0.06	0.06
SumC	5.00	5.00	5.00	4.92	4.81	5.00	5.00	4.98	5.00	5.00	5.00	4.98	5.00
BMg	0.00	0.00	0.00	0.00	0.00	0.06	0.03	0.00	0.00	0.00	0.00	0.00	0.00
BFe_2	0.09	0.05	0.07	0.00	0.00	0.04	0.00	0.00	0.00	0.00	0.06	0.00	0.00
BMn	0.01	0.01	0.01	0.00	0.00	0.00	0.00	0.00	0.00	0.00	0.11	0.00	0.04
BCa	1.89	1.87	1.90	1.85	1.96	1.85	1.82	2.00	1.96	1.98	1.91	2.00	1.92
BNa	0.12	0.04	0.02	0.02	0.03	0.03	0.09	0.09	0.09	0.02	0.02	0.03	0.04
SumB	2.00	1.95	2.00	1.87	1.99	1.98	1.94	2.09	2.05	2.00	2.00	2.03	2.00
ACa	0.00	0.00	0.00	0.00	0.00	0.00	0.00	0.04	0.00	0.00	0.00	0.04	0.00
ANa	0.12	0.00	0.02	0.00	0.00	0.00	0.00	0.02	0.03	0.04	0.02	0.03	0.04
AK	0.03	0.01	0.02	0.01	0.01	0.02	0.04	0.01	0.01	0.01	0.03	0.01	0.03
SumA	0.15	0.01	0.04	0.01	0.01	0.02	0.04	0.07	0.04	0.05	0.05	0.09	0.07
Sumcat	15.15	14.96	15.04	14.80	14.81	15.00	14.98	15.14	15.09	15.05	15.05	15.10	15.07
$Mg^{2+}/(Mg^{2+} + Fe^{2+})$	0.98	0.98	0.98	1.00	1.00	1.00	1.00	0.99	1.00	0.98	0.98	0.92	0.93

Amphibole formulae recalculated on the basis of 23 oxygen atoms; 0.00—concentration below the detection limit. T, C, B, and A represent the occupation of cations in tremolite.

Table 3. Chemical compositions of pargasite and diopside in the Yinggelike nephrite (wt.%).

Samples	H-3	H-3	H-4	H-5	H-1	H-5	H-2	H-3
Minerals	Prg	Prg	Prg	Prg	Di	Di	Di	Di
SiO_2	45.78	45.90	44.90	45.02	55.03	54.20	54.54	54.69
TiO_2	0.33	0.62	1.42	1.43	0.01	0.18	0.00	0.04
Al_2O_3	14.60	15.10	14.52	14.24	0.50	0.79	0.49	2.01
FeO	0.52	0.50	0.52	0.23	0.12	0.07	0.05	0.40
Cr_2O_3	0.00	0.03	0.00	0.00	0.04	0.00	0.01	0.03
MnO	0.03	0.00	0.00	0.04	0.00	0.04	0.04	0.07
MgO	19.36	18.99	19.39	19.76	18.46	17.68	18.47	16.50
CaO	14.50	14.37	13.96	14.38	26.13	26.94	26.23	25.93
Na_2O	2.25	1.95	1.97	1.93	0.02	0.05	0.00	0.20
K_2O	0.11	0.15	0.24	0.28	0.00	0.01	0.00	0.03
NiO	0.00	0.00	0.00	0.00	0.02	0.00	0.03	0.00
Total	97.44	97.58	97.27	97.30	100.33	99.96	99.86	99.90
Si	6.39	6.39	6.29	6.29	1.98	1.96	1.97	1.98
Ti	0.03	0.07	0.15	0.15	0.00	0.00	0.00	0.00
Cr	0.00	0.00	0.00	0.00	0.00	0.00	0.00	0.00
Al	2.40	2.48	2.40	2.34	0.02	0.03	0.02	0.09
Fe^{3+}	0.00	0.00	0.03	0.03	0.00	0.00	0.00	0.00
Mg	4.03	3.94	4.05	4.12	0.99	0.95	0.99	0.89
Fe^{2+}	0.06	0.06	0.03	0.00	0.00	0.00	0.00	0.01
Mn	0.00	0.00	0.00	0.00	0.00	0.00	0.00	0.00
Ca	2.17	2.14	2.10	2.15	1.01	1.04	1.01	1.00
Na	0.64	0.55	0.56	0.55	0.00	0.00	0.00	0.01
K	0.02	0.03	0.04	0.05	0.00	0.00	0.00	0.00
Sum	15.75	15.66	15.66	15.68	4.00	4.00	4.00	3.99
BCa	2.00	2.00	2.00	2.00	n.d.	n.d.	n.d.	n.d.
$(Na + K)_A$	0.57	0.64	0.58	0.57	n.d.	n.d.	n.d.	n.d.
$Mg^{2+}/(Mg^{2+} + Fe^{2+})$	0.99	0.99	0.99	1.00	n.d.	n.d.	n.d.	n.d.

Amphibole formulae recalculated on the basis of 23 oxygens; 0.00—concentration below the detection limit; n.d.—not detected. Diopside formulae recalculated on the basis of 4 cations.

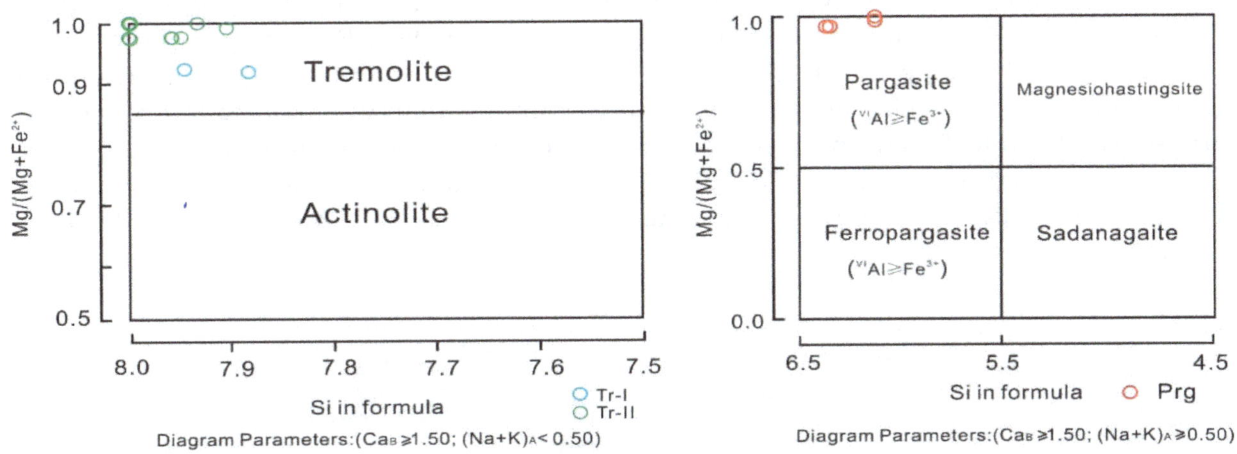

Figure 7. Amphibole classification diagram for the Yinggelike nephrite.

4.4. Bulk-Rock Chemistry

All the nephrite samples have narrow ranges of major element contents (in wt.%): SiO_2 (55.05–58.27), MgO (23.80–27.85), CaO (11.43–13.21), TFe_2O_3 (0.37–1.15), Al_2O_3 (0.76–2.64), and low TiO_2 (0.01–0.03) and MnO (0.02–0.08) (Table 4), which resemble theoretical tremolite end-member composition. This is in agreement with our petrographic observations that tremolite is the main component in the nephrite. The Fe/(Fe + Mg) ratios are 0.02 to 0.05. The transition metal contents of the nephrite purchased from the mine are listed in Table 5. As seen in Table 5, Fe^{2+}, Fe^{3+} and Mn contents are generally high in all the samples, followed by Ti, whereas the V, Cr, Co, Ni, Cu contents are too low (generally <10 ppm) to have significant impact on the color. Therefore, the Fe^{2+}, Fe^{3+}, Mn, and Ti may play an important role in Yinggelike nephrite color.

Table 4. Bulk-rock chemical compositions of the Yinggelike nephrite samples.

Samples	H-1	H-2	H-3	H-4	H-5	Q-1	Q-2	Q-3	Q-4	Q-5	Q-6	AVG
SiO_2	55.93	55.05	55.21	56.20	56.12	57.97	57.88	58.27	57.18	55.89	57.67	56.70
TiO_2	0.02	0.02	0.02	0.03	0.03	0.02	0.02	0.01	0.01	0.01	0.01	0.02
Al_2O_3	0.76	1.10	2.11	2.19	0.85	0.77	1.09	0.76	1.30	2.64	1.05	1.33
TFe_2O_3	0.56	1.15	0.50	0.78	0.37	0.78	0.66	0.74	0.60	0.48	0.57	0.65
FeO	0.40	0.83	0.33	0.49	0.26	0.61	0.49	0.58	0.53	0.41	0.50	0.49
Fe_2O_3	0.09	0.23	0.13	0.24	0.08	0.10	0.11	0.10	0.01	0.03	0.01	0.10
MnO	0.02	0.08	0.03	0.02	0.04	0.07	0.06	0.08	0.04	0.03	0.05	0.05
MgO	27.44	26.49	27.44	26.88	27.85	24.25	23.81	23.80	24.51	25.07	24.63	25.70
CaO	12.26	12.40	11.43	13.12	11.76	12.58	13.21	12.99	12.37	11.93	12.76	12.4
Na_2O	0.12	0.09	0.15	0.13	0.14	0.44	0.39	0.47	0.41	0.36	0.33	0.28
K_2O	0.08	0.09	0.22	0.14	0.07	0.20	0.11	0.11	0.61	0.39	0.41	0.22
P_2O_5	0.01	0.01	0.01	0.03	0.01	0.01	0.02	0.01	0.01	0.01	0.01	0.01
LOI	2.81	3.55	2.83	0.44	2.73	2.97	2.78	2.80	2.99	3.23	2.55	2.70
Total	100.49	101.1	100.5	100.85	100.37	99.99	99.98	99.98	99.98	99.99	99.99	100
Fe/(Fe + Mg)	0.02	0.05	0.02	0.03	0.02	0.03	0.03	0.03	0.03	0.02	0.03	0.03

T—Total; LOI—Loss on ignition.

Table 5. Transition metal concentrations (ppm) of the Yinggelike nephrite.

Samples	Fe^{3+}	Fe^{2+}	Mn	Ti	V	Cr	Co	Ni	Cu	Color
Q-1	695	4731	550	108	11.6	4.42	1.82	3.44	0.52	Greenish-yellow
Q-2	769	3824	459	118	7.42	8.81	1.87	3.51	1.04	Greenish-yellow
Q-3	733	4472	578	81	11.5	9.14	2.14	3.72	0.93	Greenish-yellow
Q-4	96	4083	341	61	1.37	8.44	1.55	4.12	1.74	Light-yellow
Q-5	184	3175	210	79	2.77	2.03	0.83	3.19	0.37	Light-yellow
Q-6	98	3889	361	88	1.49	6.03	1.56	3.46	0.41	Light-yellow

Fe^{3+} and Fe^{2+} contents were re-calculated (in ppm) from the XRF data; the remaining elements were analyzed by inductively coupled plasma-mass spectrometry (ICP-MS).

The nephrite samples have low total REE contents (2.16 to 11.25 ppm), except for sample H-5 (736.23 ppm) (Table A2; Figure 8). Photomicrographs (Figure 5g), BSE images (Figure 6e,i) and EMPA data (Table A1) suggest that mineral inclusions such as allanite and zircon in sample H-5 have generated higher trace element (including REE) contents than in the other samples (Table A2). Chondrite-normalized REE patterns of the Yinggelike nephrite are characterized by steep LREE and flat HREE patterns (Figure 8) with LREE/HREE of 5.16 to 9.05, and distinct to moderately negative Eu anomalies (Eu/Eu* = 0.04–0.67). The nephrite samples have $(La/Sm)_N$ = 1.72 to 4.04 and $(Gd/Yb)_N$ = 0.53 to 2.62 (Table A2).

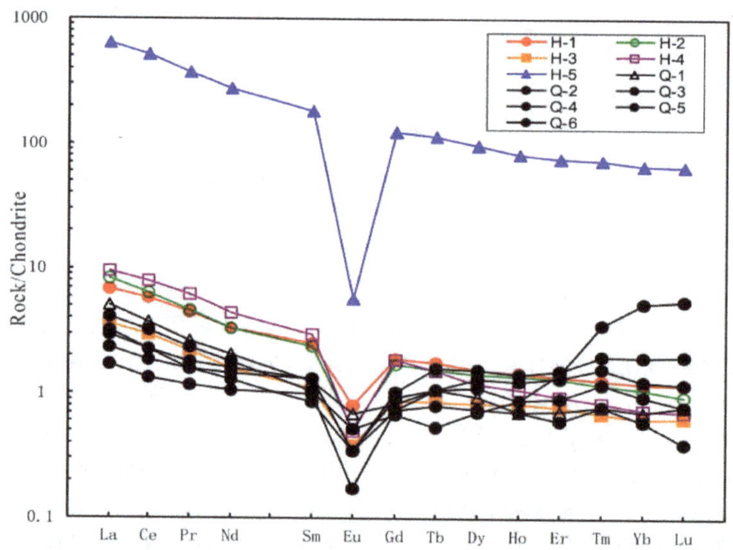

Figure 8. Chondrite-normalized REE patterns of the Yinggelike nephrite [60].

The nephrite samples contain $\delta^{18}O$ = 5.3 to 7.4‰ and δD = −74.9 to −86.7‰, exhibiting a narrow range. Hydrogen and oxygen isotope compositions in the nephrite from Yingelike and other well-known nephrite deposits worldwide are summarized in Table 6 and illustrated in Figure 9.

Table 6. Hydrogen and oxygen isotope compositions of the nephrite from Yingelike and other deposits.

Samples	δD	$\delta^{18}O$			
H-3	–74.9	5.3			
H-4	–78.0	7.4			
H-5	–86.7	5.4			
Average	–79.9	6.0			
Types	**Dolomite-related**		**Serpentinite-related**		
Deposits	**δD**	**$\delta^{18}O$**	**Deposits**	**δD**	**$\delta^{18}O$**
Kunlun, China[1]	−108~−124	0.5~2.3	Chara Jelgra River, Siberia[1]	−39	6.9
Cowell, Australia[1]	−57	3.4	Red Mountain, New Zealand[1]	–54	7.5
Wyoming, USA[1]	−56	1.5	Mt. Ogden, Canada[1]	–52	9.6
Chuncheon, Korea[1]	−118~−105	−9.9~−7.9	Shulaps Range, Canada[1]	–49	8.4
Alamas, Xinjiang, China[2]	−95~−75	3.2~6.2	Fengtien, China[1]	−68~−33	4.5~5.3
Hetian placer nephrite, China[3]	–72.4~–55.7	1.1~5.6	Qinghai, China[7]	−60~−56	8.1~8.6
Złoty Stok, SW Poland[4]	−77.2~−74.6	8.3~10.4	Pakistan[7]	−56~−43	13.2~13.4
Tashisayi, Xinjiang, China[5]	−89.3~−82.5	3.0~6.5	Jordanów, SW Poland[8]	–61	6.1
Apocarbonate nephrite, Russia[6]	−179~−119	−17~−15			

Data source: [1] From [31]; [2] From [18]; [3] from [19]; [4] from [36]; [5] from [38]; [6,7] from [61]; [8] from [32].

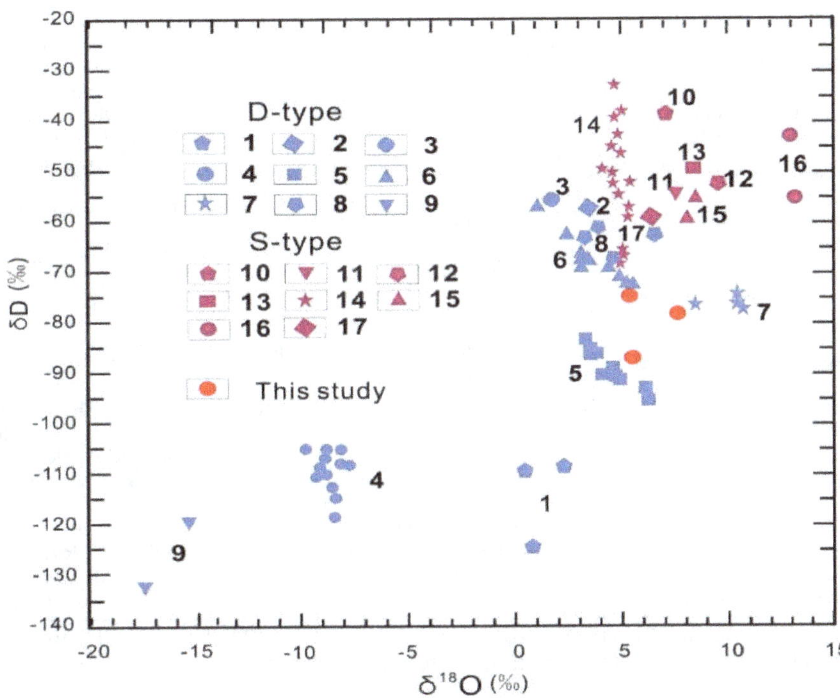

Figure 9. δD and $\delta^{18}O$ of nephrites from Yinggelike (this study) and other dolomite-related (D-type) deposits: (1) Kunlun (n = 3); (2) Cowell (n = 1); (3) Wyoming (n = 1); (4) Chuncheon (n = 12); (5) Alamas (n = 12); (6) Hetian placer nephrite (n = 10); (7) Złoty Stok (n = 4), (8) Tashisayi (n = 4); (9) Apocarbonate nephrite (n = 2), and serpentinite-related (S-type) ones: (10) Chara Jelgra River (n = 1); (11) Red Mountain (n = 1); (12) Mount Ogden (n = 1); (13) Shulaps Range (n = 1); (14) Fengtien (n = 16), (15) Qinghai (n = 2); (16) Pakistan (n = 2); (17) Jordanów, SW Poland (n = 1) (see Table 6 for details).

5. Discussion

5.1. Dolomite-Related Origin

Field investigation shows that the Yinggelike nephrite orebody occurs in the dolomitic marble near the intruding granitoids (Figure 2), which suggests that the Yinggelike nephrite deposit is of a dolomite-related type. This conclusion is supported by the mineralogy, bulk-rock chemistry, and O and H isotopes characteristics of the nephrite.

From the mineralogical characteristics of nephrites of different genetic origins around the world (summarized in Table 7), major differences are present between the dolomite-related (D-type) and serpentinite-related (S-type) nephrites: the main mineral of D-type nephrites is tremolite (>95 vol.%), although diopside exists in almost all D-type nephrites and is considered to be related to the nephrite formation [18,19,22]. Minor minerals mainly include actinolite, carbonate, epidote, clinozolisite, apatite, allanite, quartz, etc. Almost all the minerals in the D-type nephrites were iron-poor [18,19,22,37,62–65]. In contrast, S-type nephrites mainly contained tremolite-actinolite series minerals, with minor grossular, spinel, chlorite, serpentine, chromite, magnetite, magnesite, monazite, maucherite, etc., but no carbonate minerals [22,33,66–71]. Iron-rich minerals, such as magnetite and chromite, are highly common. S-type nephrite has commonly black impurities (chromite, spinel or other opaque minerals) visible to the naked eye. Our petrographic observations, whole-rock geochemical and mineral EMPA data all suggest that the Yinggelike nephrite is composed predominantly of tremolite (>98 vol.%), with minor calcite but no serpentine or Fe-rich minerals such as magnetite and chromite. The occurrence of tremolite in carbonate pseudomorphs points to metasomatic replacement processes (Figures 5b and 6a), similar to that proposed for the typical D-type Chuncheon nephrite deposit [64].

Table 7. Mineralogical characteristics of the world's major D-/S-type nephrite deposits.

Types	Deposits	Major Minerals	Minor Minerals
D-type	Złoty Stok, SW Poland[1]	Type 1: actinolite, diopside Type 2: tremolite-actinolite (>90%)	Type 1: lollingite, carbonates, quartType 2: diopside, calcite, quartz, As-minerals, apatite, scheelite, allanite
	Alamas, Xinjiang, China[2]	Tremolite (>98%)	Diopside, actinolite, calcite
	Hetian placer nephrite, China[3]	Tremolite (>95%)	Diopside, calcite, allanite, zircon, rutile
	Chuncheon, Korea[4]	Tremolite (~98%)	Diopside, serpentine, clinochlore, carbonates, apatite
	Qinghai, China[5]	Tremolite (>95%)	Diopside, calcite, epidote, clinozolisite, titanite, actinolite
	Tiantai, Xinjiang, China[6]	Tremolite (>95%)	Diopside, dolomite, titanite, epidote, magnetite, apatite, limonite
	Liyang, Jiangsu, China[7]	Tremolite (~99%)	Diopside, apatite, carbonates
S-type	Jordanów, SW Poland[8]	Tremolite (97.2–89.8%), diopside (4.0–5.7%) and chlorite (3.8–8.1%)	Grossular, prehnite, opaque spinel, antigorite, hydrogrossular, titanite, zircon, apatite, monazite
	Mount Ogden area, Central British Columbia[9]	Actinolite-tremolite (>90%), serpentine (<10%), talc (~6%)	Spinel, titanite, ilmenite, hematite, chlorite
	Qinghai, China[10]	Tremolite, minor actinolite	Diopside, chlorite, zoisite, rutile, zircon, serpentine, magnetite, magnesite, chromite, augite
	Manasi, Xinjiang, China[11]	Tremolite-actinolite (75–90%)	Chromite, kirschsteinite, chlorite
	South Island, New Zealand[12]	Mainly tremolite, less actinolite	Albite, Chlorite, diopside, Maucherite
	Hualian, Taiwan, China[13]	Tremolite-actinolite	Grossular, chromite, chlorite
	East Sayan, Siberia, Russia[14]	Tremolite	Chlorite, chromite

Data source: [1] from [33,36]; [2] from [18]; [3] from [19]; [4] from [62–64]; [5] from [22]; [6] from [37]; [7] from [65]; [8] from [32,33]; [9] from [66,67]; [10] from [22]; [11] from [35]; [12] from [68]; [13] from [69]; [14] from [70,71].

The Fe/(Fe + Mg) ratios can distinguish between D- and S-type nephrites, with those of D-type nephrites often falling below 0.06 (considerably lower than S-type) [2,72,73]. The Yinggelike nephrite has Fe/(Fe + Mg) ratios of 0.02 to 0.05, thus falling within the D-type field. In addition, the Yinggelike nephrites have very low concentrations of Cr (0.81–34.68 ppm), Co (1.10–2.91 ppm), and Ni (0.52–20.15 ppm), which are consistent with a D-type origin (Cr = 1.9–179.0 ppm, Co = 0.5–9.8 ppm, Ni = 0.05–470.7 ppm), but inconsistent with an S-type one (Cr = 719–2812 ppm, Co = 42–260 ppm, Ni = 959–1898 ppm) [32]. The low compatible trace element (Cr, Co, Ni) contents for the Yinggelike nephrite are likely inherited from its protoliths, i.e., dolomitic marble instead of serpentinized ultramafic rocks.

Previous study has summarized the REE characteristics of nephrite from different deposits [74]. REE features of nephrites are likely co-influenced by their geological environment and petrogenetic processes, and can thus be used to distinguish nephrites of different genetic origins [74]. The Yinggelike nephrite shares similar REE patterns to many D-type nephrites (e.g., Alamas and Qinghai ones; [18,22]), including moderate LREE enrichment (La_N/Sm_N = 1.72–4.04), nearly flat HREE patterns (Gd_N/Yb_N = 0.13–2.62), strong to moderate negative Eu anomalies (Eu/Eu* = 0.04–0.67), and low ΣREE contents (2.16–11.25 ppm) (Figure 8; Table A2). Meanwhile, these REE characteristics of the Yinggelike nephrite are different from many S-type nephrites, which are featured by left-inclining REE patterns, strong positive Eu anomalies, significantly elevated ΣREE (2200–3000 ppm) and LREE/HREE (151–326) that reflect a serpentinite-origin (e.g., Qinghai, Hualian, and Manasi nephrites) [74].

As summarized by the compilation of data presented in Table 6 and Figure 9. D-type nephrites tend to have relatively low $\delta^{18}O$ and δD values, and vice versa for S-type ones albeit some overlapping. The $\delta^{18}O$ (5.3–7.4‰) and δD (–74.9 to –86.7‰) values of the Yinggelike nephrite fall closer to many D-type nephrite deposits (esp. Alamas, Hetian and Złoty Stok) than the S-type ones in the $\delta^{18}O$ vs. δD diagram. The isotopic difference from some other D-type nephrite deposits (e.g., Chuncheon; Russia) may have been caused by different formation temperatures, fluid/rock ratios, and/or protolith isotope inheritance [25,31].

5.2. Metasomatic Process in Nephrite

Based on mineral paragenetic relationships obtained from this and previous studies [1,16,18,19,75], three mineral crystallization stages are recognized at Yinggelike.

Stage I is associated with granitoid intrusion (Figure 1c). Representative mineral formed during this stage is diopside (Figure 5e). Anhydrous andesine, titanite, and zircon were probably formed at this stage (Figure 6f,h,i). The prograde metasomatic process that formed the diopside can be expressed as follows:

$$CaMg(CO_3)_2 \text{ (dolomite)} + 2SiO_{2\,(aq)} \rightarrow CaMgSi_2O_6 \text{ (diopside)} + 2CO_{2\,(aq.)} \quad (1)$$

Stage II features the formation of fine and randomly oriented tremolite fibers, which contributes to the gem-quality. This stage is related to the retrograde hydrothermal alteration, during which various mineral replacement reactions had occurred. The main formation processes are as follows:

(1) The dolomitic marble was replaced by tremolite. This is a common explanation for the formation of nephrite [1].

$$5CaMg(CO_3)_2 \text{ (dolomite)} + 8SiO_{2\,(aq.)} + H_2O \rightarrow Ca_2Mg_5Si_8O_{22}(OH)_2 \text{ (tremolite)} + 3CaCO_3 \text{ (calcite)} + 7CO_{2\,(aq.)} \quad (2)$$

Both tremolite Tr-I and Tr-II can be formed in the reaction presented in Equation (2). The coarse-grained Tr-I formed were likely replaced by or recrystallized into later Tr-II (Figure 5b). The lower formation temperature (plus possible rapid cooling led by meteoric fluid incursion) likely deprived Tr-II of sufficient time to crystallize, which generated its characteristic micro-/cryptocrystalline texture.

(2) Stage I diopside was partially replaced by tremolite along its grain boundary and cracks (Figure 5e), either directly by Tr-II or by Tr-I and then Tr-II (Figure 6b). This process can be expressed in the following equation:

$$5CaMgSi_2O_6 \text{ (diopside)} + H_2O \rightarrow Ca_2Mg_5(Si_4O_{11})_2(OH)_2 \text{ (tremolite)} + 3Ca^{2+} + 6SiO_2 \quad (3)$$

Unlike most other nephrite deposits in the Hetian nephrite belt [17–21], pargasite is common in the Yinggelike nephrite. In some cases, Stage I diopside was replaced by pargasite, which was in turn replaced by Tr-II (Figure 6c). The Al and Na contents in pargasite are higher than those in diopside, which indicates that the process of diopside replacement by paragasite may have been the result of Al- and Na-rich hydrothermal alteration (Table 3).

(3) Isolated calcite grains in nephrites are suggested to be the products of Reaction (2) [1,18,19,22]. At Yinggelike, isolated calcite grains were not observed, and calcite occurs mainly as late-stage veins.

We suggest that the Stage II isolated calcite grains may have been completely taken up by the formation of fine-grained Tr-II, as expressed in the following equation:

$$2CaCO_3 \text{ (calcite)} + 5Mg^{2+} + 8SiO_2 + 6H_2O \rightarrow Ca_2Mg_5(Si_4O_{11})_2(OH)_2 \text{ (tremolite)} + 10H^+ + 2CO_2 \quad (4)$$

In the above equation, the tremolite formation requires Mg, which is likely derived from the dolomitic marble wallrocks leached by hydrothermal fluids, as first proposed by [75]. With the gradual consumption of calcite and production of H^+ ions, the hydrothermal fluid may have become more acidic. This likely led to further carbonate decomposition and produced an excess of Ca^{2+} ions, which may have formed other Ca-rich intermediate reaction products such as epidote, allanite, and prehnite (Figures 5f–g and 6e–g).

Stage III, characterized by calcite veins (Figure 5h), is related to the hydrothermal processes after nephrite formation.

5.3. Gemological Significance

The general preference of yellow color (associated with royalty) in many East Asian cultures has placed yellowish nephrite (and to a lesser extent white one, which is known "suet nephrite" in China) above nephrites of other colors [16]. Nevertheless, yellow nephrite is very rare, and is only reported in South Korea and the Qinghai and Liaoning provinces in China [22,64,76]. This makes yellow nephrite even more precious.

The factors that control the coloring of nephrites are still not fully understood, and thus we compare different yellowish-colored Yinggelike nephrite samples in this study. The colors of our samples can be broadly divided into a more greenish-yellow (Q-1, Q-2, Q-3) to a light-yellow (Q-4, Q-5, Q-6) group (Figure 4). EMPA and bulk-rock XRF/ICP-MS analyses show that all the samples were mainly composed of tremolite, and were almost free of impurities (Tables 2 and 4). Therefore, the main chromogenic mineral was tremolite. The high contents of chromogenic cations such as Fe^{2+}, Fe^{3+}, Mn and Ti suggest that they may be important in determining the Yinggelike nephrite color. Some workers have suggested that the green color of nephrite is mainly caused by the substitution of Fe^{2+} for Mg^{2+}, and the yellow color may be caused primarily by Fe^{3+} and Mn [15,18,19,77–80]. Titanium by itself is not known as a coloration factor for most terrestrial minerals, and as for nephrite, it affects the coloration of the blue-violet variety [79]. We therefore propose that Fe^{2+}, Fe^{3+}, and Mn are the main chromogenic factors for the nephrite. The greenish color of nephrite is related to Fe^{2+}, while the yellowish color is related to Fe^{3+} and Mn. In addition, the greenish-yellow group has markedly higher Fe^{3+}, and moderately higher Fe^{2+} and Mn contents than those of the light-yellow group, which indicates that Fe^{3+} is more important in giving the yellow color than Mn.

Generally, the finer the tremolite grains, the better the jade quality. Our petrographic observations indicate that the Yinggelike nephrite consists mainly of fine-grained tremolite (especially Tr-II), the aggregates of which give the nephrite a fine and compact texture (Figure 4). The randomly oriented bundles of twisted micro-fibers in the nephrite also result in its extreme toughness (Figure 5c). This toughness, combined with a relatively high Mohs hardness (5.8-6.1) (Table 1), makes the Yingglike nephrite a precious carving material.

6. Conclusions

In this paper, we presented the first systematic study of field geological occurrence, mineralogy, bulk-rock chemistry, and stable (O and H) isotope characteristics of nephrite from Yinggelike deposit in the Altyn Tagh Mountains (NW China). These data contribute to the overall understanding of the Yinggelike nephrite deposit.

The nephrite is dominated by tremolite, with minor minerals such as pargasite, diopside, epidote, allanite, prehnite, andesine, titanite, zircon, and calcite. Geochemical analyses shows that the nephrite samples have low bulk-rock Fe/(Fe + Mg) values and Cr, Co, and Ni concentrations. The chondrite-normalized REE patterns of most samples exhibit strong to moderate negative Eu anomalies with moderate LREE enrichment, nearly flat HREE, and low ΣREE contents. The mineral and geochemical characteristics of Yinggelike nephrite are consistent with those of the dolomite-related ones.

Mineral paragenetic relationships define three mineral crystallization stages for the Yinggelike nephrite: (1) diopside formed by prograde metasomatism; (2) nephrite jade formed by retrograde metasomatism and replacement of Stage I anhydrous minerals; (3) hydrothermal alteration after the nephrite formation. Features of transition metal contents indicate that the color of the Yinggelike nephrite is probably controlled by Fe^{2+}, Fe^{3+}, and Mn. Our new mineralogical and geochemical results on the Yinggelike nephrite provide better constraints on the formation of other nephrite deposits in the Altyn Tagh Mountains, and can facilitate future nephrite prospecting and research in the region.

Author Contributions: Conceptualization, Y.J. and G.S.; methodology, Y.J. and L.X.; validation, Y.J., G.S. and L.X.; formal analysis, Y.J.; investigation, Y.J., G.S. and L.X; resources, G.S. and X.L.; writing—original draft preparation, Y.J.; writing—review and editing, Y.J. and G.S; supervision, G.S.; project administration, G.S.; funding acquisition, G.S. All authors have read and agreed to the published version of the manuscript.

Acknowledgments: We appreciate Shoucheng Wang, Shiqi Wang (PKU), and Xiaojin Yu (USTB) for their kind field supports. We also thank Ji Zhang, Xu Li, and Kong Gao for the fruitful discussion. Constructive reviews by John Valley has greatly enhanced the manuscript. The Editors and two anonymous reviewers are thanked for their insightful comments, which significantly enhanced the manuscript.

Appendix A

Table A1. Chemical compositions of epidote, allanite, andesine, prehnite, titanite, and calcite in the Yinggelike nephrite, Xinjiang, NW China (wt.%).

Samples	H-5-1	H-5-2	H-3	H-3	H-3	H-3-1	H-3-2	H-1	H-3	H-5
Minerals	Allanite	Allanite	Epidote	Epidote	Andesine	Prehnite	Prehnite	Titanite	Calcite	Zircon
SiO_2	35.64	36.25	37.75	37.14	61.19	44.85	44.75	30.59	0.01	31.52
ZrO_2	n.d.	n.d.	n.d.	n.d.	n.d.	n.d.	n.d.	n.d.	n.d.	65.54
UO_3	n.d.	n.d.	n.d.	n.d.	n.d.	n.d.	n.d.	n.d.	n.d.	1.79
TiO_2	0.00	0.19	0.08	0.02	0.00	0.05	0.02	36.85	0.00	n.d.
Al_2O_3	23.22	24.73	28.05	28.30	24.32	22.86	23.53	2.13	0.00	n.d.
FeO	7.24	5.11	3.78	3.40	0.13	0.17	0.22	0.04	0.03	n.d.
Cr_2O_3	0.00	0.00	0.02	0.04	0.05	0.00	0.00	0.05	0.01	n.d.
MnO	0.32	0.15	0.05	0.00	0.00	0.00	0.01	0.02	0.06	n.d.
MgO	0.73	2.11	0.52	0.44	0.04	0.14	0.07	0.02	0.25	n.d.
CaO	20.11	18.14	25.72	25.71	7.27	27.23	27.01	29.28	55.88	n.d.
Na_2O	0.01	0.00	0.04	0.00	7.23	0.01	0.05	0.01	0.06	1.18
K_2O	0.02	0.02	0.00	0.02	0.12	0.02	0.01	0.03	0.00	n.d.
NiO	0.00	0.00	0.00	0.05	0.01	0.02	0.03	0.01	0.03	n.d.
La_2O_3	2.58	2.65	n.d.	n.d.	n.d.	n.d.	n.d.	n.d.	n.d.	n.d.
Ce_2O_3	5.85	5.96	n.d.	n.d.	n.d.	n.d.	n.d.	n.d.	n.d.	n.d.
Pr_2O_3	0.43	1.20	n.d.	n.d.	n.d.	n.d.	n.d.	n.d.	n.d.	n.d.
Nd_2O_3	2.03	2.37	n.d.	n.d.	n.d.	n.d.	n.d.	n.d.	n.d.	n.d.
Total	98.19	98.88	95.99	95.12	100.36	95.35	95.70	99.02	56.34	100.03

0.00—concentration below the detection limit; n.d.—not detected.

Table A2. Trace element compositions of the nephrite samples collected from the Yinggelike open pit, Xinjiang, NW China (ppm).

Samples	H-1	H-2	H-3	H-4	H-5	Q-1	Q-2	Q-3	Q-4	Q-5	Q-6
Cr	11.53	34.68	34.58	12.02	0.81	4.42	8.81	9.14	8.44	2.03	6.03
Co	1.45	2.91	1.82	1.10	1.48	1.82	1.87	2.14	1.55	1.83	1.56
Ni	5.15	10.29	20.15	6.00	0.52	3.44	3.51	3.72	4.12	3.19	3.46
Y	2.50	3.22	1.71	1.95	123.18	1.50	2.79	1.91	1.69	2.76	1.28
La	1.62	1.95	0.85	2.23	150.14	1.19	0.75	0.96	0.70	0.55	0.40
Ce	3.54	3.86	1.80	4.78	310.13	2.25	1.37	1.94	1.37	1.13	0.81
Pr	0.42	0.43	0.21	0.59	34.83	0.25	0.15	0.22	0.17	0.15	0.11
Nd	1.55	1.54	0.73	2.05	127.53	0.95	0.60	0.84	0.74	0.67	0.49
Sm	0.38	0.36	0.17	0.45	27.55	0.19	0.13	0.16	0.20	0.20	0.15
Eu	0.05	0.03	0.02	0.03	0.33	0.04	0.02	0.03	0.02	0.02	0.01
Gd	0.39	0.35	0.18	0.38	25.27	0.17	0.15	0.14	0.19	0.21	0.15
Tb	0.07	0.06	0.03	0.06	4.26	0.04	0.04	0.02	0.04	0.06	0.03
Dy	0.39	0.36	0.21	0.30	24.33	0.24	0.33	0.18	0.28	0.39	0.19
Ho	0.08	0.08	0.05	0.06	4.64	0.04	0.07	0.05	0.05	0.08	0.04
Er	0.22	0.22	0.13	0.16	12.43	0.12	0.22	0.24	0.15	0.25	0.10
Tm	0.03	0.03	0.02	0.02	1.87	0.02	0.04	0.09	0.03	0.05	0.02
Yb	0.20	0.18	0.11	0.12	11.28	0.12	0.21	0.88	0.16	0.33	0.10
Lu	0.03	0.02	0.02	0.02	1.64	0.02	0.03	0.14	0.02	0.05	0.01
ΣREE	8.97	9.47	4.53	11.25	736.23	5.64	4.11	5.89	4.12	4.14	2.16
ΣREE+Y	11.47	12.69	6.24	13.20	859.41	7.14	6.90	7.80	5.81	6.9	3.89
Eu/Eu*	0.36	0.25	0.39	0.21	0.04	0.67	0.44	0.60	0.31	0.30	0.20
LREE/HREE	5.34	6.31	5.16	9.05	7.59	6.32	2.77	2.39	3.48	1.92	3.08
$(La/Sm)_N$	2.75	3.50	3.23	3.20	3.52	4.04	3.72	3.87	2.26	1.78	1.72
$(Gd/Yb)_N$	1.61	1.61	1.35	2.62	1.85	1.17	0.59	0.13	0.98	0.53	1.24

$Eu/Eu^* = 2Eu_N/(Sm_N + Gd_N)$.

References

1. Harlow, G.E.; Sorensen, S.S. Jade (nephrite and jadeitite) and serpentinite: Metasomatic connections. *Int. Geol. Rev.* **2005**, *47*, 113–146. [CrossRef]
2. Wen, G.; Jing, Z.C. Chinese Neolithic Jade: A Preliminary Geoarchaeological Study. *Geoarchaeology* **1992**, *7*, 251–275. [CrossRef]
3. D'Amico, C.; Starnini, E.; Gasparotto, G.; Ghedini, M. Eclogites, jades and other HP-metaophiolites employed for prehistoric polished stone implements in Italy and Europe. *Period. Mineral.* **2004**, *73*, 17–42.
4. Middleton, A. Jade–geology and mineralogy. In *Gems*; O'Donoghue, M., Ed.; Elsevier Ltd.: Oxford, UK, 2006; pp. 332–354.
5. Losey, R.J.; Bazaliiskii, V.I.; Garvie-Lok, S.; Germonpré, M.; Leonard, J.A.; Allen, A.L.; Anne Katzenberg, M.; Sablin, M.V. Canids as persons: Early Neolithic dog and wolf burials, Cis-Baikal, Siberia. *J. Anthropol. Archaeol.* **2011**, *30*, 174–189. [CrossRef]
6. Tsydenova, N.; Morozov, M.V.; Rampilova, M.V.; Vasil'ev, Y.A.; Matveeva, O.P.; Konovalov, P.B. Chemical and spectroscopic study of nephrite artifacts from Transbaikalia, Russia: Geological sources and possible transportation routes. *Quat. Int.* **2015**, *355*, 114–125. [CrossRef]
7. Kostov, R.I. Gemmological significance of the prehistoric Balkan "nephrite culture" (cases from Bulgaria). *Annu. Univ. Min. Geol.* **2005**, *48*, 91–94.
8. Kostov, R.I.; Protochristov, C.; Stoyanov, C.; Csedreki, L.; Simon, A.; Szikszai, Z.; Uzonyi, I.; Gaydarska, B.; Chapman, J. Micro-PIXE geochemical fingerprinting of nephrite neolithic artifacts from Southwest Bulgaria. *Geoarchaeology* **2012**, *27*, 457–469. [CrossRef]
9. Péterdi, B.; Szakmány, G.; Judik, K.; Dobosi, G.; Kasztovszky, Z.; Szilágyi, V.; Maróti, B.; Bendõ, Z.; Gil, G. Petrographic and geochemical investigation of a stone adze made of nephrite from the balatonoszöd-Temetoi dulo site (Hungary), with a review of the nephrite occurrences in Europe (especially in Switzerland and in the Bohemian Massif). *Geol. Q.* **2014**, *58*, 181–192. [CrossRef]
10. Chen, T.-H.; Calligaro, T.; Pagès-Camagna, S.; Menu, M. Investigation of Chinese archaic jade by PIXE and μRaman spectrometry. *Appl. Phys. A* **2004**, *79*, 177–180. [CrossRef]
11. Sax, M.; Meeks, N.D.; Michaelson, C.; Middleton, A.P. The identification of carving techniques on Chinese Jade. *J. Archaeol. Sci.* **2004**, *31*, 1413–1428. [CrossRef]

12. Derevianko, A.P.; Olsen, J.W.; Tseveendorj, D.; Gladyshev, S.A.; Nokhrina, T.I.; Tabarev, A.V. New insights into the archaeological record at Chikhen Agui Rockshelter (Mongolia). *Archaeol. Ethnol. Anthropol. Eurasia* **2008**, *34*, 2–12. [CrossRef]
13. Fournelle, J.; Law, R.; Konishi, H. A nephrite jade amulet from Harappa: Implications for long-distance contacts in the Harappan Period. In Proceedings of the 39th Annual Conference on South Asia, New York, NY, USA, 14–17 October 2010.
14. Hung, H.-C.; Iizuka, Y.; Bellwood, P.; Nguyen, K.D.; Bellina, B.; Silapanth, P.; Dizon, E.; Santiago, R.; Datan, I.; Manton, J.H. Ancient jades map 3,000 years of prehistoric exchange in Southeast Asia. *Proc. Natl. Acad. Sci. USA* **2007**, *104*, 19745–19750. [CrossRef] [PubMed]
15. Wilkins, C.J.; Tennant, W.C.; Williamson, B.E.; McCammon, C.A. Spectroscopic and related evidence on the coloring and constitution of New Zealand jade. *Am. Mineral.* **2003**, *88*, 1336–1344. [CrossRef]
16. Tang, Y.L.; Chen, B.Z.; Jiang, R.H. *Chinese Hetian Nephrite*; Xinjiang People's Publishing House: Xinjiang, China, 1994; pp. 103–206. (In Chinese with English abstract)
17. Liu, Y.; Deng, J.; Shi, G.H.; Lu, T.J.; He, H.Y.; Ng, Y.-N.; Shen, C.H.; Yang, L.Q.; Wang, Q.F. Chemical Zone of Nephrite in Alamas, Xinjiang, China. *Resour. Geol.* **2010**, *60*, 249–259. [CrossRef]
18. Liu, Y.; Deng, J.; Shi, G.H.; Yui, T.-F.; Zhang, G.B.; Abuduwayiti, M.; Yang, L.Q.; Sun, X. Geochemistry and petrology of nephrite from Alamas, Xinjiang, NW China. *J. Asian Earth Sci.* **2011**, *42*, 440–451. [CrossRef]
19. Liu, Y.; Deng, J.; Shi, G.H.; Sun, X.; Yang, L.Q. Geochemistry and petrogenesis of placer nephrite from Hetian, Xinjiang, Northwest China. *Ore Geol. Rev.* **2011**, *41*, 122–132. [CrossRef]
20. Liu, Y.; Zhang, R.Q.; Abuduwayiti, M.; Wang, C.; Zhang, S.P.; Shen, C.H.; Zhang, Z.Y.; He, M.Y.; Zhang, Y.; Yang, X.D. SHRIMP U–Pb zircon ages, mineral compositions and geochemistry of placer nephrite in the Yurungkash and Karakash River deposits, West Kunlun, Xinjiang, Northwest China: Implication for a magnesium skarn. *Ore Geol. Rev.* **2016**, *72*, 699–727. [CrossRef]
21. Liu, Y.; Zhang, R.Q.; Zhang, Z.Y.; Shi, G.H.; Zhang, Q.C.; Abuduwayiti, M.; Liu, J.H. Mineral inclusions and SHRIMP U–Pb dating of zircons from the Alamas nephrite and granodiorite: Implications for the genesis of a magnesian skarn deposit. *Lithos* **2015**, *212–215*, 128–144. [CrossRef]
22. Yu, H.Y.; Wang, R.C.; Guo, J.C.; Li, J.G.; Yang, X.W. Study of the minerogenetic mechanism and origin of Qinghai nephrite from Golmud, Qinghai, Northwest China. *Sci. China Earth Sci.* **2016**, *59*, 1597–1609. [CrossRef]
23. Yin, Z.W.; Jiang, C.; Santosh, M.; Chen, Y.M.; Bao, Y.; Chen, Q.L. Nephrite Jade from Guangxi Province, China. *Gems Gemol.* **2014**, *50*, 228–235. [CrossRef]
24. Zhang, C.; Yu, X.; Jiang, T. Mineral association and graphite inclusions in nephrite jade from Liaoning, northeast China: Implications for metamorphic conditions and ore genesis. *Geosci. Front.* **2019**, *10*, 425–437. [CrossRef]
25. Burtseva, M.V.; Ripp, G.S.; Posokhov, V.F.; Zyablitsev, Y.A.; Murzintseva, A.E. The sources of fluids for the formation of nephritic rocks of the southern folded belt of the Siberian. *Craton. Dokl. Earth Sci.* **2015**, *460*, 82–86. [CrossRef]
26. Burtseva, M.V.; Ripp, G.S.; Posokhov, V.F.; Murzintseva, A.E. Nephrites of East Siberia: Geochemical features and problems of genesis. *Russ. Geol. Geophys.* **2015**, *56*, 402–410. [CrossRef]
27. Leaming, S.F. *Jade in Canada*; Geological Survey of Canada: Ottawa, ON, Canada, 1978; pp. 1–59.
28. Adams, C.J.; Beck, R.J.; Campbell, H.J. Characterisation and origin of New Zealand nephrite jade using its strontium isotopic signature. *Lithos* **2007**, *97*, 307–322. [CrossRef]
29. Cox, S.C.; Nibourel, L. Bedload composition, transport and modification in rivers of Westland, New Zealand, with implications for the distribution of alluvial pounamu (jade). *N. Z. J. Geol. Geophys.* **2015**, *58*, 154–175. [CrossRef]
30. Cooper, A.F. Nephrite and Metagabbro in the Haast Schist at Muddy Creek, Northwest Otago, New Zealand. *N. Z. J. Geol. Geophys.* **1995**, *38*, 325–332. [CrossRef]
31. Yui, T.-F.; Kwon, S.-T. Origin of a Dolomite-Related Jade Deposit at Chuncheon, Korea. *Econ. Geol.* **2002**, *97*, 593–601. [CrossRef]
32. Gil, G.; Barnes, J.D.; Boschi, C.; Gunia, P.; Szakmány, G.; Bendő, Z.; Raczynski, P.; Petérdi, B. Origin of serpentinite-related nephrite from Jordanów and adjacent areas (SW Poland) and its comparison with selected nephrite occurrences. *Geol. Q.* **2015**, *59*, 457–472. [CrossRef]

33. Gil, G. Petrographic and microprobe study of nephrites from Lower Silesia (SW Poland). *Geol. Q.* **2013**, *57*, 395–404. [CrossRef]
34. Yui, T.-F.; Yeh, H.-W.; Lee, C.W. Stable isotope studies of nephrite deposits from Fengtien, Taiwan. *Geochim. Cosmochim. Acta* **1988**, *52*, 593–602. [CrossRef]
35. Tian, S. Gemological Characteristics and Mineral Typomorphic Characteristics of Green Xinjiang Manasi Nephrite. Master's Thesis, Shijiazhuang University of Economics, Shijiazhuang, China, 2014.
36. Gil, G.; Barnes, J.D.; Boschi, C.; Gunia, P.; Raczynski, P.; Szakmány, G.; Bendő, Z.; Petérdi, B. Nephrite from Złoty Stok (Sudetes, SW Poland): Petrological, geochemical, and isotopic evidence for a dolomite-related origin. *Can. Mineral.* **2015**, *53*, 533–556. [CrossRef]
37. Wu, L.J. Study on Gemological and Mineralogical Characteristics and Genesis of Nephrite in Tiantai, Qiemo, Xinjiang. Master's Thesis, China University of Geosciences (Beijing), Beijing, China, 2016.
38. Gao, K. Study on the Metallogenic Mechanism of Tashisayi Nephrite from Xinjiang. Unpublished. Ph.D. Thesis, China University of Geosciences (Beijing), Beijing, China, 2018.
39. Gao, K.; Shi, G.H.; Wang, M.L.; Xie, G.; Wang, J.; Zhang, X.C.; Fang, T.; Lei, W.Y.; Liu, Y. The Tashisayi nephrite deposit from South Altyn Tagh, Xinjiang, northwest China. *Geosci. Front.* **2019**, *10*, 1597–1612. [CrossRef]
40. Han, H.W.; Zhang, C.J.; Wang, H.; Mou, L.X.; Wei, Y.; Bai, D. Metallogenic geological characteristics and prospecting prospects of nephrite deposits in the middle and western Altyn Tagh. *Acta Mineral. Sin.* **2011**, *31*, 952–953. (In Chinese)
41. Local Chronicle Compilation Committee. *Ruoqiang County Local Chronicle*; Chinese Literature and History Press: Ruoqiang, China, 2017. (In Chinese)
42. Yu, M. *Chinese Jade Yearbook*; Beijing Science Press: Beijing, China, 2017. (In Chinese)
43. Guangxi Regional Geological Survey Institute. *Geological Map of the Washixia, Xinjiang, China, Scale 1: 250,000*; Guangxi Regional Geological Survey Institute Press: Guilin, China, 2003. (In Chinese)
44. Xinjiang Geology and Mineral Bureau third Geology Brigade. *Geological Survey of Yingelike Nephrite Mine, Ruoqiang County, Xinjiang, China*; Xinjiang Geology and Mineral Bureau Third Geology Brigade Press: Kuerle, China, 2002. (In Chinese)
45. Wang, C.; Liu, L.; Xiao, P.X.; Cao, Y.T.; Yu, H.Y.; Meert, J.G.; Liang, W.T. Geochemical and geochronologic constraints for Paleozoic magmatism related to the orogenic collapse in the Qimantagh-South Altyn region, northwestern China. *Lithos* **2014**, *202–203*, 1–20. [CrossRef]
46. Wang, C.; Liu, L.; Yang, W.Q.; Zhu, X.H.; Cao, Y.T.; Kang, L.; Chen, S.F.; Li, R.S.; He, S.P. Provenance and ages of the Altyn Complex in Altyn Tagh: Implications for the early Neoproterozoic evolution of northwestern China. *Precambrian Res.* **2013**, *230*, 193–208. [CrossRef]
47. Liu, L.; Sun, Y.; Xiao, P.X.; Che, Z.C.; Luo, J.H.; Chen, D.L.; Wang, Y.; Zhang, A.D.; Chen, L.; Wang, Y.H. Discovery of ultrahigh-pressure magnesite-bearing garnet lherzolite (>3.8 GPa) in the Altyn Tagh, Northwest China. *Chin. Sci. Bull.* **2002**, *47*, 881–886. [CrossRef]
48. Cowgill, E.; Yin, A.; Harrison, T.M.; Wang, X.F. Reconstruction of the Altyn Tagh fault based on U-Pb geochronology: Role of back thrusts, mantle sutures, and heterogeneous crustal strength in forming the Tibetan Plateau. *J. Geophys. Res. Solid Earth* **2003**, *108*, B7. [CrossRef]
49. Zhang, J.X.; Zhang, Z.M.; Xu, Z.Q.; Yang, J.S.; Cui, J.W. Petrology and geochronology of eclogites from the western segment of the Altyn Tagh, Northwestern China. *Lithos* **2001**, *56*, 187–206. [CrossRef]
50. Mohsen, M.-D. Distant vision method. In *Dictionary of Gems and Gemology*, 3rd ed.; Springer: Berlin/Heidelberg, Germany, 2009; p. 271. [CrossRef]
51. Mohsen, M.-D. Hydrostatic weighing of specific gravity. In *Dictionary of Gems and Gemology*, 3rd ed.; Springer: Berlin/Heidelberg, Germany, 2009; p. 440. [CrossRef]
52. Meng, X.Z.; Zhao, M.F. The legal measurement unit of Vickers hardness and its conversion. *J. Gems Gemmol.* **2007**, *2*, 52.
53. GB/T 14506. *30-2010. National Standards of the People's Republic of China: Methods for Chemical Analysis of Silicate Rocks-Part 30: Determination of 44 Elements*; Standardization Administration of China Press: Beijing, China, 2010. (In Chinese)
54. Clayton, R.N.; Mayeda, T.K. The use of bromine pentafluoride in the extraction of oxygen from oxides and silicates for isotopic analysis. *Geochim. Cosmochim. Acta* **1963**, *27*, 43–52. [CrossRef]
55. Liu, H.B.; Jin, G.S.; Li, J.J.; Han, J.; Zhang, J.F.; Zhang, J.; Zhong, F.W.; Guo, D.Q. Determination of stable

isotope composition in uranium geological samles. *World Nucl. Geosci.* **2013**, *3*, 174–179. (In Chinese with English abstract) [CrossRef]

56. Friedman, I. Deuterium content of natural waters and other substances. *Geochim. Cosmochim. Acta* **1953**, *4*, 89–103. [CrossRef]
57. Adamo, I.; Bocchio, R. Nephrite Jade from Val Malenco, Italy: Review and Update. *Gems Gemol.* **2013**, *49*, 98–106. [CrossRef]
58. Leake, B.E.; Woolley, A.R.; Arpes, C.E.S.; Birch, W.D.; Gilbert, M.C.; Grice, J.D.; Hawthorne, F.C.; Kato, A.; Kisch, H.J.; Krivovichev, V.G.; et al. Nomenclature of Amphiboles; Report of the Subcommittee on Amphiboles of the International Mineralogical Association Commission on New Minerals and Mineral Names. *Mineral. Mag.* **1997**, *61*, 295–310. [CrossRef]
59. Morimoto, N.; Fabries, J.; Ferguson, A.K.; Ginzburg, I.V.; Ross, M.; Seifert, F.A.; Zussman, J.; Aoki, K.; Gottardi, G. Nomenclature of pyroxenes. *Mineral. Petrol.* **1988**, *39*, 55–76. [CrossRef]
60. Sun, S.-S.; McDonough, W.F. Chemical and isotopic systematics of oceanic basalts: Implications for mantle composition and processes. *Geol. Soc. Spec. Publ.* **1989**, *42*, 313–345. [CrossRef]
61. Liu, X.F.; Zhang, H.Q.; Liu, Y.; Zhang, Y.; Li, Z.J.; Zhang, J.H.; Zheng, F. Mineralogical Characteristics and Genesis of Green Nephrite from the World. *Rock Miner. Anal.* **2018**, *23*, 479–489. (In Chinese) [CrossRef]
62. Kim, S.J.; Lee, D.J.; Chang, S. A mineralogical and gemological characterization of the Korean jade from Chuncheon, Korea. *J. Geol. Soc. Korea* **1986**, *22*, 278–288. (In Korean)
63. Noh, J.H.; Yu, J.-Y.; Choi, J.B. Genesis of nephrite and associated calc-silicate minerals in Chuncheon area. *J. Geol. Soc. Korea* **1993**, *29*, 199–224. (In Korean)
64. Pei, X.X. Study on the Chuncheon Nephrite Deposit, Korea. Master's Thesis, China University of Geosciences (Beijing), Beijing, China, 2012.
65. Li, H.J.; Cai, Y.T. Study on Characteristics of Nephrite from Liyang, Jiangsu Province. *J. Gems Gemmol.* **2008**, *10*, 16–19. (In Chinese with English abstract) [CrossRef]
66. Simandl, G.J.; Riveros, C.P.; Schiarizza, P. Nephrite (Jade) Deposits, Mount Ogden Area, Central British Columbia (NTS 093N 13W). *Geol. Fieldwork* **2000**, *1999*, 339–347.
67. Mi, L.L. Green Nephrite Jade from Canada. *J. Gems Gemmol.* **2003**, *1*, 10–13, (In Chinese with English abstract). [CrossRef]
68. Zhao, Y.Y. Study of Gemological Mineralogy of Green Nephrite from New Zealand. Master's Thesis, China University of Geosciences (Beijing), Beijing, China, 2015.
69. Chen, C.X.; Chen, S.Y.; Li, G.G.; Liu, D.Y. Comparative research on gemological characteristics of Hualian nephrite from Taiwan. *Acta Petrol. Mineral.* **2014**, *S2*, 35–40. (In Chinese with English abstract)
70. Yuan, M.; Wu, R.H.; Zhang, J.H. A study of gemological and color influencing ions of green nephrite from Ospinsk No. 7 mining area, Russia. *Acta Petrol. Mineral.* **2014**, *S1*, 48–54. (In Chinese with English abstract)
71. Prokhor, S.A. The genesis of nephrite and emplacement of the nephrite bearing ultramafic complexes of East Sayan. *Int. Geol. Rev.* **1991**, *33*, 290–300. [CrossRef]
72. Zhang, Z.W.; Gan, F.X.; Cheng, H.S. PIXE analysis of nephrite minerals from different deposits. *Nucl. Instrum. Methods Phys. Res., Sect. B* **2011**, *269*, 460–465. [CrossRef]
73. Siqin, B.; Qian, R.; Zhou, S.J.; Gan, F.X.; Dong, M.; Hua, Y.F. Glow discharge mass spectrometry studies on nephrite minerals formed by different metallogenic mechanisms and geological environments. *Int. J. Mass Spectrom.* **2012**, *309*, 206–211. [CrossRef]
74. Siqin, B.; Qian, R.; Zhuo, S.J.; Gao, J.; Jin, J.; Wen, Z.Y. Studies of rare earth elements to distinguish nephrite samples from different deposits using direct current glow discharge mass spectrometry. *J. Anal. At. Spectrom.* **2014**, *29*, 2064–2071. [CrossRef]
75. Men, Y. The Causes and Genesis Study of Nephrite in Alamas, Yutian, Xinjiang, China. Master's Thesis, China University of Geosciences (Beijing), Beijing, China, 2014.
76. Zhou, Z.H.; Feng, J.R. A petrological and mineralogical comparison between Xinjiang nephrite and Xiuyan nephrite. *Acta Petrol. Mineral.* **2010**, *29*, 331–340. (In Chinese with English abstract)
77. Feng, X.Y.; Zhang, Y.; Lu, T.J.; Zhang, H. Characterization of Mg and Fe contents in nephrite using Raman spectroscopy. *Gems Gemol.* **2017**, *53*, 204–212. [CrossRef]

78. Fritsch, E.; Rossman, G.R. An update on color in gems. Part 1: Introduction and colors caused by dispersed metal ions. *Gems Gemol.* **1987**, *23*, 126–139. [CrossRef]
79. Yu, H.Y.; Wang, R.C.; Guo, J.C.; Li, J.G.; Yang, X.W. Color–inducing elements and mechanisms in nephrites from Golmud, Qinghai, NW China: Insights from spectroscopic and compositional analyses. *J. Mineral. Petrol. Sci.* **2016**, *111*, 313–325. [CrossRef]
80. Guo, L.H.; Han, J.Y. The IR analyses of M1 and M3 cation occupation of Hetian jade, Manasi green jade and Xiuyanold jade. *Acta Petrol. Mineral.* **2002**, *21*, 68–71. (In Chinese)

2

Crystallinity and Play-of-Colour in Gem Opal with Digit Patterns from Wegel Tena, Ethiopia

Kehan Zhao and Feng Bai *

School of Gemmology, China University of Geosciences (Beijing), Beijing 100083, China; 2109180006@cugb.edu.cn

* Correspondence: baifeng@cugb.edu.cn

Abstract: A typical feature of Wegel Tena opal is the "digit pattern". This pattern consists of two parts, columns and matrix, with different colours, transparency or play-of-colour effect, which is still unexplained. This study aims at investigating the various parts of the digit pattern using different spectroscopic methods, and scanning and transmission electron microscopy (SEM and TEM). The band at 780 cm^{-1} on the Fourier transform infrared (FTIR) spectrum is correlated to the symmetric stretching vibration of Si–O. The bands at 1085, 895, 785 and 3600 cm^{-1} on Raman spectra indicate that Wegel Tena opal is opal-CT. Comparison of the relative intensity of the Raman signals around 360 cm^{-1} indicates that the microcrystalline opal on the top of the sample contains a higher amount of tridymite-like structural units, and the tridymite-type regions in the matrix contain a higher degree of structural defects. Silica spheres in the columns tend to be smaller and better ordered than in the matrix. The diameter of the silica spheres (d = 80–500 nm) or agglomerates (d = 200–580 nm) in Wegel Tena opal satisfies the conditions of diffraction of visible light, and the thickness of the silica layer (h = 120–200 nm) satisfies the conditions for film interference.

Keywords: Ethiopia; opal; digit patterns; play-of-colour; FTIR; Raman spectroscopy; SEM; TEM; opal-CT

---

1. Introduction

Opal is an amorphous hydrated silica to micro-crystalline silica mineral, with a chemical formula of $SiO_2 \cdot nH_2O$ [1,2]. According to Jones and Segnit, opals are divided into opal-C, opal-CT and opal-A, based on X-ray diffraction (XRD). Opal-C is a relatively well-ordered α-cristobalite with minor evidence of tridymite; opal-CT is a disordered α-cristobalite with α-tridymite type stacking; opal-A is amorphous. In recent years, however, this classification has been questioned, and the nature of opal-CT remains far from conclusive [1]. Wilson explained the opal-CT structure by a tridymite-dominant paracrystalline model [3]. Recent work by Curtis et al. [4] found that opal-CT appears to be a paracrystalline form of silica that has some structural characteristics of cristobalite and tridymite.

Other techniques, such as infrared spectroscopy [5–7] and Raman spectroscopy [8–14], have also been used to distinguish different types of opal. Furthermore, Raman spectroscopy has been considered a method to distinguish the complex structural relationships between different types of opal [4,8,10,13].

Silica spheres can be observed in some of the opals using an electron microscope. When the size and arrangement of the spheres meet certain conditions, a special optical effect, play-of-colour, can appear [1,2,13]. Gem opal with play-of-colour is called "precious opal"; conversely, opal without play-of-colour is called "common opal". Play-of-colour is independent of the structural features revealed by XRD. The atomic structure of the silica in the spheres of opal-A and opal-CT is clearly different, although both of them can show play-of-colour [4]. Precious opals are mostly mined in Australia, Mexico, Brazil and Ethiopia. In the early 1990s, opal was discovered in Mezezo,

Shewa Province, Ethiopia. The body colour was mostly orange to red or "chocolate" brown [15–18]. In 2008, high-quality precious opal was discovered in Wegel Tena, in the northeast of Wollo Province, Ethiopia [19–21]. The deposit occurs in an Oligocene volcano-sedimentary sequence of alternating basalt and rhyolitic ignimbrite layers [22]. Opals mined at Wegel Tena [18,21] and Mezezo [16,17,21] always show a "digit pattern", a macroscopic finger-like structure. This pattern consists of almost parallel vertical columns, separated by a homogeneous matrix of a different colour, transparency or play-of-colour [21,23]. Rondeau et al. [23] built a model of digit formation based on observation containing four basic steps: (1) The occurrence of sedimentation and dehydration of silica spheres, which take on a columnar structure; (2) polygonization of the columns; (3) water inflowing, altering the columns by erosion along the polygonal boundaries; (4) new silica-rich fluid filling the interstice and settling, the formation of silica lepispheres and dehydration.

This study analyses the crystallinity and microstructure in opals from Wegel Tena, finds the difference between columns and matrix, and explains the cause of play-of-colour in Wegel Tena opal by linking the microstructure with the various effects.

2. Materials and Methods

We studied 11 opals from Wegel Tena, Wollo Province, Ethiopia. Six of them were cabochon (Figure 1), while five were rough samples with surrounding rocks (Figure 2). Each sample showed play-of-colour. Opals 01, 02, 03, 04, 05, 06, 09 and 11 showed digit patterns. All cabochon opals had polished surfaces, and polishing had no effect on the spectral data.

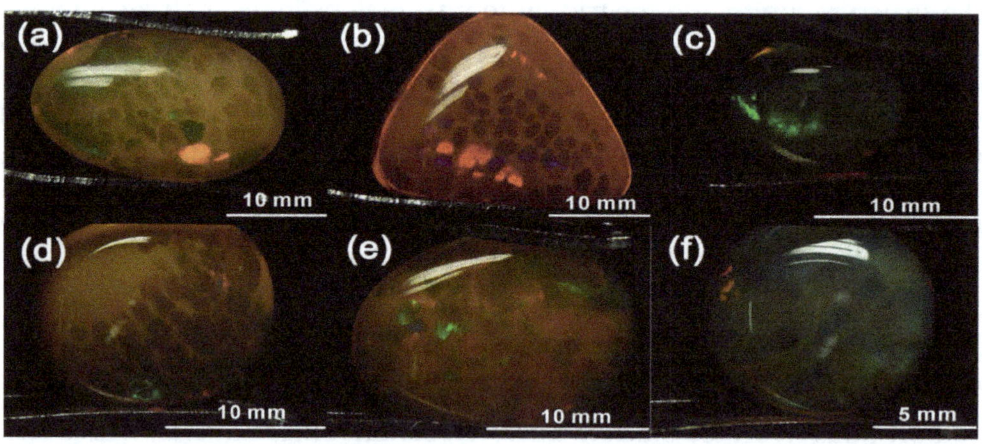

Figure 1. Cabochon samples. (**a**) Opal-01; (**b**) Opal-02; (**c**) Opal-03; (**d**) Opal-04; (**e**) Opal-05; (**f**) Opal-06.

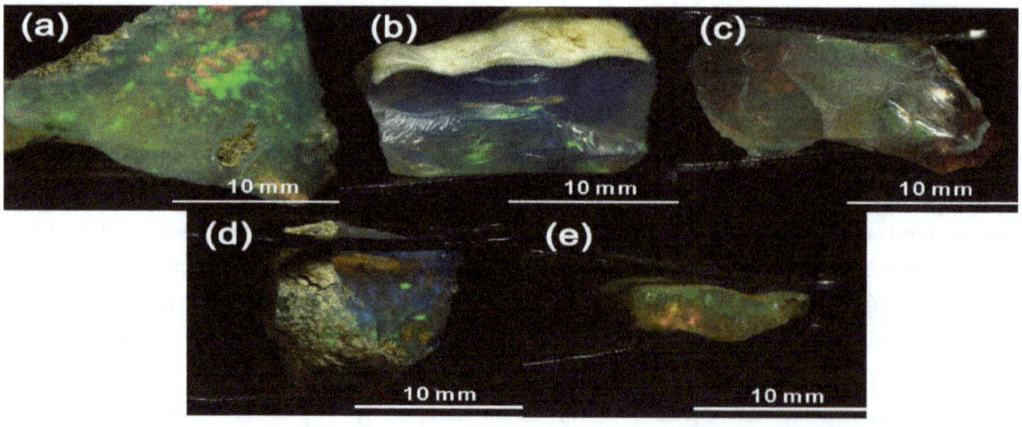

Figure 2. Rough samples with surrounding rock. (**a**) Opal-07; (**b**) Opal-08; (**c**) Opal-09; (**d**) Opal-10; (**e**) Opal-11.

We used a refractometer to measure the index of refraction (RI) with an optical contact liquid of 1.78 RI. Ultraviolet (UV) fluorescence was observed using a UV 240 shortwave (254 nm) and longwave (365 nm) UV lamp. Samples were observed using a gemmological microscope with a maximum magnification of 40, using reflected and transmitted light.

FTIR and Raman spectroscopy were used to identify the silica forms. These are non-destructive and do not require preparation of the samples [24–26]. The work was carried out at the Experimental Teaching Center of Jewellery and Mineral Materials, China University of Geosciences, Beijing.

Infrared spectra in the mid-infrared range (2000 to 400 cm^{-1}) were recorded for all 11 samples on a polished surface in reflection mode, using a Tensor 27 FTIR (Bruker, Ettlingen, Baden-Württemberg, Germany). Thirty scans were collected at a resolution of 4 cm^{-1}.

Micro-Raman spectra were measured on seven samples (Opals 01, 02, 03, 04, 05, 06 and 08) and collected in the 100 to 4000 wavenumber (cm^{-1}) region using a LabRAM HR Evolution confocal Raman spectrometer (Horiba, Kyoto, Kyoto-fu Japan), with a 600 lines/mm grating (500 nm blaze) and an excitation wavelength of 532 nm at a power of 30 mW. Each spectrum was scanned two times for 5 s, with a 100 μm slit and a 100 μm laser hole. ICS correction (instrument specific intensity correction) was on. Samples were observed using a 50x VIS LWD (visible and long working distance) objective (Horiba, Kyoto, Kyoto-fu, Japan). Two to four points were selected for each sample. For the samples with strong transparency contrast, the Raman spectra were collected in the matrix and columns respectively. All spectra were baseline corrected by LabSpec software (version 5). Baseline correction did not affect band positions and their relative intensity ratios. The broad band around 360 cm^{-1} was selected from several spectra, and fitted with Voigt functions using Origin 9 software (version 9.0.0.45).

The micromorphology of opals was observed on surfaces and fresh-fractured material. Samples were coated with a thin Pt film. High-resolution images of six samples (Opals 01, 03, 04, 07, 08 and 09) were obtained with a SUPRA 55 Sapphire scanning electron microscope (SEM) at the Institute of Earth Science in China University of Geosciences (Beijing), equipped with a field-effect electron gun, at an acceleration voltage of 15 kV. The test temperature was 22 °C, and the humidity was 45%.

Three samples (Opal-01, 05, 06) were prepared into flakes by mechanical grinding and ion thinning to electron transparency. The test surface of the flake was vertical to the length of columns. The microstructure was investigated on an H-8100 transmission electron microscope (TEM) provided by the School of Earth Sciences and Resources of China University of Geosciences (Beijing). The accelerating voltage was 200 kV, and the observation magnification was from 900 to 10,000.

3. Results

3.1. Appearance and Gemmological Properties

The appearance and gemmological properties of the samples are presented in Table 1. The body colours are brown, yellowish-brown, orange-brown, yellow and colourless. The digit patterns can appear on samples of different colours and sizes. The columns and matrix are of similar hue on the same sample, but the transparency is different. Columns are transparent, and the matrices are translucent or semi-translucent. The refractive index (RI) values of the samples range from 1.430 to 1.465 and are typical for opal, but slightly higher than the data measured by Rondeau et al. [18]. Fluorescence under UV is white or bluish-white luminescence, and fluorescence under LWUV (longwave ultraviolet) is stronger than under SWUV (shortwave ultraviolet).

The cross-sectional diameter of columns is about 0.2 to 5 mm, and the shape is from angular polygonal to oval or round. The cross-sectional shape of columns tends to be similar on the same sample (e.g., Opal-02, see Figure 3), but large variations also occur (e.g., Opal-05, see Figure 4). In the samples with the characteristic digit patterns (e.g., Opal-02, as shown in Figure 3), the columns are parallel to each other, and the ends of the columns are rounded. According to the model of digit formation [23], the honeycomb-like side of Opal-02 is the bottom, and the round-end of the column covered by common opal is the top. In the sample Opal-09 with weak digit patterns (Figure 5), the matrix is dendritic.

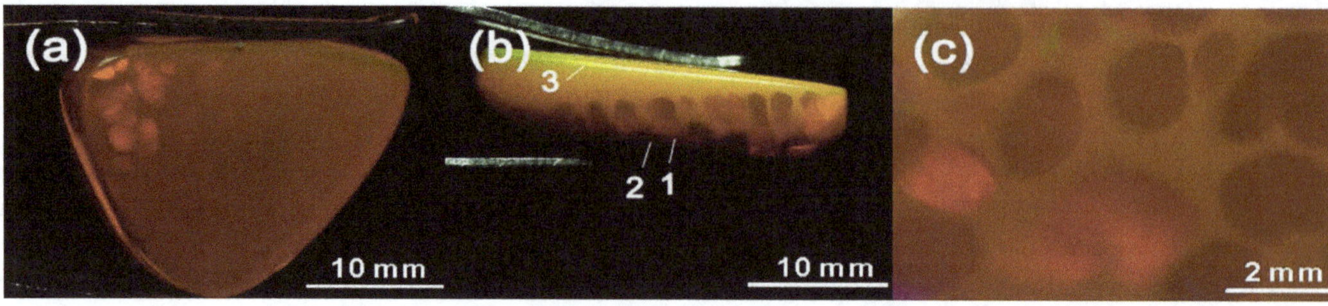

Figure 3. Opal-02 showing the characteristic digit pattern. Columns (1) are transparent with a vivid play-of-colour, separated by common opal matrix (2); (**a**) The top of the sample is covered by common opal (3); (**b**) When viewed from the perpendicular direction, the surface shows a finger-light structure. Transparent areas are the vertical sections of parallel columns and rounded at one end; (**c**) When viewed from the bottom of the sample, the surface shows a mosaic pattern. The transparent rounded patches are cross sections of columns, separated by an opaque matrix composed of common opal [23].

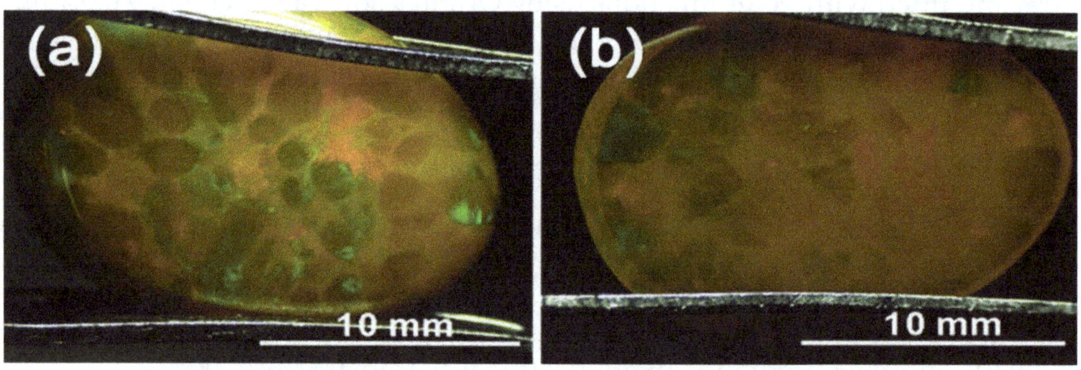

Figure 4. Play-of-colour is shown on both columns and matrix in Opal-05. The shape of the digits is very different. On the one side, (**a**) the cross section of the columns is near-circular, and on the other side (**b**) they are angular polygons.

Table 1. Descriptions of appearance and gemmological properties. Opals 07, 08 and 10 do not show a digit pattern. No visible inclusions are observed under the gemmological microscope.

Sample	W [1] (ct)	Colour of Columns	Transparency of Columns	Colour of Matrix	Transparency of Matrix	RI [2]	LWUV [3] Fluorescence	SWUV [4] Fluorescence
Opal-01	3.369	Pale-yellowish-brown	Transparent	Pale-brown	Translucent	1.430	White, moderate	White, weak
Opal-02	4.144	Pale-orange-brown	Transparent	Orange-brown	Semi-translucent	1.465	White, weak	White, very weak
Opal-03	0.537	Yellow	Transparent	Brown	Translucent	1.464	White, weak	White, very weak
Opal-04	1.017	Pale-yellowish-brown	Transparent	Pale-brown	Semi-translucent	1.459	Inert	Inert
Opal-05	0.909	Pale-yellowish-brown	Transparent	Pale-brown	Translucent	1.464	White, weak	White, very weak
Opal-06	0.659	Pale-brown	Transparent	Brown	Translucent	1.460	White, weak	White, very weak
Opal-07	2.473	Colourless	Transparent	-	-	N.M. [5]	Bluish-white, moderate	Bluish-white, weak
Opal-08	1.214	Colourless	Transparent	-	-	N.M.	Bluish-white, moderate	Bluish-white, weak
Opal-09	2.898	Pale-yellow	Transparent	Pale-brown	Translucent	N.M.	Inert	Inert
Opal-10	2.093	Pale-yellow	Transparent	-	-	N.M.	Inert	Inert
Opal-11	0.625	Pale-brown	Transparent	Pale-brown	Translucent	N.M.	Inert	Inert

[1] Weight. [2] Index of refraction. [3] Longwave ultraviolet radiation. [4] Short wave ultraviolet radiation. [5] Not measured.

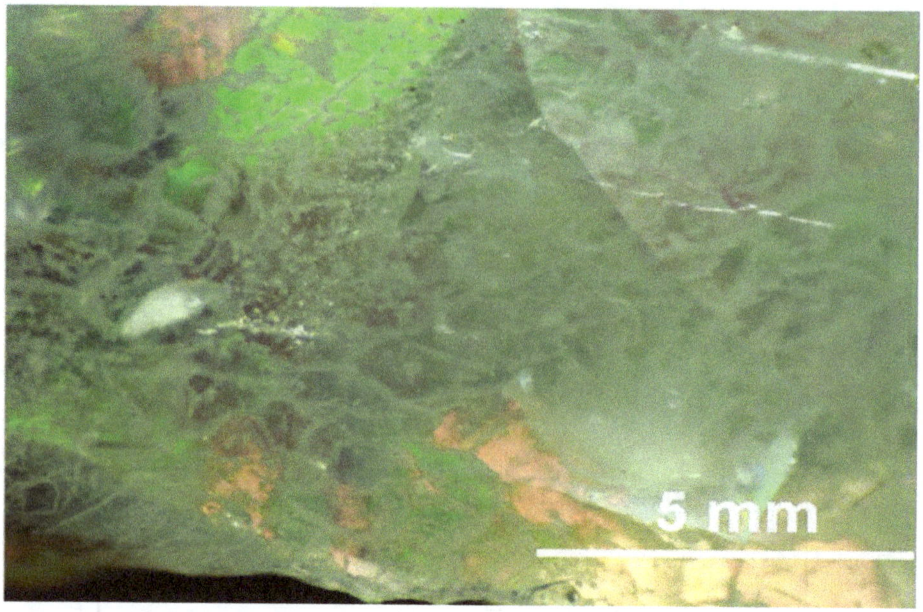

Figure 5. The digit pattern in Opal-09 is undeveloped and has a dendritic matrix.

The play-of-colour patches are bright and two-dimensional silk-like, flaming, flaky, or band-shaped, and show bright and dark stripes parallel to each other (Figure 6). A single patch usually consists of only one colour in one direction of observation, and a continuous spectral colour also appears. There are two kinds of spectral colours, one being the pure spectral colour with clear boundaries (Figure 7a,b), and the other being the spectral gradient colour with blurred boundaries (Figure 7c). In most samples (e.g., Opal-02, see Figure 3), play-of-colour appears only in columns, and the matrix is common opal. A patch can be divided into parts by the matrix, and on adjacent columns, the shape and texture of the colour patch are continuous (Figure 3). Sometimes, play-of-colour can be seen on both columns and matrix (Figure 4). A patch on columns can be divided by the matrix, and a patch on the matrix is divided by columns. The patches on these two types of areas are not connected to each other.

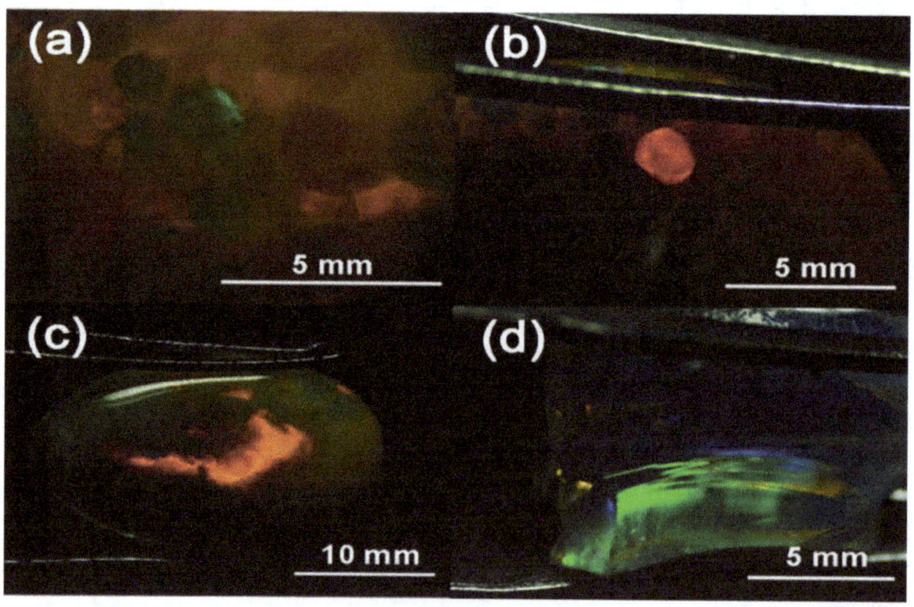

Figure 6. Different shaped play-of-colour patches in Ethiopian opals. (**a**) Silk-like (Opal-05); (**b**) Flaky (Opal-04); (**c**) Flaming (Opal-01); (**d**) Band-shaped (Opal-08).

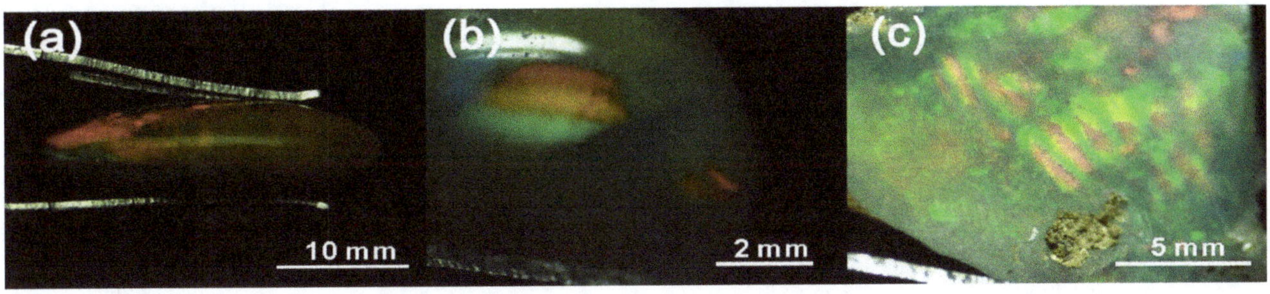

Figure 7. Spectral colour in Ethiopian opals. (**a**) Pure spectral colour (Opal-01); (**b**) Pure spectral colour (Opal-06); (**c**) Gradient spectral colour (Opal-07).

3.2. FTIR Spectroscopy

The infrared reflectance spectra of all samples are similar (Figure 8), with three strong bands around 474, 780 and 1100 cm^{-1}, relating to the fundamental Si–O vibrations. An inflexion point appears around 1245 cm^{-1}. The 1100 cm^{-1} band is caused by the antisymmetric stretching vibration of Si–O $_{as}$ (Si–O)), and the 780 cm^{-1} band is caused by the symmetric stretching vibration of Si–O (ν_s (Si–O)). The 474 cm^{-1} band is related to O–Si–O bending vibration δ (O–Si–O) [5,27–33]. The transition from the amorphous to the crystalline form is evident in the FTIR spectra. One characteristic of this transition is a significant inflexion point at 1245 cm^{-1} [34,35]. Opal-C can be identified by the absorption appearing at 480, 620, 675, 800 and 1110 cm^{-1}, especially the intense bands at about 630 cm^{-1} [5–7]. The infrared spectra of our samples are obviously different from those of opal-C, and more similar to the spectra of opal-A and opal-CT. The symmetric stretching vibration of the Si–O absorption band in opal-CT is in the range of 788 to 792 cm^{-1}, while this band is at 796 to 800 cm^{-1} in opal-A [5]. The band measured in this study is around 780 cm^{-1}, which is the closest to the opal-CT.

However, it is not convincing enough to rely on the small difference of band position at around 780 cm^{-1}. Due to the limitations of the FTIR test conditions (the size of the infrared spot is larger than most columns), the columns and matrix could not be tested separately as well. Therefore, we used a micro-Raman spectrometer to further study the fine-scale structure of samples.

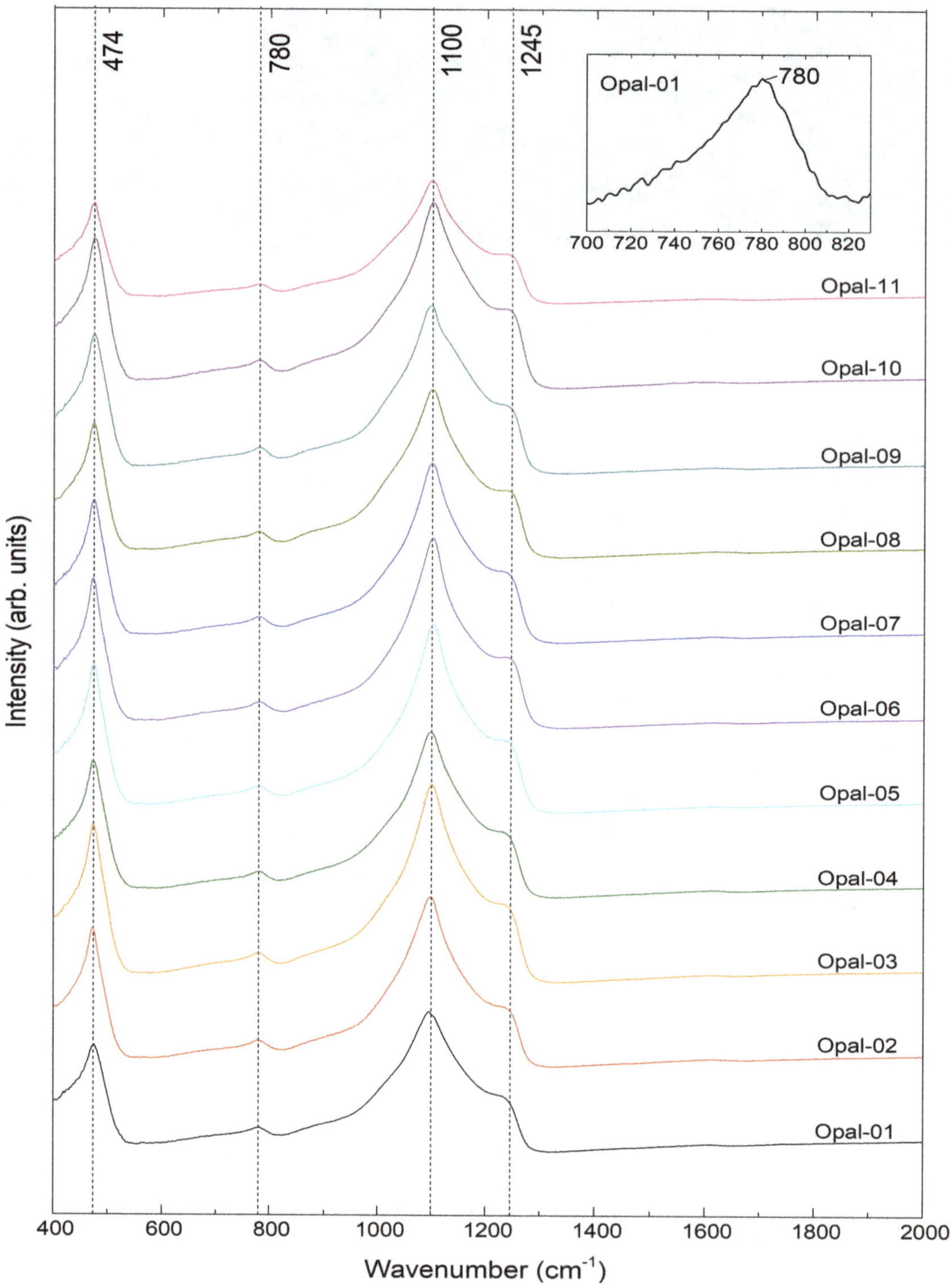

Figure 8. Infrared spectra of all 11 samples, with three bands around 474, 780 and 1100 cm^{-1}, and an inflexion point at 1245 cm^{-1}.

3.3. Micro-Raman Spectroscopy

Each sample shows a high Raman background, which is most likely related to the occurrence of crystalline defects interacting with light or the presence of water [36]. Raman spectra were baseline corrected to eliminate the influence. All samples show the following Raman shifts (Figure 9): broad bands around 360 cm^{-1}, a series of weaker bands in the range of 750 to 1650 cm^{-1}, as well as broad bands around 2945 cm^{-1} and 3550 cm^{-1}. Antisymmetric Si–O–Si stretching vibration bands are located around 1210 and 1085 cm^{-1}; the Si–O stretching vibration band is located around 895 cm^{-1}; the band located around 785 cm^{-1} is assigned to symmetric Si–O–Si stretching vibration [10,13,37,38]. The bands from 1600 to 3600 cm^{-1} are related to water: the band around 2945 cm^{-1} indicates "cristobalite" water [39,40]; the band centred at 1610 cm^{-1} and the broad band around 3550 cm^{-1} are caused by physiosorbed water molecules [12,18,41]. Bands at 1085, 895, 785 and 360 cm^{-1} are characteristic of opal-CT [8–13].

The broad band centred around 360 cm^{-1} is mainly composed of five signals: cristobalite-type atomic arrangements at 230 and 420 cm^{-1}, tridymite-type nanoscale spatial regions at 300 and 355 cm^{-1}, and a shoulder around 490 cm^{-1} representing vibration of Si–O–Si rings with three to four terms [12,13]. By analysing this broad band, the fine-scale structure of opal can be obtained.

Sharp bands at 148 cm^{-1} are observed in Opals 01 to 04, which are attributed to anatase [42–44]. Anatase also has bands at 397, 516 and 640 cm^{-1}, which can affect the intensity and shape of the broad band at 360 cm^{-1}. A small amount of dark anatase inclusions can be observed in those samples with the microscope equipped with a Raman spectrometer. The size of anatase inclusions in our samples is about 1 μm. We subtracted the spectrum of anatase [44] in proportion, and concluded that the impact of anatase on this broad band was very limited. Since our aim was to compare the spectra of different regions on the same sample, and the presence of anatase was not sufficient to affect the comparison of the relative intensity of tridymite and cristobalite type regions, we did not take account of the influence of anatase in the subsequent data analysis.

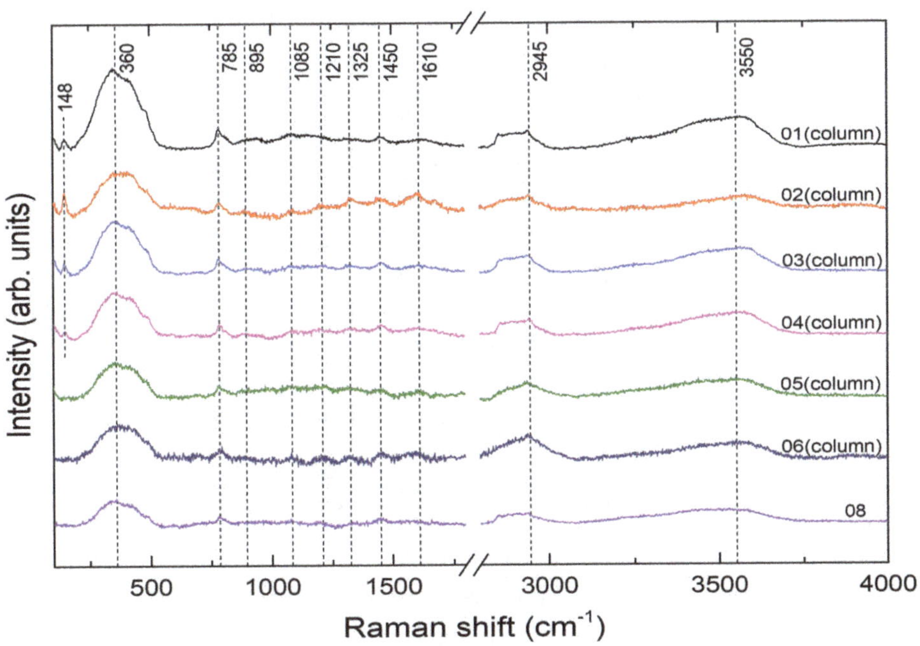

Figure 9. Raman spectra in the range of 100–4000 cm^{-1} of opals from Wegel Tena, Ethiopia. Each sample shows a main band at around 360 cm^{-1}, several small bands in the range of 750–1650 cm^{-1}, as well as broad bands around 2945 cm^{-1} and 3550 cm^{-1}. Opals 01 to 04 show a sharp band at 148 cm^{-1}, which is related to anatase.

We chose Raman spectra in different areas on the sample for comparison, including (1) The column and matrix at the bottom, and the common opal on the top of Opal-02 (respectively corresponding to "1", "2" and "3" in Figure 3); (2) The column and matrix on Opal-04. Raman spectra data in the range of 170 to 600 cm^{-1} were fitted with five Voigt functions to understand the structure state of the samples (Figure 10). Band parameters after fitting are listed in Table 2.

Table 2. Band parameters after fitting the Raman spectra.

Test Position	R1 [1]			R2			R3			R4			R5			η [5]	SE [6] $_\eta$	ξ [7]	SE_ξ
	ω [2]	I [3]	Γ [4]	ω	I	Γ	ω	I	Γ	ω	I	Γ	ω	I	Γ				
02 column (bottom)	230.43	4514	76.12	300.15	65,467	82.96	349.95	69,756	74.91	415.73	83,556	90.42	486.38	37,986	70.49	0.61	0.14	0.83	0.17
02 matrix (bottom)	233.20	3445	44.26	302.20	62,437	79.54	355.50	64,916	70.84	418.00	76,706	76.50	484.40	29,522	63.40	0.61	0.17	0.93	0.20
02 matrix (top)	230.44	4537	124.55	307.72	36,506	83.80	356.94	30,301	64.80	419.37	33,122	74.42	486.38	20,970	58.32	0.64	0.25	0.87	0.20
04 column	237.41	12,841	64.33	302.43	56,312	74.68	353.50	59,810	64.00	416.36	91,546	88.38	486.38	31,884	56.43	0.53	0.09	0.72	0.09
04 matrix	231.13	1862	28.77	298.39	49,522	76.80	351.78	67,882	72.95	421.38	3543	90.00	490.28	15,814	38.66	0.61	0.06	0.81	0.06

[1] R1, R2, R3, R4 and R5 refer to the bands at 230, 300, 355, 420 and 490 cm^{-1}, respectively. [2] Band position (cm^{-1}). [3] Integrated intensity. [4] Full width at half maximum (cm^{-1}). [5] Equation (1). [6] Standard Error. [7] Equation (2).

The value of η is used to estimate the tridymitic fraction in microcrystalline opal [12], and the calculation formula (Equation (1)) is as follows:

$$\eta = \frac{I(R2) + I(R3)}{I(R1) + I(R2) + I(R3) + I(R4)}, \quad (1)$$

where I is the integrated intensity, and R1, R2, R3, R4 and R5 represent the Raman signals near 230, 300, 355, 420 and 490 cm^{-1}, respectively.

For Opal-02, η of the matrix (common opal) on the top is larger than the η of the column and the matrix at the bottom. Similarly, η of the matrix is larger than the columns in Opal-04. This means the tridymite-like structural categories have a larger proportion in the matrix on top compared to their proportion in the matrix and column on the bottom.

The parameter (Equation (2)):

$$\xi = \frac{\Gamma(R3)}{\Gamma(R4)}, \quad (2)$$

where Γ is the full width at half maximum (FWHM), and can offer additional information in the structural state. When ξ is closer to 1, the degree of defectiveness in tridymitic and cristobalitic nanoregions will be closer to each other. From column to matrix, the structural defects of tridymite increase relative to cristobalite nanoregions.

The value of η is related to the integrated intensity of R1, R2, R3 and R4, and the value of ξ is related to the FWHM of R3 and R4. The estimated standard error calculation formula (Equations (S1) and (S2)) [45] and error bars of η and ξ (Figure S1) are shown in the supplementary materials. In this study, each Raman spectrum was only scanned two times for 5 seconds, resulting in a high noise of the spectra, thus affecting the accuracy of the values of η and ξ. However, the trends are observed, which can support the conclusion. If the acquisition time is prolonged, the spectra with higher quality will be obtained, and the standard error of the values of η and ξ will be smaller, which can be more statistically significant and better support the conclusions.

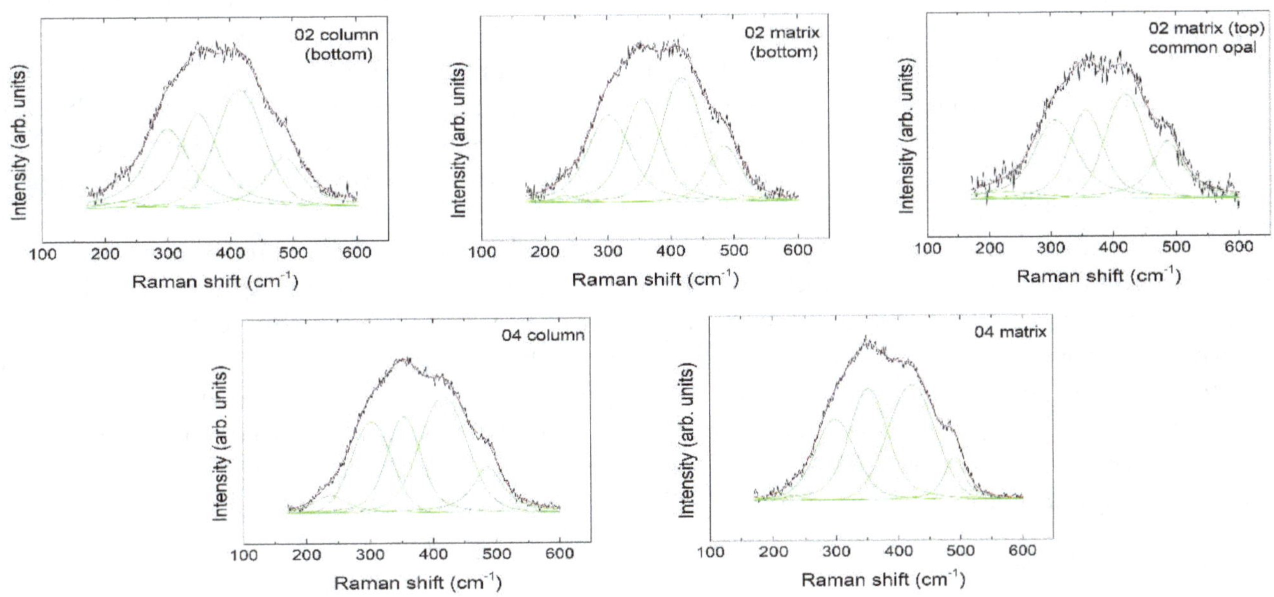

Figure 10. Non-linear curve fitting of Raman spectra of different areas in Opal-02 and 04.

3.4. Microstructure

3.4.1. SEM Observations

All samples consist of aggregates with variable shape, size, ordering and degree of cementation. Silica sphere is a special form of aggregate with a near-spherical shape. On Opal-01, 03 and 04, a layer formed by the combination of silica spheres is observed. Silica spheres are tight-bound, and the gap between them is almost non-existent. The layers are arranged parallel to each other (Figure 11a). The agglomerates formed by the silica spheres are scattered in the layers, with a diameter of about 290 to 350 nm.

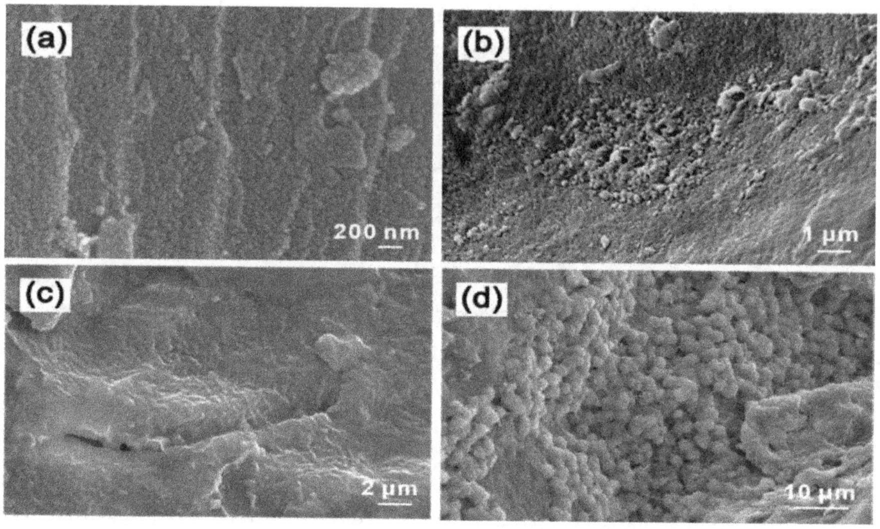

Figure 11. Layers and silica aggregates observed by SEM. (**a**) Opal-01 shows a layered structure; (**b**) Silica aggregates of various shapes and agglomerates on Opal-08 (diameters 130–500 nm); (**c**) Blurred boundary of Opal-09; (**d**) Close-packed small spheres on the surrounding rock of Opal-09 with diameters of 2.2–3.8 μm (2200–3800 nm).

Silica aggregates of various shapes (oval, spherical, long stick, etc.; diameter 80 to 500 nm) and a disorderly accumulation are observed on Opal-07 and 08 (Figure 11b). The silica aggregates can make up irregular-shaped agglomerates, making it difficult to observe the gap between the aggregates. Spheres of about 2.2 to 3.8 μm (2200 to 3800 nm) in diameter are observed on the surrounding rock of Opal-07, which is mainly composed of opaque common opal (Figure 11d). We can see a "bridge" of silica linking two neighbouring spheres. This structure was described by Roudeau et al. [14], who proposed that it could be the result of a capillarity phenomenon involving silica deposition from silica-rich water circulation.

In sample Opal-09, where the digit patterns are not well-formed, the boundary between the silica spheres or between the layers is blurred (Figure 11c). The irregularity of the spheres is related to the instability of the geological environment, and most natural opals have this feature. When the hydrothermal environment is affected by high temperatures, the silica spheres are easily agglomerated; when the temperature and pressure are stable, the silica spheres are regularly arranged [46]. We speculate that the environment of Opal-09 was relatively stable at the initial stage of formation, and the silica spheres were arranged in layers. After that, the environment changed, and the original stacked agglomerates or spheres slowly blended together, but the process stopped before it was completed.

3.4.2. TEM Observations

Three different types of structural forms are observed in Opal-05 and Opal-06 by TEM. Apparent structure is not observed in Opal-01, because the thickness of the sample does not reach the optimal thickness for a standard TEM observation. The most typical structure is composed of silica spheres in an approximately hexagonal close-packed arrangement, which can be observed in both the columns and matrix, while the dimensions of the spheres in each area are different (e.g., Opal 05, see Figure 12a,b). The diameters of the spheres in the columns vary from 210 to 240 nm (Figure 12a), while they are about 230 to 260 nm in the matrix (Figure 12b). The electron diffraction pattern is diffuse (Figure 12b), indicating that the spheres are amorphous. Furthermore, two different structures in Opal-06 can also be detected in Figure 12. In the columns, the silica particles are arranged in oriented pyramids of different sizes (Figure 12c), with a width of 840 to 900 nm and a height of 780 to 800 nm. This form is similar to the wedge-shaped twin crystal of tridymite described by Dong [47]. The matrix consists of spheres of varying sizes (250 to 580 nm in diameter) with a distinct boundary (Figure 12d). Parts of the spheres are fused into multiple grains, and the diameter of the grains can be as small as 50 nm.

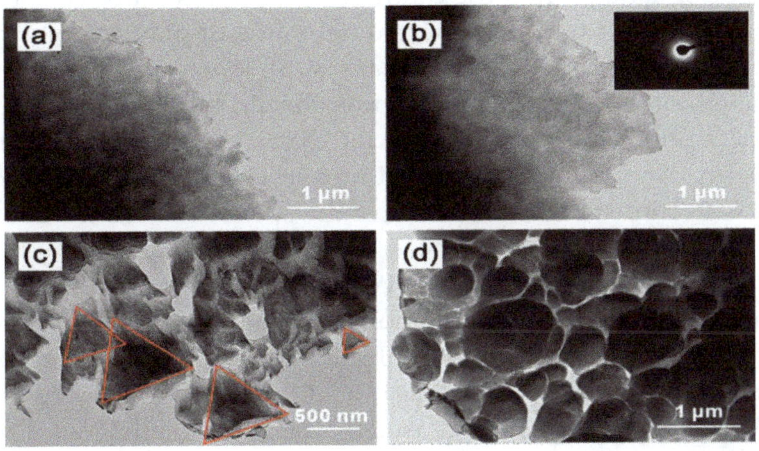

Figure 12. Opal observed by TEM. (**a**) Silica spheres in columns (d = 210–240 nm) from Opal-05; (**b**) Silica spheres in the matrix (d = 230–260 nm) from Opal-05; (**c**) The silica particles in columns from Opal-06 are arranged in pyramids, the width of the cone is about 840–900 nm, and the height is about 780–800 nm; (**d**) Silica spheres in the matrix (d = 250–580 nm) with a clear boundary from Opal-06, composed of small spheres, the diameters of which can be as small as 50 nm.

4. Discussion

4.1. Differences Between Columns and Matrix in Microcrystalline Opal, and Steps of Digit Formation

The Raman spectra indicate that the matrix on the top has a higher tridymite/cristobalite ratio compared to the columns and matrix at the bottom, that means more tridymite structure components will occur closer to the top. In addition, columns have a smaller ξ value than matrix, which means that the degree of order of tridymite structural units increases from the matrix to the columns. At the bottom of the sample, a larger proportion of "tridymite" transforms into "cristobalite", tridymite nanoregions are better ordered in columns, and this state is closer to the end stage of opal-CT formation [12,25,48]. Therefore, we confirm that columns should represent an earlier stage of diagenetic alternation as compared with the matrix, which is consistent with the steps of digit formation [23]. If we consider Opal-02 (Figure 3), the digit pattern is formed in the following steps: (1) Silica spheres undergo sedimentation from the bottom to the top, and then are divided into columns after dendritic; (2) water alters the columns by progressive erosion; (3) new interstitial silica-rich filler settles; (4) additional deposition is covered on top.

4.2. The Cause of Play-of-Colour and Iridescence

The consensus for the formation of play-of-colour is that the internal structure of opal is mainly composed of amorphous and microcrystalline silica spheres with similar diameters, which are close-arranged and regular layered [2,13,49]. The distance between the silica spheres is close to the wavelength of visible light, which produces diffraction. There is a direct correlation between the colour, the diameter of the spheres and the distance between them [2,49]. Opal-A always shows a "smooth sphere" structure, while opal-CT shows a "lepisphere" structure [50,51]. Although neither SEM or TEM observation helped us determine which structure our sample belonged to, the regular arrangement of the spheres was still sufficient to form a diffraction grating that produced play-of-colour.

According to Sanders [51], the wavelength of visible light and the radius of the spheres satisfy the following Equation (3):

$$\lambda_{\max} = 2\mu r\sqrt{3}, \quad (3)$$

where λ_{max} is the maximum wavelength, μ the refractive index of opal (1.46) and r the radius of the spheres. The calculated results show that the condition of white light (wavelength ranging from 400 to 760 nm) diffraction can be satisfied when the diameter of spheres is above 158 nm.

If we consider Opal-05 as an example, different networks of play-of-colour can be observed in the columns and matrix (Figure 4). According to Equation (3), since green play-of-colour is observed in the columns, the diameters of spheres in the columns should be in the range of 198 to 228 nm, while when red play-of-colour is observed in the matrix, the diameters of spheres in the matrix should be in the range of 245 to 300 nm. According to the TEM measurements, the diameter of the spheres in the columns is about 210 to 240 nm, and the diameter of the spheres in the matrix is about 230 to 260 nm, which agrees with the calculated result.

The diameter of the silica spheres from the Wegel Tena opal we observed are between 120 and 500 nm, and most of them conform to the calculations of Equation (3). Although the diameters of some silica spheres are extremely small, the diameter of the agglomerates formed by them is about 200 to 580 nm, which conforms to the conditions needed for diffraction, so play-of-colour is possible. Thus, the play-of-colour in Wegel Tena opal can be caused by a three-dimensional diffraction grating formed by silica spheres, including monochromatic diffraction and a layering of play-of-colour that can be shifted from red to yellow to green. The layering of play-of-colour is caused by the gradual change in the silica sphere size—the diameter of the sphere in the red region is larger, and the diameter of the sphere in the green region is smaller. According to [23], the red layer corresponds to the bottom and the green layer to the top. We observed this phenomenon in Opal-01 (Figure 7a). The sedimentation sequence of silica spheres is controlled by gravity, which leads to the deposition of larger spheres first, followed by progressively smaller spheres [23].

In addition to the above two phenomena, Wegel Tena opal can also show rainbow-like continuous spectral colours (Figure 7c), similar to iridescence. Visible interference can occur between the silica layer and the gap between layers. The colour is related to the incident angle of light and the thickness of the silica layer. The interference colour formula for thin-film interference is determined by the following Equation (4) [52]:

$$2\mu h\cos\theta = k\lambda, \quad (4)$$

where μ is the refractive index of opal (1.46), h is the layer thickness, θ is the refraction angle, λ is the wavelength of visible light (400 to 760 nm) and $k = 1$. $\cos\theta$ decreases in the range of 0° to 90°, so when we take the refraction angle as 0°, we can get the minimum layer thickness of each wavelength in visible light to produce the interference effect. When $\theta = 0°$, the calculated h of 137 to 260 nm can be substituted into the formula. In Opal-07, the diameter of the silica spheres or agglomerates is 120 to 200 nm, and a thin layer (120 to 200 nm thick) fused from a single layer of silica spheres meets the conditions which produce iridescence. As the layer thickness gradually becomes larger or smaller within a small range, monochromatic light of different wavelengths is sequentially strengthened, thereby forming a gradually-changing rainbow-like colour patch observed by the naked eye.

5. Conclusions

The opal from Wegel Tena is classified as opal-CT type. The maturity of different regions of the digit pattern can be distinguished by comparing the relative intensity of the Raman signals at around

360 cm^{-1}. On the bottom of the sample, a larger proportion of tridymite-type regions transform into cristobalite in the matrix than on the top; while the tridymite-type regions in the matrix contain a higher degree of structural defects than in the columns. In the same digit pattern opal, columns formed earlier than the matrix. The difference in microstructure between the matrix and columns is mainly based on the size or morphology of the silica spheres. The diameter and regular arrangement of the silica spheres in Wegel Tena opals satisfies the conditions for diffraction; as a consequence, the play-of-colour can take place. The silica layer may cause interference of light, resulting in continuous spectral colour patches. Two problems remain to be solved: (1) Whether the opals of the columns and matrices are from the same source; and (2) Determining the elemental differences between the columns and the matrix, and whether this influences the differences in their appearance.

Author Contributions: Conceptualisation, K.Z.; methodology, K.Z. and F.B.; validation, K.Z.; formal analysis, K.Z.; investigation, K.Z.; resources, K.Z. and F.B.; data curation, K.Z.; writing—original draft preparation, K.Z.; writing—review and editing, F.B.; visualisation, K.Z.; supervision, F.B.; project administration, F.B. All authors have read and agreed to the published version of the manuscript.

Acknowledgments: We thank Opal Style for providing samples from reliable sources. Thanks to the School of Gemmology, Institute of Earth Science, School of Earth Sciences and Resources of the China University of Geosciences, Beijing, and the School of Materials Science and Engineering of the University of Science and Technology, Beijing for the experimental equipment. Thanks to Li Zhechen, Hu Zhikang, Zhang Yu and Sun Zhulin for their help with instrument operation.

References

1. Jones, J.B.; Segnit, E.R. The nature of opal I. nomenclature and constituent phases. *J. Geol. Soc. Austral.* **1971**, *18*, 57–68. [CrossRef]
2. Yu, X.Y. *Colored Gemmology*, 2nd ed.; Geological Publishing House: Beijing, China, 2009; p. 10.
3. Wilson, J. The structure of opal-CT revisited. *J. Non-Cryst. Solids* **2014**, *405*, 68–75. [CrossRef]
4. Curtis, N.J.; Gascooke, J.R.; Johnston, M.R.; Pring, A. A review of the classification of opal with reference to recent new localities. *Mineral* **2019**, *9*, 299. [CrossRef]
5. Adamo, I.; Ghisoli, C.; Caucia, F. A contribution to the study of ftir spectra of opals. *Neues Jahrb. Mineral. Abh. J. Mineral. Geochem.* **2010**, *187*, 63–68. [CrossRef]
6. Graetsch, H.; Flörke, O.W.; Miehe, G. The nature of water in chalcedony and opal-c from brazilian agate geodes. *Phys. Chem. Miner.* **1985**, *12*, 300–306. [CrossRef]
7. Awadh, S.M.; Yaseen, Z.M. Investigation of silica polymorphs stratified in siliceous geode using ftir and xrd methods. *Mater. Chem. Phys.* **2019**, *228*, 45–50. [CrossRef]
8. Smallwood, A.; Thomas, P.S.; Ray, A.S. Characterisation of sedimentary opals by fourier transform raman spectroscopy. *Spectrochim. Acta Part. A Mol. Biomol. Spectrosc.* **1997**, *53*, 2341–2345. [CrossRef]
9. Fritsch, E.; Rondeau, B.; Ostrooumov, M.; Lasnier, B.; Marie, A.M.; Barreau, A.; Wery, J.; Connoué, J.; Lefrant, S. Découvertes récentes sur l'opale [recent discoveries about opal]. *Rev. Gemmol. AFG* **1999**, *138/139*, 34–40.
10. Ostrooumov, M.; Fritsch, E.; Lasnier, B.; Lefrant, S. Spectres raman des opales: Aspect diagnostic et aide à la classification. *Eur. J. Miner.* **1999**, *11*, 899–908. [CrossRef]
11. Rodgers, K.A.; Hampton, W.A. Laser raman identification of silica phases comprising microtextural components of sinters. *Miner. Mag.* **2003**, *67*, 1–13. [CrossRef]
12. Ilieva, A.; Mihailova, B.; Tsintsov, Z.; Petrov, O. Structural state of microcrystalline opals: A Raman spectroscopic study. *Am. Miner.* **2007**, *92*, 1325–1333. [CrossRef]
13. Sodo, A.; Municchia, A.C.; Barucca, S.; Bellatreccia, F.; Della-Ventura, G.; Butini, F.; Ricci, M.A. Raman, FT-IR and XRD investigation of natural opals. *J. Raman Spectrosc.* **2016**, *47*, 1444–1451. [CrossRef]
14. Rondeau, B.; Fritsch, E.; Guiraud, M.; Renac, C. Opals from slovakia ("hungarian" opals): A re-assessment of the conditions of formation. *Eur. J. Miner.* **2004**, *16*, 789–799. [CrossRef]

15. Koivula, J.I.; Kammerling, R.C.; Fritsch, E. Gem news: Opal from ethiopia. *Gems Gemol.* **1994**, *30*, 52–53.
16. Johnson, M.L.; Kammerling, R.C.; DeGhionno, D.G.; Koivula, J.I. Opal from the shewa province, ethiopia. *Gems Gemol.* **1996**, *32*, 112–120. [CrossRef]
17. Gauthier, J.P.; Mazzero, F.; Mandaba, Y.; Fritsch, E. Opal from ethiopia: Usual gemology and unusual characteristics. *Rev. Gemmol. AFG* **2004**, *149*, 15–23.
18. Rondeau, B.; Fritsch, E.; Mazzero, F.; Gauthier, J.P.; Cenki-Tok, B.; Bekele, E.; Gaillou, E. Play-of-color opal from wegel tena, wollo province, ethiopia. *Gems Gemol.* **2010**, *46*, 90–105. [CrossRef]
19. Fritsch, E.; Rondeau, B. Gem news international: Rare optical phenomenon in play-of-color opal. *Gems Gemol.* **2009**, *45*, 147–148.
20. Mazzero, F.; Gauthier, J.P.; Rondeau, B.; Fritsch, E.; Bekele, E. Nouveau gisement d'opales d'ethiopie dans la province du welo: Premières informations [New deposit of Ethiopian opals in Welo Province: Early information]. *Rev. Gemmol. AFG* **2009**, *167*, 4–5.
21. Rondeau, B.; Mazzero, F.; Bekele, E.; Gauthier, F.P.; Fritsch, E. Gem news international: New play-of-color opal from welo, ethiopia. *Gems Gemol.* **2009**, *45*, 59–60.
22. Rondeau, B.; Cenki-Tok, B.; Fritsch, E.; Mazzero, F.; Gauthier, J.P.; Bodeur, Y.; Bekele, E.; Gaillou, E.; Ayalew, D. Geochemical and petrological characterization of gem opals from wegel tena, wollo, ethiopia: Opal formation in an oligocene soil. *Geochem. Explor. Environ. Anal.* **2012**, *12*, 93–104. [CrossRef]
23. Rondeau, B.; Gauthier, J.P.; Mazzero, F.; Bodeur, Y.; Chauvire, B. On the origin of digit patterns in gem opal. *Gems Gemol.* **2013**, *49*, 138–146. [CrossRef]
24. Lippincott, E.R.; Vanvalkenburg, A.; Weir, C.E.; Bunting, E.N. Infrared studies on polymorphs of silicon dioxide and germanium dioxide. *J. Res. Natl. Bur. Stand.* **1958**, *61*, 61–70. [CrossRef]
25. Rice, S.B.; Freund, H.; Huang, W.L.; Clouse, J.A.; Isaacs, C.M. Application of fourier transform infrared spectroscopy to silica diagenesis: The opal-A to opal-CT transformation. *J. Sediment. Res.* **1995**, *65*, 639–647. [CrossRef]
26. Schmidt, P.; Fröhlich, F. Temperature dependent crystallographic transformations in chalcedony, SiO_2, assessed in mid infrared spectroscopy. *Spectrochim. Acta Part. A Mol. Biomol. Spectrosc.* **2011**, *78*, 1476–1481. [CrossRef]
27. Moenke, H.H.W. Silica, the three-dimensional silicates, borosilicates and beryllium silicates. In *The Infrared Spectra of Minerals*; Farmer, V.C., Ed.; Science Press: Beijing, China, 1982; pp. 289–303.
28. Wu, J.G. *Fourier Transform. Infrared Spectroscopy Technology and Application*; Scientific and Technology Documentation Press: Beijing, China, 1994; Volume 2, p. 722.
29. Hu, Y.; Fan, L.W.; Xue, Q.F. Study on filling treatment and identification characteristics of opal from ethiopian. *J. Gems Gemmol.* **2013**, *15*, 32–37. [CrossRef]
30. Zhong, Q.; Qi, L.J.; Shao, H.J.; Zhou, Y. Gemmological characteristics of green opals from brazil. *J. Gems Gemmol.* **2013**, *15*, 29–36. [CrossRef]
31. Yan, J.; Hu, X.C.; Fang, B.; Tao, J.B.; Peng, Q.J.; Zhang, J. Study on the mineralogical and optical characteristics of blue opal from lishui investigated by XRF-SEM-XRD-FTIR. *J. Rock Miner. Anal.* **2014**, *33*, 795–801. [CrossRef]
32. Eckert, J.; Gourdon, O.; Jacob, D.E.; Meral, C.; Monteiro, P.J.M.; Vogel, S.C.; Wirth, R.; Wenk, H. Ordering of water in opals with different microstructures. *Eur. J. Miner.* **2015**, *27*, 203–213. [CrossRef]
33. Ren, H.J.; Li, L.P. Chemistry of opal from ethiopia. *J. Gems Gemmol.* **2015**, *17*, 23–28. [CrossRef]
34. Li, J.J.; Liu, X.W.; Wang, Y.; Fan, C.X.; Ye, H.; Cheng, Y.F.; Liu, D.Z. Infrared spectral features of SiO_2 with different crystallinity and their implications. *Infrared* **2010**, *31*, 31–35. [CrossRef]
35. Chen, H.S.; Sun, Z.Y.; Shao, J.C. Investigation on FT-IR spectroscopy for eight different sources of SiO_2. *Bull. Chin. Ceram. Soc.* **2011**, *30*, 934–937. [CrossRef]
36. Splett, A.; Splett, C.; Pilz, W. Dynamics of the raman background decay. *J. Raman Spectrosc.* **1997**, *28*, 481–485. [CrossRef]
37. Zu, E.D.; Li, M.C.; Zhang, P.X. Study on jades of SiO_2 by Raman spectroscopy. *J. Kunming Univ. Sci. Technol. Nat. Sci. Ed.* **2000**, *35*, 77–78. [CrossRef]
38. Zarubin, D.P. The two-component bands at about 4500 and 800 cm^{-1} in infrared spectra of hydroxyl-containing silicas. Interpretation in terms of fermi resonance. *J. Non-Cryst. Solids* **2001**, *286*, 80–88. [CrossRef]
39. Gaillou, E.; Mocquet, B.; Fritsch, E. Gem news international: A new gem material from Madagascar: A mixture of cristobalite and opal. *Gems Gemol.* **2004**, *40*, 339–340.

40. Aguilar-Reyes, B.O.; Ostrooumov, M.; Fritsch, E. Estudio mineralógico de la desestabilización de ópalos Mexicanos. *Rev. Mex. Cienc. Geol.* **2005**, *22*, 391–400.
41. Jiang, Y.; Cao, Q. Raman spectroscopic investigation of magnetized water, raw water and distilled water. *Chin. J. Light Scatting* **1992**, *2*, 102–106.
42. Zhang, W.F.; He, Y.L.; Zhang, M.S.; Yin, Z.; Chen, Q. Raman scattering study on anatase TiO_2 nanocrystals. *J. Phys. D Appl. Phys.* **2000**, *33*, 912–916. [CrossRef]
43. Li, Y.; Duan, Y.; Li, W. Study on raman spectrum of nanophase anatase. *Spectrosc. Spectr. Anal.* **2000**, *6*, 699–702. [CrossRef]
44. RRUFF. Anatase R070582. Available online: Rruff.info/R070582 (accessed on 11 April 2020).
45. Ku, H.H. Notes on the Use of Propagation of Error Formulas. *J. Res. Nat. Bur. Stand. Sec. C: Eng. Inst.* **1966**, *70*, 263–273. [CrossRef]
46. Luo, J.; Liu, X.F.; Yan, X.X. Comparative analysis of the characteristics of australian opal and ethiopian opal. *Superhard Mater. Eng.* **2016**, *28*, 50–57. [CrossRef]
47. Dong, Z.C. Discussion on the morphology of cristobalite and tridymite. *Bull. Chin. Ceram. Soc.* **1984**, *6*, 45–50. [CrossRef]
48. Liesegang, M.; Tomaschek, F. Tracing the continental diagenetic loop of the opal-A to opal-CT transformation with X-ray diffraction. *Sediment. Geol.* **2020**, *398*, 105603. [CrossRef]
49. Sanders, J.V. Diffraction of light by opals. *Acta Cryst. Sect. A Found. Cryst.* **1968**, *24*, 427–434. [CrossRef]
50. Elzea, J.M.; Rice, S.B. TEM and X-ray diffraction evidence for cristobalite and tridymite stacking sequences in opal. *Clays Clay Min.* **1996**, *44*, 492–500. [CrossRef]
51. Gaillou, E.; Fritsch, E.; Rondeau, B.; Aguilar-Reyes, B.; Barreau, A.; Ostroumov, M.; Post, J. Common gem opal: An investigation of micro- to nano-structure. *Am. Miner.* **2008**, *93*, 1865–1873. [CrossRef]
52. Liang, Q. *Physical Optics*, 1st ed.; China Machine Press: Beijing, China, 1987; p. 345.

3

An Evaluation of the Potential for Determination of the Geographic Origin of Ruby and Sapphire using an Expanded Trace Element Suite Plus Sr–Pb Isotope Compositions

Mandy Y. Krebs [1,2,*], Matthew F. Hardman [1], David G. Pearson [1], Yan Luo [1], Andrew J. Fagan [3] and Chiranjeeb Sarkar [1]

1 Earth and Atmospheric Sciences Department, University of Alberta, 1-26 Earth Sciences Building, Edmonton, AB T6G 2E3, Canada; mhardman@ualberta.ca (M.F.H.); gdpearso@ualberta.ca (D.G.P.); yluo4@ualberta.ca (Y.L.); chiranjeeb.sarkar@ualberta.ca (C.S.)

2 Gemological Institute of America, 5355 Armada Drive, Carlsbad, CA 92008, USA

3 Hummingbird Geological Services, Vancouver, BC V5Z 1R9, Canada; ajf.geoconsulting@gmail.com

* Correspondence: mandy.krebs@nau.edu

Abstract: The geographic origin of gem corundum has emerged as one of its major value factors. Combined with gemological observations, trace element analysis is a powerful tool for the determination of corundum provenance. However, owing to similar properties and features of gem corundum from different localities, but similar geological settings, and very low levels of many trace elements in gem corundum, the determination of geographic origin remains challenging. In this study, we present trace elements compositions determined by laser ablation inductively coupled plasma mass spectrometry (LA-ICP-MS) for rubies and blue sapphires from several different localities of geologically similar deposits: high-Fe amphibolite-type rubies, low-Fe marble-type rubies, and metamorphic blue sapphires. In addition, we determined Sr and Pb isotopic ratios by offline laser ablation sampling followed by thermal ionization mass spectroscopy (TIMS). By applying new and existing elemental discrimination schemes and the multivariate statistical method linear discriminant analysis (LDA), we show that, in addition to the commonly used discriminators Mg, Fe, V, Ti, and Ga, the elements Ni, Zr, Cr, and Zn show potential for geographic origin determination. Amphibolite-type rubies from different localities can be discriminated using Sr and Pb isotope ratios, whereas the discrimination of marble-type ruby and metamorphic blue sapphires is limited. Our results re-emphasize the challenge of geographic origin determination and the need for a more powerful discriminatory tool.

Keywords: geographic origin determination; ruby; sapphire; trace elements; radiogenic isotopes; linear discriminant analysis

---

1. Introduction

Gem corundum, especially the red ruby and blue sapphire varieties, are among the most valuable gemstones. The commercial value of a gem primarily depends on its size, its color, and its clarity, however, the geographic origin has recently emerged as a major factor affecting their valuation on the market [1]. This is in large part owing to the prestige attributed to certain regions (e.g., rubies from Myanmar, formerly Burma, sapphires from Kashmir), but also because of political, environmental, and ethical considerations [2,3]. Therefore, trying to identify the geographic origin of gem corundum has developed into one of the main tasks for gem-testing laboratories, driving technological approaches to establishing accurate methods of identifying the geographic origin of gemstones.

Trace element analysis is a particularly powerful tool for the determination of gemstone provenance. Most gemological laboratories today use trace elements combined with gemological observations (e.g., inclusion characteristics and growth features) and spectroscopic analysis (Raman, Fourier transform infrared (FTIR) spectroscopy, and ultraviolet-visible-near-infrared (UV/vis-NIR) spectroscopy) to determine the country of origin [4–6]. Another technique that has proven useful for determining gem corundum provenance is the application of oxygen isotopes [7]. These approaches have proven successful in distinguishing between rubies and blue sapphires of different geological origin, however, they are not yet able to reliably distinguish between gems from different geographic regions with a similar geological formation environment [4,5,8–11]. To date, no unique fingerprinting method exists, illustrating the need for a more powerful tool for provenance determination. Determining the provenance of marble-type rubies and metamorphic blue sapphires remains particularly challenging.

Only a small number of trace elements substitute for Al_{3+} into the crystal structure: Cr, Fe, V, Ti, Mg, and Ga. All other trace elements (including Sr and Pb) in gem corundum (chemical formula Al_2O_3) are present at ultra-trace levels and are hosted in micron to sub-micron inclusions. These micron to sub-micron inclusions may be derived during the original formation of the corundum, or from later metasomatic re-crystallization, and are either protogenetic, syngenetic, or epigenetic. In either case, the micron to sub-micron inclusion assemblage is unique for each individual deposit and as such, ultra-trace element data may provide additional discriminatory power for geographic origin determination. A recent study [12] developed a novel laser ablation inductively coupled plasma mass spectrometry (LA-ICP-MS) method for the determination of a broader spectrum of trace element compositions at quantitative levels than has been published previously. In addition to expanding the narrow range on quantifiable trace elements in ruby, Krebs et al. [12] measured Sr and Pb radiogenic isotope compositions in ruby for the first time using an offline laser ablation sampling technique followed by thermal ionization mass spectrometry (TIMS). Radiogenic isotope tracers are an attractive approach to the determination of geographic provenance because they remain largely unaffected by crystal fractionation [13].

Corundum forms in mafic and siliceous geological environments, always associated with rocks depleted in silica and enriched in alumina, because in the presence of silica, Al will be preferentially incorporated into aluminosilicate minerals such as feldspars and micas [14]. Gem corundum is rare because it also requires the presence of Cr, Fe, and Ti to substitute for Al in the structure [9,15], and thermobarometric conditions favorable for its crystallization and stability [16,17].

Two major geological environments have been found to be favorable for the generation of gem-quality corundum; amphibolite- to medium pressure granulite-facies metamorphic belts and alkaline basaltic volcanism in continental rifting environments. The global distribution of corundum deposits is closely linked to collision, rift, and subduction geodynamics, and three main periods of corundum formation are recognized worldwide [11,18,19]: (1) The Pan-African orogeny (750–450 Ma), (2) The Cenozoic Himalayan orogeny (45 Ma–Quaternary), and (3) The Cenozoic alkali basalt extrusions (65 Ma–Quaternary). Gem corundum deposits formed during these periods are often referred to as classic sources and include (1) most deposits in Southeast and Central Asia (e.g., Myanmar, Vietnam, and Afghanistan) formed during the Himalayan orogeny for marble-hosted ruby deposits;(2) Mozambique and Madagascar, among others, for metamorphic or metasomatic rubies; (3) Sri Lanka, Myanmar, and Kashmir, as well as the more modern source of Madagascar, for the metamorphic blue sapphire group; and (4) Australia, Thailand, and Cambodia for basalt-related rubies and sapphires [4,5,20]. Gem corundum deposits that formed unrelated to these periods, such as the Yogo sapphire deposit in Montana, USA (360 to 325 Ma, Ar–Ar phlogopite [21]), are referred to as "non-classical" [4]. For these types of deposits, geographic origin determination is often easier, as they can be characterized by distinct geochemical characteristics that distinguish them from other localities. Corundum deposits can be classified into primary, where the corundum is either hosted in the rock where it crystallized, or in the rock that carried it from the zone of crystallization in the crust or mantle to the Earth's surface, and secondary, where the corundum is of detrital origin, that is, it formed in a different petrogenetic setting as where it is deposited [17]. Many of the classical gem corundum deposits are of the second type, which adds a further layer of difficulty to origin determination efforts.

In this study, we analyzed rubies and blue sapphires from several different localities of three geologically similar deposit types, amphibolite-type and marble-type rubies and metamorphic blue sapphires, using the LA-ICP-MS method of [12], to identify previously unquantifiable trace elements that show potential for geographic origin determination. In addition, Sr and Pb isotope compositions were measured for a subset of samples, and the potential usefulness of radiogenic isotopes for geographic origin determination is explored. While both LA-ICP-MS and isotope measurement approaches are currently too "destructive", that is, visible damage and measurable weight loss are incurred, to be used on high-value client stones, our results provide further insight into gem corundum geochemistry as well as directions of focus for future method development for gemological laboratories.

2. Materials and Methods

2.1. Materials

A total of 135 gem corundum grains from 11 different locations are analyzed in this study (Figure 1, Table 1). The quality of origin determination studies heavily depends on the completeness and reliability of the reference samples used. Ideally, the samples should be collected in the deposits or mines, however, this is not always feasible because some mines are no longer active, while others are inaccessible or off limits to foreigners [22]. In this study, the samples were classified using a classification scheme developed by the Gemological Institute of America (GIA), which reflects the degree of confidence for origin determination (Table S1; the classification codes can be found in Supplementary Material S6).

The samples are divided into three groups according to their variety and known or suspected geological origin.

2.1.1. Amphibolite-Type Ruby

This group comprises samples from four different localities, including the Namahaca deposit at Montepuez in Mozambique [23], the Winza deposit in Tanzania [24,25], and a deposit in the Zahamena National Park located on the Eastern coast of Madagascar [26,27]. For conventional LA-ICP-MS, a sample set provided by the American Museum of Natural History—thought to represent a deposit near Ampanihy, Toliara Province in southwestern Madagascar (George Harlow, pers. communication)—was also analyzed. However, the exact origin of this sample set could not be confirmed, and it was thus excluded from radiogenic isotope analysis. Previously published data from the Aappaluttoq deposit in SW Greenland [12] are included in the discussion, adding a fifth locality to this group.

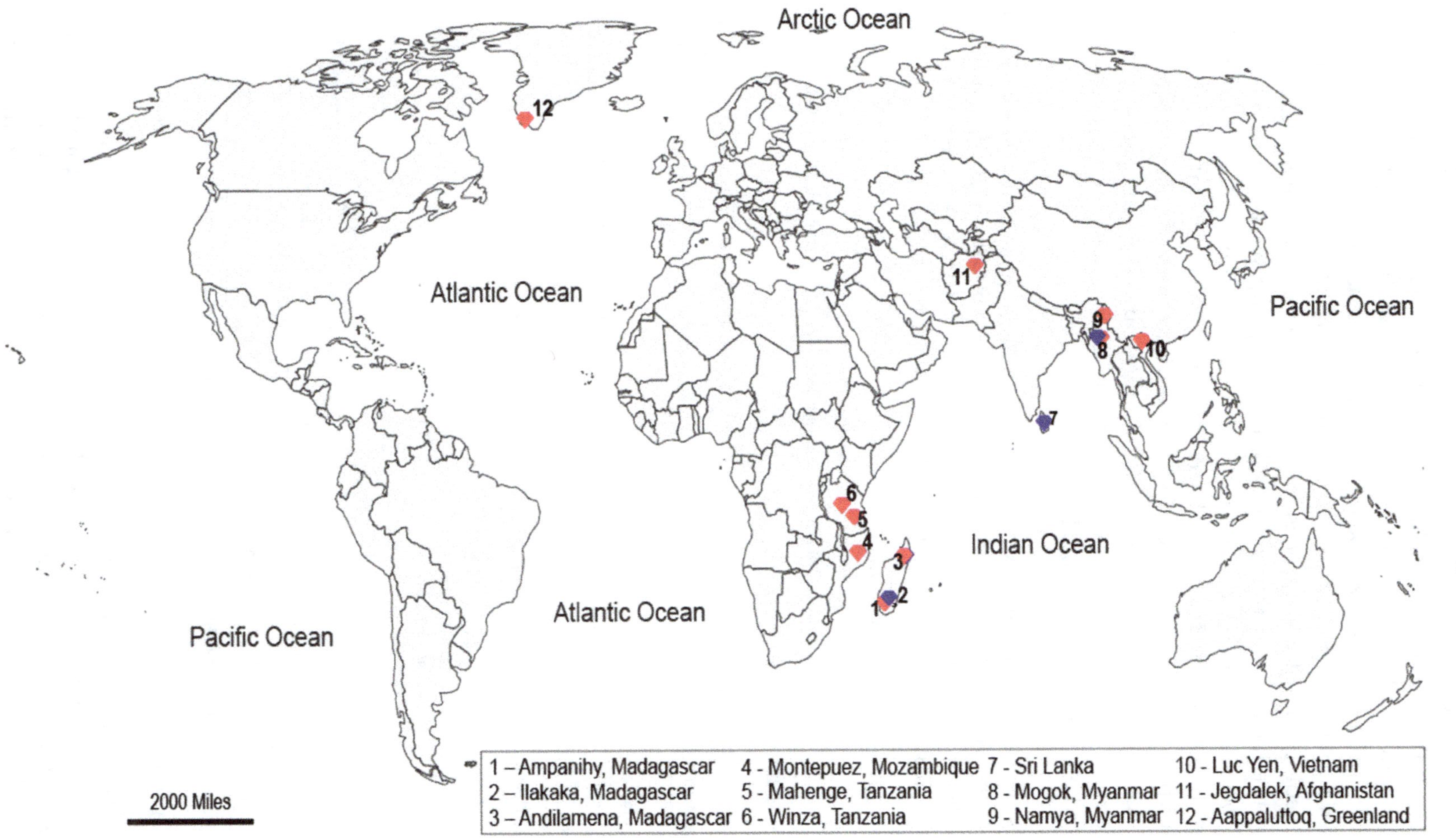

Figure 1. World map showing the 11 localities of origin for ruby (pink symbols) and blue sapphire (blue symbols) analyzed in this study. Also shown is a 12th locality, Greenland. Data of rubies from this locality [12] are included in the discussion.

The localities were selected based on shared features commonly found in rubies from these sources as detailed in the literature: (1) Moderate (as defined by [20]) Fe contents (in this study, between 500 to 5000 ppm); (2) Common inclusions are amphibole, mica, rutile (crystals and needles), and zircon. One exception is Winza rubies, which contain mainly amphibole, garnet, and spinel inclusions [23,25–28]; (3) The oxygen isotope ($\delta^{18}O$) values overlap the mantle value, falling into the range determined for rubies formed in mafic-ultramafic rocks ($0.25 < \delta^{18}O < 6.8‰$ [20]) [18,25,29,30]. Information on the geology of ruby host rocks remains sparse for some deposits, but, where reported, primary deposits at the localities of the amphibolite-type group were found to be associated with mafic or ultramafic rocks [25,27,30,31]; rubies in this group are thought to have formed in a desilicified host with abundant metamorphic fluids, at high pressure and temperature (at least amphibolite facies). The scope of this study is the evaluation of the usefulness of trace element and radiogenic isotope characteristics for geographic origin determination, and thus the geology and formation of amphibolite-type ruby deposits will not be discussed further. Details on current knowledge of these deposits can be found elsewhere [20,27,30,31].

Table 1. Samples analyzed in this study. MozX_X samples were provided by A. Fagan, Hummingbird consulting, Vancouver. All ADXXX and four-digit samples are part of Gemological Institute of America's (GIA) research collection. All samples named A_XXX, V_XXX and M_XXX were provided by George Harlow from the American Museum of Natural History, New York.

Sample ID	Country of Origin	Location	Color	*N*
		Amphibolite-type ruby deposits		
MozX_X	Mozambique (MOZ)	Montepuez, Namahaca Deposit	Pale pink–red	27
AD402, AD403, AD406-408, AD410, AD413, AD432, AD435, AD442	Tanzania (TZA)	Winza	red	10
1968, 1986, 1994, 1996, 2006, 2008, 2048, 2049	Madagascar (MDG)	Andilamena/Zahamena mining area	Red–purplish red	8
A-37145a–c, A-36928a–b, A-91729	Madagascar (MDG)	Ampanihy, Toliara	red	6
		Marble-hosted ruby deposits		
M_112701-1–4 [1]; M_110599-1–4 [2]; (M_42103-1–2, 0469, 2306, 2307, 9233, 9239) [3]; M_107637 [4]; M_109274-1–4 [5]	Myanmar (MMR)	Mandalay Region: [1] Sagyin, [2] Wet Loo, [3] Mogok Stone Tract, [4] Dattaw [5] Kachin State: Namya	Pink–red	21
V-109310a–c, V-106502, V-109086, 0402, 1402, 2002, 2502, 3102, 4902, 6502, 7802, 8102, 8302	Vietnam (VNM)	Luc Yen	Pink–red	15
0902, 1059, 1064, 1088, 1117, 1902, 2707, 3202, 3802, 5402, 5602	Afghanistan (AFG)	Jegdalek	Pinkish red–purplish red	11
3748, 3751, 3753, 3755, 3760, 3797, 5202, 6402, 7902	Tanzania (TZA)	Mahenge, Morogoro region	Red	9
		Blue sapphire deposits		
(AD279, AD283, AD292, AD296) [1], (AD019, AD026, AD027, 4900) [2]	Madagascar (MDG)	[1] Andilamena/Andrebabe (NE), [2] Ilakaka (S)	Blue–dark blue	8
0098, (5210,5159) [1], (AD721, AD773) [2], (AD725, AD727-730) [3]	Myanmar (MMR)	Mogok Stone Tract; [1] Baw Mar mine, [2] Kyatpyn, [3] Kin	Pale blue–blue	10
1320, 1388, 6702, AD030, AD036, AD039, AD044, AD049, AD095	Sri Lanka (LKA)	Elahera	Greyish blue–blue	9

[1–5]: Superscript numbers match samples to their corresponding locality; *N* = number of samples analyzed.

2.1.2. Marble-Hosted Ruby

Marble-hosted rubies from four different localities, Myanmar (Mandalay Region and Namya), Vietnam (Luc Yen), Afghanistan (Jegdalek), and Tanzania (Mahenge Mountains, Morogoro Region), were selected for this study. Three of the localities, Myanmar (Mandalay Region and Namya), Vietnam (Luc Yen), and Afghanistan (Jegdalek), are from deposits where the corundum crystallized in marble as a result of retrograde isochemical reactions mainly in a closed system. Several hypotheses on the genesis of these ruby deposits have been advanced and are summarized in [20]. Rubies from Myanmar in this study are from several different areas, most of which are in the Mandalay region and include the famous Mogok Stone

Tract, Sagyin, Wet Loo, where rubies are found coating painite crystals, and Dattaw. Four samples originate from the Namya mines, located in the Kachin state, approximately 300 km north of Mogok (Figure 2).

Figure 2. A political map showing Myanmar. Regions of origin for rubies analyzed in this study are marked by stars.

In the course of this study, no systematic differences between the different localities were found and the samples were thus grouped together. The localities were chosen because rubies from these deposits share the following common features: (1) Low Fe contents (in this study, between ~20 and 800 ppm); (2) The most common inclusions include silk (rutile needles; to a varying degree), calcite, dolomite, apatite, spinel, zircon, and mica. In addition, salts are found as solid inclusions [20,32–35]; (3) The oxygen isotope ($\delta^{18}O$) values for rubies from these localities are high, consistent with previously published values for marble-hosted ruby (16.3‰ $< \delta^{18}O <$ 23‰ [7]) [35,36].

The marble-hosted ruby deposit in the Mahenge Mountains, Morogoro Region, Tanzania, is of a different type of ruby deposit associated with marble, where rubies form at the peak of prograde metamorphism in impure marble containing gneiss and silicate layers [20]. Rubies from Morogoro share similar features with those found in rubies from other marble-type deposits (e.g., Myanmar), including low Fe contents and rutile silk, spinel, zircon, and apatite inclusions [37,38]. Details on the geology and formation of marble-hosted ruby deposits have been described previously [20,38].

2.1.3. Metamorphic Blue Sapphires

Blue sapphires are broadly separated into two groups—"metamorphic" and "magmatic"—based on their geological origin [4,10]. Magmatic sapphires are primarily sapphires found as xenocrysts in alkali basalts, while metamorphic sapphires are primarily found in amphibolite- to medium pressure granulite-facies metamorphic belts. The majority of gem-quality metamorphic blue sapphire is found in secondary deposits, where it is of detrital origin, however, some primary deposits have been identified and are generally associated with desilicated pegmatites; skarn-related granites; marbles; gneisses; aluminous shale; and, in the case of Myanmar, syenite-like rocks associated with high-grade metamorphic rocks [5,17,20]. Metamorphic and magmatic sapphires can be separated using UV/vis-NIR spectroscopy. The spectra of the respective groups show many similarities; a major difference, however, is the presence of a broad band around 880 nm in magmatic basalt-type sapphires, which is always more intense than the 580 nm absorption band [4]. Blue sapphires from the Baw Mar mine in Myanmar have slightly different UV/vis spectra than those from other Mogok mines; however, they are also lacking the 880 nm band visible in basalt-type sapphires [39]. This classification is not always straight-forward, however, because a recent study on blue sapphires from Madagascar has shown that heat treatment—heating the sapphires to 800–1100 °C to improve the color—may shift the spectrum from primarily having peaks between 550 and 600 nm before heating, to having a strong peak at 880 nm after heating [40].

The blue sapphire samples analyzed in this study were classified as metamorphic based on UV/vis-NIR spectroscopy and inclusion features and are from three different countries: Sri Lanka, Myanmar, and Madagascar. Samples from Sri Lanka originate from the Elahera gem field in Central Sri Lanka [41]; sapphires from Myanmar from the Mogok Stone Tract [42,43]; and sapphires from Madagascar originate from two different areas, including the secondary Ilakaka deposit in the south [44] and Andrebabe in the northeast [45].

The localities for metamorphic blue sapphires in this study were chosen because they share several common features as published in previous studies: (1) lower Fe and Ga and higher Mg than magmatic blue sapphires, that is, sapphire xenocrysts within basalts [10]; (2) common inclusions are mica (phlogopite, muscovite, biotite), rutile crystals and needles, zircon, apatite, and spinel [27,41,42,44,46–48]; and (3) high $\delta^{18}O$ values (>7.7‰) [7].

Details on the geology and formation of metamorphic blue sapphire deposits can be found elsewhere [7,11,17].

2.2. Methods

2.2.1. Trace Element Analysis

Trace element and radiogenic isotope compositions were determined using novel methods developed for analysis of low abundance elements in ruby at the Arctic Resources Geochemistry Laboratory, University of Alberta (Edmonton, AB, Canada) [12]. Individual grains were mounted on one inch epoxy grain mounts and polished with diamond to obtain a flat surface.

Trace element compositions were determined via LA-ICP-MS using a RESOlution M-50-LR (Resonetics-Australian Scientific Instruments, Canberra, Australia) 193 nm ArF Excimer laser (CompexPro 102, Coherent, Santa Clara, CA, USA) coupled to a sector-field Thermo Scientific Element XR2 mass spectrometer (Thermo Fisher Scientific, Waltham, MA, USA) operated in low mass resolution mode ($M/\Delta M$ = ca. 300). NIST612 was used as the primary standard. For quality control, NIST614 and NIST616 were measured every 8–10 samples (Table S2) and used as secondary standards to monitor overall instrument drift. Because the primary and secondary glass standards ablate differently from corundum, that is, they are not matrix-matched, repeat measurements were performed on corundum standard 02-1032 [49] to determine precision and, for some elements, accuracy (Table S2). We note that, at this stage, this reference material (RM) is not an internationally recognised reference material, but it is currently the only matrix-matched RM. We chose not to use it as a primary standard owing to the very limited number of elements for which it is characterised. Samples and standards were ablated with a laser spot size of 285 μm, at a repetition rate of 20 Hz and with a fluence of ~6.6 J/cm^2. These parameters result in clean, circular ablation pits with precise geometry, conducive to low levels of inter-element fractionation, as shown in Figure 2 of [12]. Helium served as the carrier gas. Each analysis consisted of 60 s gas blank acquisition, followed by 60 s ablation time and a 40 s delay for washout. Data were acquired for ^{25}Mg, ^{29}Si, ^{39}K, ^{43}Ca, ^{49}Ti, ^{51}V, ^{52}Cr, ^{57}Fe, ^{60}Ni, ^{64}Zn, ^{65}Cu, ^{71}Ga, ^{88}Sr, ^{90}Zr, ^{93}Nb, ^{120}Sn, ^{139}La, ^{140}Ce, ^{141}Pr, ^{146}Nd, ^{165}Ho, ^{178}Hf, ^{181}Ta, ^{184}W, ^{208}Pb, ^{232}Th, and ^{238}U, with a dwell time of 10 ms for Mg, Si, K, Ca, Ti, V, Cr, Fe, Zn, Ga, and W; a dwell time of 20 ms for Ni, Cu, Zr, Nb, La, Ce, Pr, Nd, Ho, Hf, Ta, Th, and U; and a dwell time of 30 ms for Sr, Sn, and Pb. Two spots were measured on each grain. The analyzed areas were selected randomly, the only selection criterion was the absence of visible solid inclusions, other impurities, and fractures at the 10 μm scale. Data reduction and processing was conducted offline using Iolite v. 3.3.2 [50] with ^{27}Al (52.92 wt%) as the internal standard element. Careful screening of the time-resolved display and segment-picking for data integration in Iolite permits the identification of signal spikes from prominent inclusions, and were thus excluded during data reduction, resulting in only a single analytical run for some samples. Limits of quantification (LOQ) for all analysed elements are calculated as 3 × LOD and are listed in Table 2. Typical precision of the analyses ranges from 1% to 10% for the secondary standards (2σ) and 10% to 20% (2σ) for the corundum RM 02-1032. As expected, elements close to the LOQ have lower precision. The accuracy for corundum, using NIST 612 as the primary standard, could be determined for the elements Mg, Ti, V, Cr, Fe, and Ga, and is between 5% and 20% for the elements Ti, V, Cr, and Ga. For Mg and Fe, however, the accuracy is less satisfactory, with ~75% and ~35%, respectively, and illustrates the importance of using matrix-matched standards for LA-ICP-MS where possible. It is important to note that the isotopes ^{43}Ca and ^{29}Si have isobaric interferences, ^{27}Al^{16}O and ^{28}SiH, respectively, which are not reflected in the LOQ calculation and make accurate quantification of low abundances difficult. This is also reflected by the low precision of these elements for corundum RM 02-1032 (Table S2); therefore, the reported concentrations should be viewed as quantitative and are used here in a strictly comparative manner. The interference of ^{64}Ni on ^{64}Zn was determined to be negligible and was not corrected for.

Table 2. Median limits of quantification (LOQ − 3 × LOD) obtained for 26 elements in ppb.

Element	LOQ	Element	LOQ	Element	LOQ
^{25}Mg	378	^{64}Zn	136	^{141}Pr	0.59
^{29}Si	16266	^{65}Cu	78	^{146}Nd	2.55
^{39}K	322	^{71}Ga	20	^{165}Ho	0.37
^{43}Ca	34500	^{88}Sr	5.5	^{178}Hf	1.4
^{49}Ti	253	^{90}Zr	5.1	^{181}Ta	0.95
^{51}V	14	^{93}Nb	1.37	^{184}W	1.04
^{52}Cr	299	^{120}Sn	9.9	^{208}Pb	3.9
^{57}Fe	2530	^{139}La	0.81	^{232}Th	0.4
^{60}Ni	112	^{140}Ce	0.56	^{238}U	0.18

2.2.2. Radiogenic Isotope Analysis

Of each corundum group, the cleanest samples were selected for isotope analysis—64 grains in total. Radiogenic isotope analyses were carried out using an offline laser ablation sampling technique, originally developed for the ablation of diamond [51,52], followed by thermal ionization mass spectrometry (TIMS). The technique utilizes a closed-system, custom-designed teflon laser ablation cell in which a sample is ablated, and the products are trapped, allowing the accumulation of higher volumes of analyte needed for isotope analysis. Within corundum, only few elements substitute into the crystal lattice, and Sr, Rb, Pb, Th, or U are hosted in micron to sub-micron mineral and fluid inclusions trapped within the corundum lattice instead. This technique allows us to bulk sample the micron to sub-micron mineral assemblages, and thus obtain representative isotopic ratios. Offline ablation and TIMS measurements were carried out following the protocols outlined in [12,53] and briefly summarized here.

The corundum samples were leached in 16 N Seastar ultra purity acid (UPA)-grade HNO_3 on a hotplate at 100 °C for 24 h and then rinsed in 18.2 Ω MQ H_2O. Subsequently, samples were leached in 6 N HCl for another 24 h at 100 °C followed by another rinse in MQ H_2O, and finally dried at 100 °C for at least 1 h. All samples were weighed prior to the ablation. Ablations were performed using a RESOlution M-50HR (Resonetics) 193 nm ArF Excimer (CompexPro 110, Coherent, Santa Clara, CA, USA) laser ablation system, in a raster pattern with the following ablation conditions: spot size 90 μm, repetition rate 100 Hz, and fluence 4–7 J/cm^2. Areas selected for ablation are free of visible inclusions and fractures, however, depending on the clarity of the samples, fractures are occasionally difficult to avoid, and larger solid inclusions may not be visible beneath the surface. Ablation times varied from 3 to 6 h.

Following ablation, 2 mL of 6 N HCl ultra purity acid (UPA) was added to the sample beaker and a Teflon cap replaced the laser window. The sealed cell was then placed in an ultrasonic bath for 40 min to loosen the material from the sample. The fluid was then transferred using a pipette from the ablation cell to a 7 mL Teflon beaker and dried on a hotplate at 100 °C. To determine the average size of the blank contribution, two to four total procedural blanks (TPBs) were collected for each batch of samples processed, using the same ablation cells and reagents used for the samples. The only step omitted was the step of ablating a solid because including an ablation step would require a solid that is essentially devoid of all trace elements or has trace element concentrations below the LOD of our method. Such a matrix has not yet been found, though we have evaluated Si-wafers for this purpose. Samples were rinsed in MQ H_2O and dried before being reweighed, and the weight loss (0.02–0.42 mg) resulting from the ablation was then calculated. The dried ablation products were dissolved in 200 μL of 3 N HNO_3 and placed on a hot plate for approximately 2 h to homogenize. After cooling, a ~20% aliquot by volume was transferred into a pre-leached 7 mL Teflon beaker for trace element analysis and the remaining sample was processed for isotopic analysis.

Strontium was separated using Sr spec resin following procedures after [54,55]. For the Mozambique sample suite, Pb was eluted after Sr in the same column. However, semi-transparent, sticky jam-like residues in the Pb fraction lead to loading difficulties, resulting in low intensities of Pb for TIMS analysis. Therefore, for all other samples, the Pb fraction was separated in separate columns using BioRad AG1-X8 anion exchange resin (100–200 μm mesh) [12].

Total procedural blanks (TPBs) and aliquot sample solutions were analyzed for trace elements using a Thermo Scientific Element XR2 magnetic sector ICPMS (Thermo Fisher Scientific, Waltham, MA, USA) coupled with an APEX-Q high-efficiency sample introduction system (Elemental Scientific Inc., Omaha, NE, USA) fitted with a membrane desolvation module (ACM—actively cooled membrane). Each dried sample was taken up in 1 mL 3% UPA HNO_3, with 500 ppt In added as internal standard. Instrumental conditions were similar to those described by [53]. The isotopes measured were as follows: ^{55}Mn, ^{56}Fe, ^{59}Co, ^{63}Cu, ^{64}Zn, ^{69}Ga, ^{85}Rb, ^{88}Sr, ^{89}Y, ^{90}Zr, ^{93}Nb, ^{98}Mo, ^{133}Cs, ^{137}Ba, ^{139}La, ^{140}Ce, ^{141}Pr, ^{143}Nd, ^{147}Sm, ^{151}Eu, ^{157}Gd, ^{159}Tb, ^{161}Dy, ^{167}Er, ^{173}Yb, ^{175}Lu, ^{178}Hf, ^{181}Ta, ^{208}Pb, ^{232}Th, and ^{238}U. Solution concentrations were measured against a five-point calibration line derived from dilutions of a synthetic rock multi-element standard solution, MES-0314-01. The standard was diluted 50,000, 100,000, 250,000, and 500,000 times, yielding concentrations in the range of 1 ppt to 35 ppb for different elements and different dilutions, thus providing an appropriate matrix-match for the samples so calibration lines did not require extrapolation into the region of sample analyte concentrations. All concentrations were normalized to the sample weight loss.

The LOQ for trace element analysis were calculated following the same procedures detailed in [53]. A total of 57 TPBs were performed for the corundum ablations, resulting in the following LOQs: the elements Cs, Pr, Sm, Eu, Gd, Tb, Dy, Ho, Er, Yb, Lu, and U are less than 1 pg; the elements Rb, Y, Nb, La, Nd, and Th are less than 5 pg; the elements Co, Ce, Hf, and Pb are less than 20 pg; the elements Sr and Ba are less than 70 pg; the elements Mn, Cu, Zr, and Mo are less than 350 pg; and the elements Fe and Zn are less than 10.4 ng.

A purified TaF_5 activator was used to load Sr samples to enhance Sr ionisation. Sr isotope ratios were measured on Thermo Scientific Triton Plus TIMS, equipped with nine movable Faraday cups, in static mode using 10^{11} Ω and 10^{12} Ω amplifiers. Multiple loads (n = 21) of NBS987 standards of 5 and 10 ng size gave an average value of 0.710270 ± 92 (2σ; n = 21) and compares well with long-term data published in other laboratories (Thirlwall, 1991), with data for similar sized loads, and with values measured in the Arctic Resources laboratory for similar sized standards. Eleven loads of dissolved BHVO-2 of 6 ng in size gave an average value of 0.70352 ± 17 (2σ; n = 11), slightly higher than, but within uncertainty of, the reported average value of 0.70348 ± 7 (GeoREM Database, version 27).

Lead samples were loaded with a mixture of 3.5 μL of silicic acid activator and 0.5 μL 0.1 N H_3PO_4, following a similar procedure as for Sr. Pb isotopic ratios were analyzed on a Triton Plus TIMS (Thermo Fisher Scientific, Waltham, MA, USA) using a single SEM in peak-hopping mode using the method described in [56]. The average $^{206}Pb/^{204}Pb$, $^{207}Pb/^{204}Pb$, and $^{208}Pb/^{204}Pb$ values for NBS981 were 16.950 ± 0.015, 15.456 ± 0.013, and 36.856 ± 0.027, respectively, based on 21 analyses, which are within uncertainty of the values published for double spike analyses of the reference material by Todt et al. [57] of $^{206}Pb/^{204}Pb$ = 16.937 ± 0.0022, $^{207}Pb/^{204}Pb$ = 15.492 ± 0.0025, and $^{206}Pb/^{204}Pb$ = 36.701 ± 0.011, respectively.

3. Results

3.1. Trace Elements

3.1.1. LA-ICP-MS

The trace element results for spot analyses on polished grains are listed in Table S1. The most abundant trace elements are Fe; Ti; Ca; Mg; Ga; and, with the exception of blue sapphires, Cr. Trace elements also mostly above the LOQ (defined here as 10 standard deviations (σ) above the gas blank) are V, Zn, Ni, and Pb. Other elements that are habitually above the LOQ for at least one of the groups are Si, K, Cu, Sr, Zr, Sn, La, Ce, Nd, Pr, Th, and U. Among the three gem corundum groups analyzed, marble-hosted rubies are the purest, that is, the number of elements >LOQ is the most limited. In Table 3, the trace elements relevant or with the potential to be useful for origin determination are summarized.

Table 3. Trace element comparative ranges with medians in parentheses. All data are in ppm (μg/g).

Sample	^{25}Mg	^{49}Ti	^{51}V	^{52}Cr	^{57}Fe	^{60}Ni	^{64}Zn	^{71}Ga
Amphibolite-type ruby deposits								
MOZ n = 54	18–67 (38)	21–104 (42)	2.9–8.5 (4.1)	1119–9190 (1812)	688–1522 (1001)	0.4–3.2 (1)	0.3–5.6 (0.8)	17–36 (22)
TZA, Winza n = 19	6.4–128 (91)	16–584 (99)	0.9–8 (3.1)	526–2850 (991)	1648–3937 (3204)	0.4–9.7 (6.6)	0.2–1 (0.6)	16–44 (34)
MDG, Zahamena n = 13	48–103 (78)	73–149 (109)	29–59 (37)	1782–8770 (2839)	1878–3521 (2940)	0.3–3.5 (0.8)	0.3–1 (0.4)	49–101 (64)
MDG, Ampanihy n = 12	13–40 (27)	21–86 (42)	26–49 (32)	1704–2679 (2189)	2720–3560 (3095)	0.5–2.7 (0.8)	0.4–1.4 (0.8)	84–112 (97)
Marble-hosted ruby deposits								
MMR n = 40	13–129 (61)	35–207 (109)	46–1074 (310)	772–6850 (2352)	23–267 (101)	0.2–2.6 (0.6)	0.2–1.8 (0.5)	12–192 (131)
VNM n = 39	12–161 (39)	36–627 (126)	8–480 (63)	242–8190 (875)	20–718 (122)	0.1–3.9 (0.4)	0.2–1.3 (0.4)	9.3–133 (58)
AFG n = 23	26–144 (59)	47–367 (111)	16–210 (82)	1267–9507 (3145)	41–679 (81)	0.07–3.2 (0.6)	0.2–1.9 (0.5)	33–92 (48)
TZA, Morogoro n = 19	42–162 (71)	84–360 (126)	15–212 (34)	462–2319 (1207)	128–1250 (449)	0.08–1.7 (0.2)	0.2–0.6 (0.4)	42–117 (60)
Blue Sapphire deposits								
MDG, Andrebabe n = 7	6.4–44 (19)	22–35 (26)	1.8–7 (3.6)	2.7–11 (5.4)	2218–4975 (3460)	0.2–4.2 (1.9)	0.3–3.1 (1.3)	125–216 (159)
MDG, Ilakaka n = 6	21–133 (63)	41–841 (164)	12–34 (23)	2.2–20 (6)	804–2136 (924)	0.09–0.4 (0.23)	0.25–0.75 (0.5)	50–233 (63)
LKA n = 18	28–146 (63)	50–283 (138)	3.8–62 (9.4)	0.9–4.9 (1.8)	497–2402 (997)	0.8–1.9 (0.33)	0.1–0.9 (0.3)	46–131 (97)
MMR n = 20	5.8–77 (25)	8.7–275 (59)	0.5–27 (1.5)	1.6–9.8 (2.6)	1186–5617 (2040)	0.4–3.2 (1)	0.1–1.3 (0.5)	51–193 (89)

The trace element characteristics of the two ruby groups are consistent with those previously reported in the literature [20]: moderate contents of Cr (<1 wt%), low Ga (<130 ppm), and low Ti (~16–630 ppm) for all rubies, moderate contents of Fe for amphibolite-type rubies (~700–4000 ppm), low Fe for marble-hosted rubies (~20–1250 ppm), and low V for amphibolite-type rubies (~1–60 ppm) compared with elevated V for marble-hosted rubies (~8–1074 ppm). Among amphibolite-type rubies, some differences in trace element abundance are observed, including higher median V and Ga contents for rubies from Madagascar (medians of V = 37 and Ga = 64 ppm in the north, and V = 32 and Ga = 97 ppm in the south) than for Mozambique (medians of V = 4.1 and Ga = 22 ppm) and Tanzania (medians of V = 3.1 and Ga = 34 ppm) rubies. Rubies from Tanzania and Zahamena, MDG have approximately twice as much Mg and Ti as those from Mozambique and Ampanihy, MDG. In addition, Mozambique rubies analyzed in this study have lower Fe (median = 1001 ppm) than rubies from the other amphibolite-type deposits (median ~3000 ppm), while Winza, TZA rubies are characterized by higher Ni contents (median = 6.6 ppm vs. max ~1 ppm for the other the deposits). Among the marble-hosted rubies, rubies from Myanmar have higher V (median = 310 ppm) and Ga (median = 131 ppm) than rubies from the other localities, while rubies from Morogoro, TZA have the lowest V contents (median = 34 ppm), which overlap with the highest V values in the amphibolite-type deposits, and are also characterized by higher Fe contents (median = 449 vs. median = 81–122 ppm for the other deposits). Differences in trace elements of blue sapphires from different deposits are as follows: sapphires from Andrebabe, MDG, and Myanmar have lower Mg, Ti, and V, and higher Fe and Ni contents than Ilakaka, MDG and Sri Lanka sapphires. Both Madagascar sample suites have higher Cr contents than Sri Lanka and Myanmar samples. Blue sapphires from Andrebabe, MDG are further characterized by higher Ga contents. Overall, the trace element contents of the blue sapphire samples analyzed in this study exhibit trace element characteristics typical for metamorphic blue sapphires reported by [10], including low Ga/Mg ratios (<10).

3.1.2. Solution ICP-MS

For 64 rubies and sapphires, a fraction of the offline ablated material was analyzed for trace elements using solution ICP-MS. This comprised 17 amphibolite-type rubies (five from Zahamena, MDG; eight from Mozambique; and four from Winza, TZA), 24 marble-type rubies (10 from Myanmar; five from Vietnam; five from Afghanistan; and four from Morogoro, TZA) and 22 metamorphic blue sapphires (seven from Sri Lanka; eight from Myanmar; four from Ilakaka, MDG; and three from Andrebabe, MDG). For five samples, multiple offline ablations were performed to examine homogeneity and look for isotopic variations within single crystals.

The elemental abundances of all analyzed samples for rare earth elements (REEs), high field strength elements (HFSEs), and large ion lithophile elements (LILEs) range from low ppb levels up to 100s of ppm in some cases (Table S3). When normalized to primitive mantle values [58], most samples show enrichment of LILE and HFSE over LREE and Nb, a positive Pb_N anomaly (relative to Ce_N and Pr_N) and LREE enrichment relative to HREE (La_N/Yb_N ~1.5–991) (Figure 3).

Except for the Mozambique and Zahamena, MDG suites, all Y_N anomalies (relative to Dy_N and Ho_N) are not real, but are caused by single extremely high values in the Vietnam and Ilakaka, MDG datasets (Figure 3c,d). Eu anomalies ($Eu/Eu^* = 2 \times Eu_N/(Sm + Gd)_N$) are mostly negative, with only rubies from Mozambique showing $Eu/Eu^* > 1$ (Figure 4a), while Ce anomalies are typically ≥1 (Figure 4d). Sr anomalies for rubies are mostly positive—exceptions are the majority of the amphibolite-type Mozambique and some marble-type Myanmar rubies, while all blue sapphires have $Sr/Sr^* < 1$ (Figure 4c).

It is also noteworthy that Sr is <LOQ in all rubies from Zahamena, MDG. Ta/Nb ratios are predominantly >1 for all deposit groups (Figure 5a–c), while the U/Pb ratios range are mostly <1 for amphibolite-type rubies, >1 for marble-type rubies, and range from U/Pb_N = 0.01 to 20 for metamorphic blue sapphires (Figure 5d–f). Rubies mostly have Th/U < 1, with the exception of amphibolite-type Mozambique rubies, most of which have Th/U < 1 (Figure 5g–i).

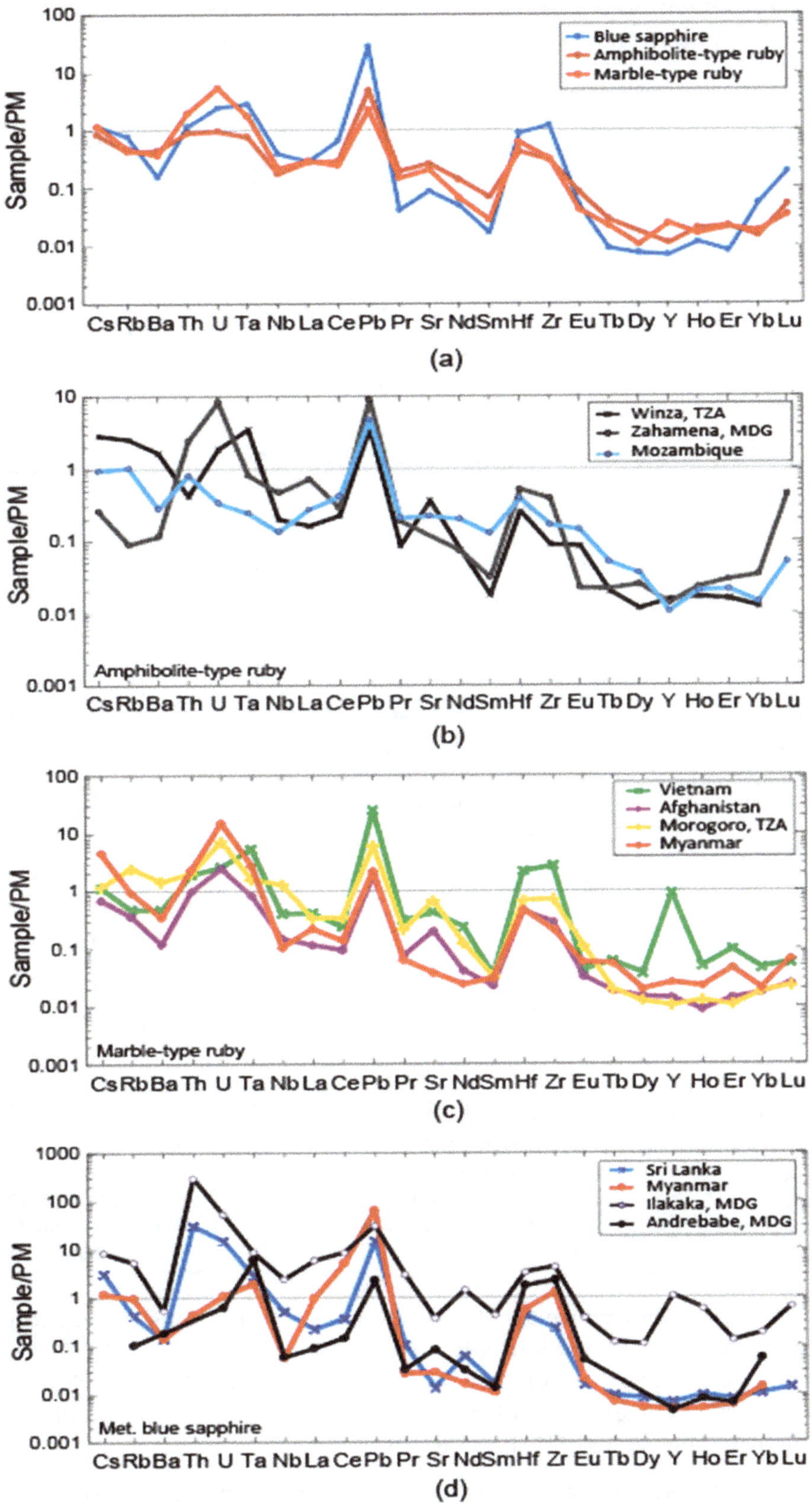

Figure 3. Primitive mantle (PM)-normalized [58] median trace element concentrations in (**a**) the three different deposit groups, (**b**) the amphibolite-type ruby localities, (**c**) the marble-type ruby localities, and (**d**) the localities of metamorphic blue sapphire analyzed by offline ablation followed by solution inductively coupled plasma mass spectrometry (ICPMS) in this study. All data plotted are >limit of quantification (LOQ). TZA, Tanzania; MDG, Madagascar.

Most blue sapphires have Th < LOQ, but where Th > LOQ, Th/U is predominantly <1. Zr and Hf, more often >LOQ in ruby than in blue sapphire, are enriched with Zr/Hf less than or greater than 1 (Figure 5j–l).

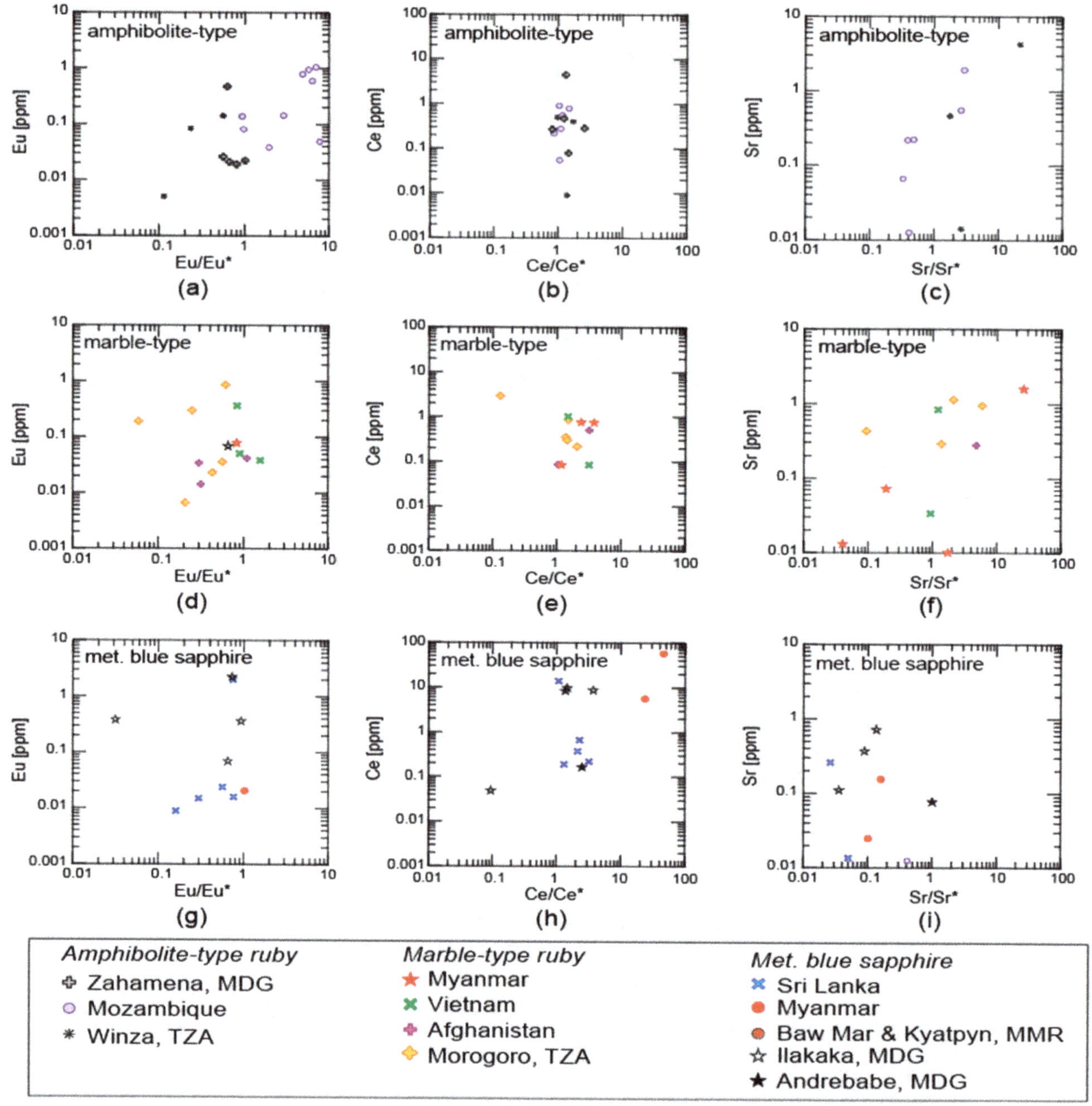

Figure 4. Primitive mantle-normalized [58] trace element compositions in gem corundum ablated offine in this study. Points represent individual analyses. (**a**–**c**) Eu versus Eu/Eu*, (**d**–**f**) Ce versus Ce/Ce*, and (**g**–**i**) Sr versus Sr/Sr*. Anomalies were calculated using $Eu/Eu^* = 2 \times Eu_N/(Sm + Gd)_N$, $Ce/Ce^* = Ce_N/((La_N)^{0.667} \times (Nd_N)^{0.333})$, and $Sr/Sr^* = 2 \times Sr/(Ce + Nd)$, where N indicates normalization to average primitive mantle values of [58]. MNR, Myanmar.

Some differences between the different localities within each group are apparent. Among amphibolite-type ruby deposits, only the Mozambique suite is characterized by Eu/Eu* > 1 (Figure 4a) and Th/U ratios > 1 (Figure 5g). In addition, Zr and Hf in Mozambique rubies are predominately <LOQ—a characteristic also seen in the LA-ICP-MS data (Figure A2). Furthermore, rubies from Winza, TZA are characterized by positive Gd anomalies; rubies from Zahamena, MDG are characterized by the absence of Sr; and both the Zahamena, MDG and Mozambique suites show Lu enrichment (Figure 3b).

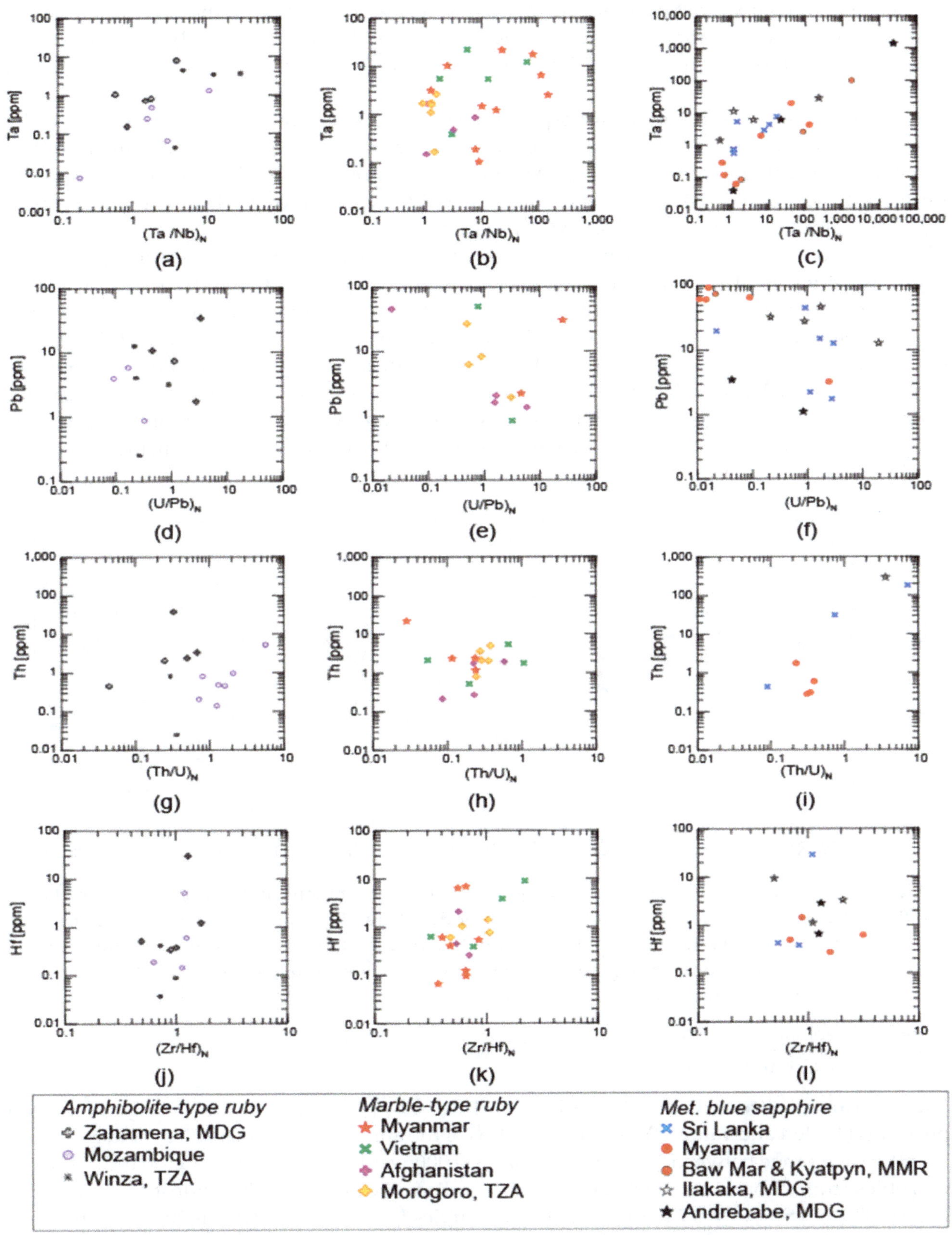

Figure 5. Primitive mantle-normalized [58] trace element compositions in offline ablated gem corundum analyzed in this study. Points represent individual analyses. **(a–c)** Ta versus $(Ta/Nb)_N$; **(d–f)** Pb versus $(U/Pb)_N$, **(g–i)** Th versus $(Th/U)_N$, and **(j–l)** Hf versus $(Zr/Hf)_N$ for amphibole-type ruby, marble-type ruby, and met. blue sapphire, respectively.

Trace element characteristics in offline-ablated marble-type rubies are mostly similar. However, different characteristics are apparent and include mostly very high Ta/Nb for Myanmar and Vietnam rubies versus Ta/Nb ~1 for Morogoro, TZA rubies, and Sr/Sr* < 1 for Myanmar rubies (Figures 3, 4c and 5b).

Among the different suites of metamorphic blue sapphires, several differences are observed. Blue sapphires from Sri Lanka and Ilakaka, MDG are very similar to each other, with U/Pb mostly >1 and U (and Th where present) enrichment over Ta. Myanmar blue sapphires, on the other hand, show U/Pb < 1 (Figure 5f). In addition, in Myanmar and Andrebabe, MDG sapphires, Cs is mostly <LOQ, whereas in sapphires from Sri Lanka and Ilakaka, MDG, Cs is >LOQ (Table S3).

Figure 6 shows the primitive mantle-normalized trace element patterns of the samples for which two separate offline ablations were performed to investigate the reproducibility of the data.

The trace element patterns of the two repeat analyses are remarkably similar for all samples, however, in some samples, variations for single elements are observed. For ruby moz2_2, for instance, Sr and Ba vary, while for blue sapphire 0098 from Myanmar, Sr and Ta of the two analyses are different, and one analysis of ruby moz1_4 shows a negative Tb anomaly. The observed variations in concentrations probably reflect different densities of inclusions, while differences in the trace element patterns of the replicates indicate that the composition of the micron to submicron inclusion assemblage varied in the ablated areas.

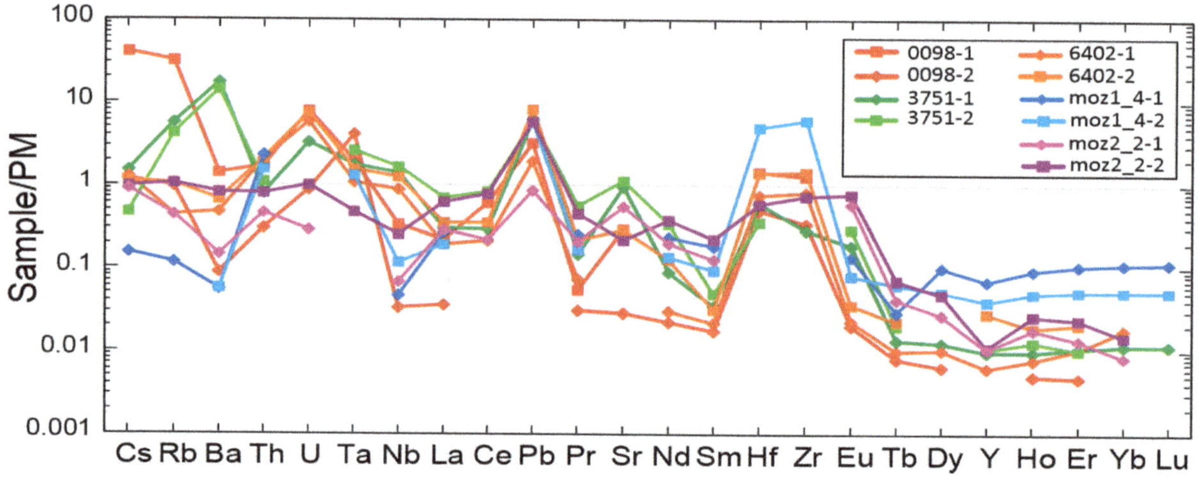

Figure 6. Primitive mantle-normalized [58] trace element concentrations of duplicate offline-ablations for five samples in this study. Lines of the same color denote replicate analyses of the same sample in a different area of the crystal. Sample 0098 is from the Mogok Stone Tract, Myanmar; samples 3751 and 6402 are from the Morogoro region in Tanzania; and samples moz1_4 and moz2_2 are from the Montepuez area in Mozambique.

3.2. *Radiogenic Isotopes*

3.2.1. Sr Isotopic Composition

Gem corundum from different locations and deposit types span a wide range of Sr isotopic compositions (Table 4 and Table S3). Sr isotopic ratios of the amphibolite-type rubies (21 analyses—three duplicate) span a range between $^{87}Sr/^{86}Sr = 0.7071 \pm 0.0005$ and $^{87}Sr/^{86}Sr = 0.7463 \pm 0.0042$. Within this group, rubies from Zahamena, MDG are the most radiogenic and have the widest range ($^{87}Sr/^{86}Sr = 0.7115 \pm 0.0035$ to $^{87}Sr/^{86}Sr = 0.7463 \pm 0.0042$), while rubies from Winza, TZA are the most unradiogenic and define a narrow range between $^{87}Sr/^{86}Sr = 0.7071 \pm 0.0006$ and $^{87}Sr/^{86}Sr = 0.7101 \pm 0.0014$ with rubies from Mozambique falling in between (Figure 7b, Table 4). For three of the Mozambique rubies, two replicate analyses were conducted, with Sr isotopic compositions for moz2_2 of $^{87}Sr/^{86}Sr = 0.7089 \pm 0.0030$ and $^{87}Sr/^{86}Sr = 0.7075 \pm 0.0006$, for moz2_6 of $^{87}Sr/^{86}Sr = 0.7148 \pm 0.0009$ and $^{87}Sr/^{86}Sr = 0.7108 \pm 0.0007$, and for moz2_8 of $^{87}Sr/^{86}Sr = 0.7164 \pm 0.0005$ and $^{87}Sr/^{86}Sr = 0.7112 \pm 0.0009$. For sample moz2_2, the analyses are within error of one another. For samples moz2_6 and moz2_8, however, the $^{87}Sr/^{86}Sr$ ratios of the duplicate ablations are outside of their respective uncertainty intervals, indicating Sr isotopic heterogeneity that is likely controlled by the ablation of larger inclusions [12].

The range in $^{87}Sr/^{86}Sr$ isotope ratios for the marble-type rubies ($^{87}Sr/^{86}Sr = 0.7059 \pm 0.0007$ to $^{87}Sr/^{86}Sr = 0.7182 \pm 0.0021$, median $^{87}Sr/^{86}Sr$ ~0.7082) is much narrower and more unradiogenic than that of both the amphibolite-type rubies (median $^{87}Sr/^{86}Sr$ ~0.7113) and the metamorphic blue sapphires (median $^{87}Sr/^{86}Sr$ ~0.7156) (Figure 7a). Among the different localities of marble-hosted ruby, rubies from Myanmar have a more radiogenic distribution (median $^{87}Sr/^{86}Sr$ ~0.7099) and the widest range, while including both the most and least radiogenic marble-hosted ruby samples (Figure 7c). Rubies from Afghanistan; Morogoro, TZA; and Vietnam are less radiogenic, with median $^{87}Sr/^{86}Sr$ of ~0.7083, ~0.7070, and ~0.7063, respectively. For two rubies from Morogoro, TZA, two ablations and subsequent Sr isotope analyses were performed, with Sr isotopic ratios of $^{87}Sr/^{86}Sr = 0.7081 \pm 0.0005$ and $^{87}Sr/^{86}Sr = 0.7063 \pm 0.0030$ for sample 3751 and $^{87}Sr/^{86}Sr = 0.7072 \pm 0.0029$ and $^{87}Sr/^{86}Sr = 0.7093 \pm 0.0047$ for sample 6402. For both samples, the duplicate analyses are within uncertainty of each other.

Table 4. Ranges in Sr and Pb isotopic ratios for the different localities analyzed in this study. Number of analyses in parentheses.

Sample	$^{87}Sr/^{86}Sr$	$^{206}Pb/^{204}Pb$	$^{207}Pb/^{204}Pb$	$^{207}Pb/^{206}Pb$	$^{208}Pb/^{206}Pb$
		Amphibolite-type deposits			
MOZ	0.7075–0.7291 ($n = 12$)	17.86–19.00 ($n = 4$)	15.33–15.60 ($n = 4$)	0.8305–0.8655 ($n = 4$)	2.0292–2.0955 ($n = 4$)
TZA, Winza	0.7071–0.7101 ($n = 4$)	16.88–17.57 ($n = 3$)	15.57–15.62 ($n = 3$)	0.8880–0.9223 ($n = 4$)	2.1290–2.1786 ($n = 3$)
MDG, Zahamena	0.7115–0.7463 ($n = 4$)	15.92–24.64 ($n = 4$)	15.21–15.99 ($n = 4$)	0.6488–0.9539 ($n = 4$)	1.5866–2.8679 ($n = 4$)
		Marble-hosted deposits			
MMR	0.7059–0.7182 ($n = 8$)	18.02–19.75 ($n = 9$)	15.55–15.72 ($n = 9$)	0.7959–0.8636 ($n = 9$)	1.9643–2.1037 ($n = 9$)
VNM	0.7062–0.7066 ($n = 3$)	18.11–19.08 ($n = 3$)	15.50–15.64 ($n = 3$)	0.8177–0.8558 ($n = 3$)	2.0223–2.0819 ($n = 3$)
AFG	0.7059–0.7089 ($n = 5$)	17.77–19.13 ($n = 5$)	15.48–15.67 ($n = 5$)	0.8196–0.9721 ($n = 5$)	2.0090–2.1081 ($n = 5$)
TZA, Morogoro	0.7063–0.7093 ($n = 5$)	17.38–35.86 ($n = 4$)	15.39–16.57 ($n = 4$)	0.4616–0.8858 ($n = 4$)	1.0561–2.1320 ($n = 4$)
		Blue sapphire deposits			
MDG, Andrebabe	0.7066–0.7087 ($n = 2$)	17.08–18.30 ($n = 3$)	15.47–15.57 ($n = 3$)	0.8505–0.9056 ($n = 3$)	2.1113–2.1484 ($n = 3$)
MDG, Ilakaka	0.7156–0.7316 ($n = 4$)	17.45–20.01 ($n = 4$)	15.54–15.80 ($n = 4$)	0.7895–0.8899 ($n = 4$)	2.0243–2.5975 ($n = 4$)
LKA	0.7081–0.7316 ($n = 7$)	17.24–20.69 ($n = 7$)	15.49–15.95 ($n = 7$)	0.7706–0.8974 ($n = 7$)	1.8330–3.2644 ($n = 7$)
MMR	0.7098–0.7152 ($n = 7$)	17.86–18.65 ($n = 3$)	15.40–15.65 ($n = 3$)	0.8381–0.8634 ($n = 3$)	2.0629–2.0856 ($n = 3$)

Metamorphic blue sapphires have Sr isotopic compositions ranging from $^{87}Sr/^{86}Sr = 0.7066 \pm 0.0020$ to $^{87}Sr/^{86}Sr = 0.7316 \pm 0.0008$. Among the different localities, Ilakaka, MDG has blue sapphires with the most radiogenic Sr isotopic composition (median $^{87}Sr/^{86}Sr$ ~0.7206), followed by Sri Lanka and Myanmar with median $^{87}Sr/^{86}Sr$ ~0.7141 and $^{87}Sr/^{86}Sr$ ~0.7109, respectively (Figure 7d, Table 4). Only two blue sapphires from Andrebabe, MDG were analyzed, and represent the most unradiogenic suite with Sr isotopic ratios of 0.7066 ± 0.0020 and 0.7087 ± 0.0024. For one blue sapphire from Myanmar (sample 0098), two Sr isotope analyses were carried out, giving two different $^{87}Sr/^{86}Sr$ ratios: $^{87}Sr/^{86}Sr = 0.7152 \pm 0.0008$ and $^{87}Sr/^{86}Sr = 0.7130 \pm 0.0010$.

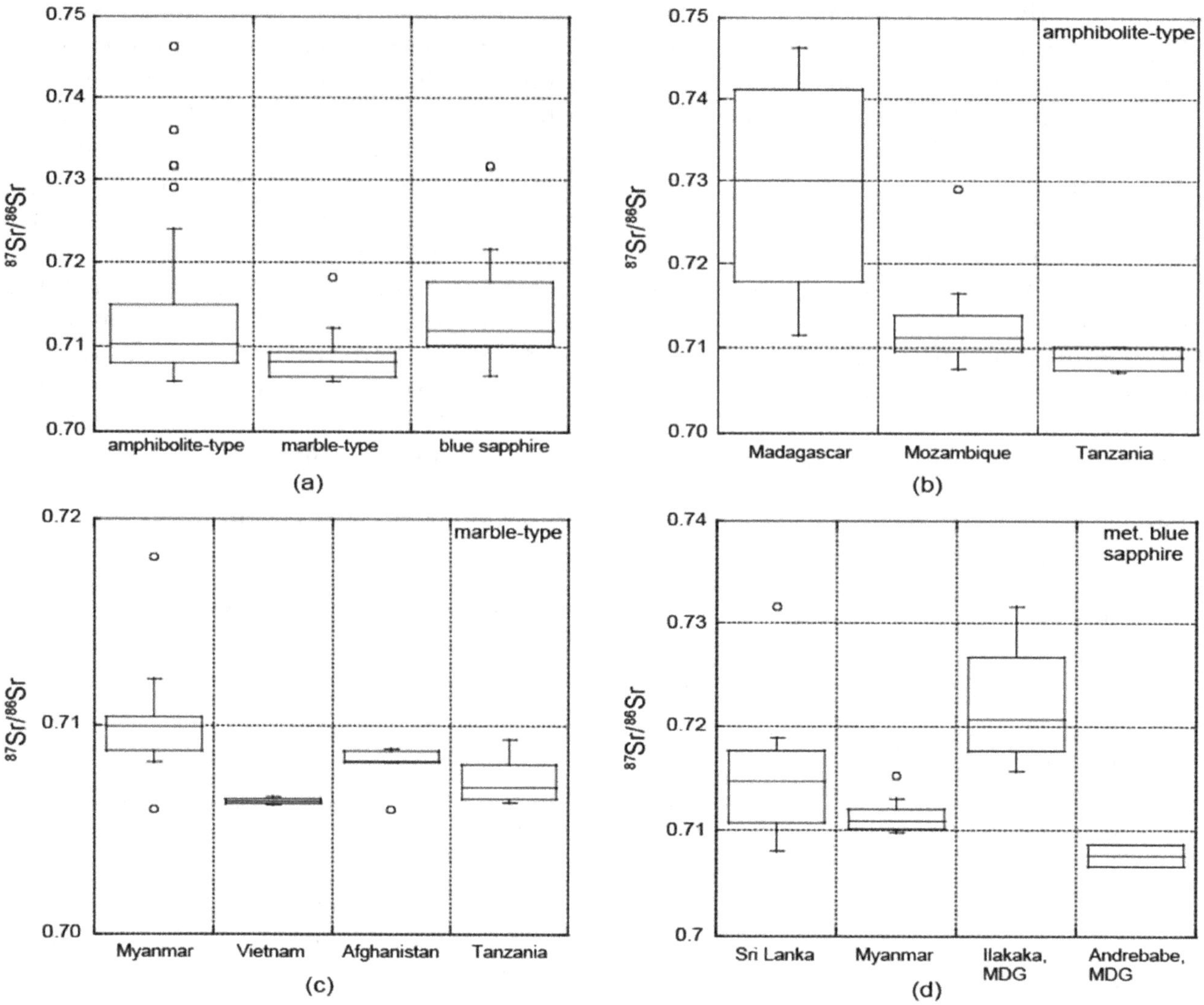

Figure 7. Box and whisker plots displaying $^{87}Sr/^{86}Sr$ isotopic compositions of gem corundum from (**a**) the three different deposit groups, (**b**) the amphibolite-type ruby localities, (**c**) the marble-type ruby localities, and (**d**) the localities of metamorphic blue sapphire measured in this study. Individual data points are outliers whose value is either greater than UQ + 1.5 × IQD or less than LQ − 1.5 × IQD (where UQ = upper quartile, LQ = lower quartile, and IQD = inter quartile distance).

3.2.2. Pb Isotopic Composition

The relative variations in Pb isotopic compositions in corundum of the different groups are minor (Table 4, Figure 8a–d); the amphibolite-type rubies have slightly more unradiogenic compositions (median $^{206}Pb/^{204}Pb$ ~17.93, median $^{207}Pb/^{204}Pb$ ~15.58, median $^{207}Pb/^{206}Pb$ ~0.8655, and median $^{208}Pb/^{206}Pb$ ~2.1290) than the marble-type rubies (median $^{206}Pb/^{204}Pb$ ~18.63, median $^{207}Pb/^{204}Pb$ ~15.61, median $^{207}Pb/^{206}Pb$ ~0.8308, and median $^{208}Pb/^{206}Pb$ ~2.0483) and the metamorphic blue sapphires (median $^{206}Pb/^{204}Pb$ ~18.50, median $^{207}Pb/^{204}Pb$ ~15.58, median $^{207}Pb/^{206}Pb$ ~0.8436, and median $^{208}Pb/^{206}Pb$ ~2.0851).

Among the different localities of amphibolite-type ruby, Pb isotopic ratios in samples from Zahamena, MDG span a wide range (e.g., $^{207}Pb/^{206}Pb$ = 0.6488 ± 0.0003 to 0.9539 ± 0.0005, Figure 9a), while the ranges for samples from Mozambique and Winza, TZA are much more narrow, $^{207}Pb/^{206}Pb$ = 0.8305 ± 0.0026 to 0.8655 ± 0.0016 and $^{207}Pb/^{206}Pb$ = 0.8880 ± 0.0003 to 0.9223 ± 0.0002, respectively. For one ruby (sample moz2_6), two Pb isotope analyses were carried out, giving two different $^{207}Pb/^{206}Pb$ ratios, 0.8655 ± 0.0016 and 0.8305 ± 0.0026.

Pb isotopic ratios of rubies from different marble-type localities overlap significantly, with median $^{207}Pb/^{206}Pb$ values of ~0.8280, ~0.8400, ~0.8489, and ~0.8328 for rubies from Myanmar; Vietnam; Afghanistan; and Morogoro, TZA, respectively (Figure 9b). Sample 5202 from Morogoro, TZA is much more radiogenic ($^{207}Pb/^{206}Pb$ = 0.4615 ± 0.0002) than any other sample in this group, resulting in a wide range of Pb isotopic compositions for this locality.

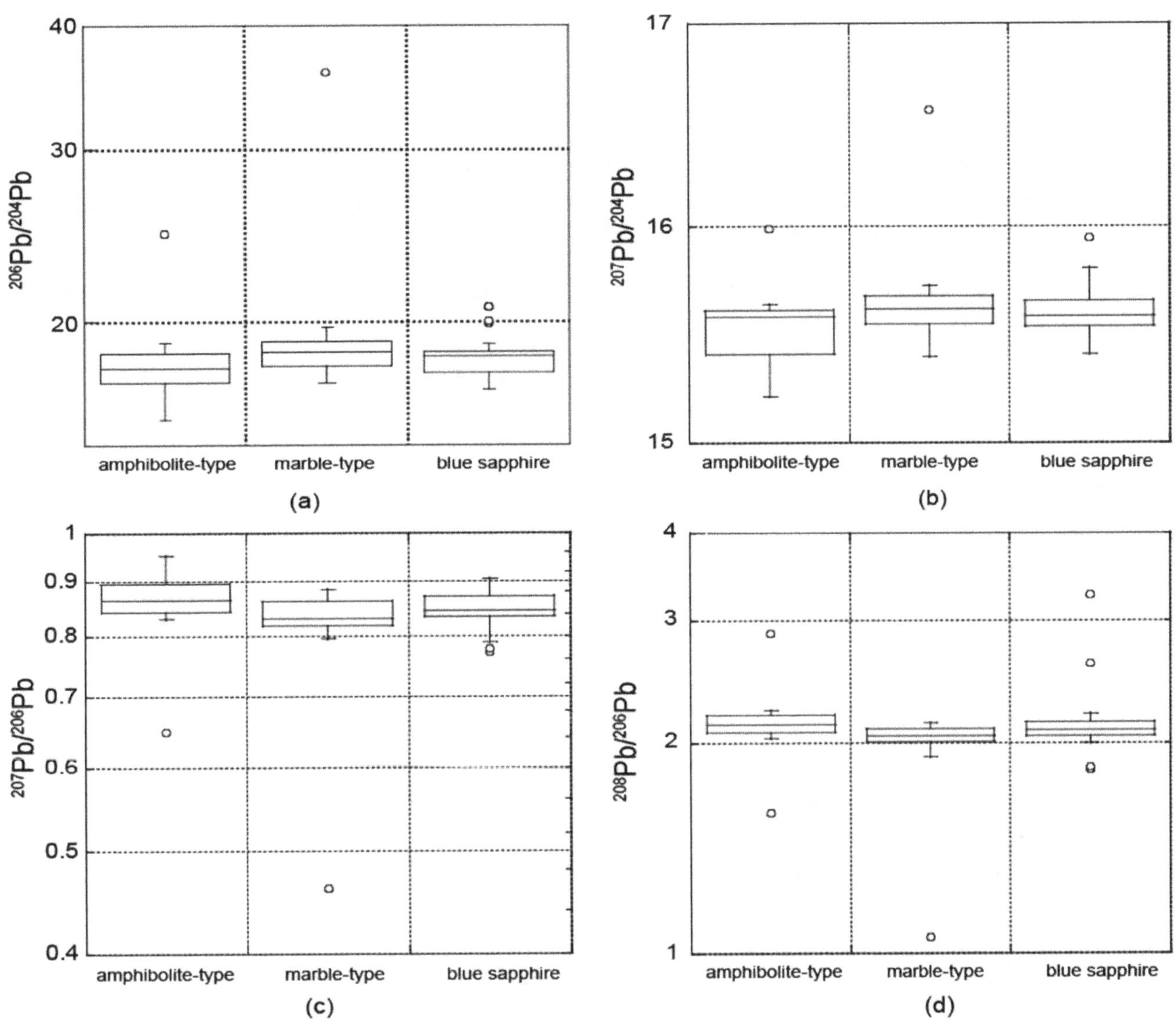

Figure 8. Box and whisker plots displaying (**a**) $^{206}Pb/^{204}Pb$, (**b**) $^{207}Pb/^{204}Pb$, (**c**) $^{207}Pb/^{206}Pb$, and (**d**) $^{208}Pb/^{206}Pb$ isotopic compositions of gem corundum from the three different deposit groups measured in this study. Individual data points are outliers whose value is either greater than UQ + 1.5 × IQD or less than LQ − 1.5 × IQD.

Similar to the amphibolite-type and marble-type rubies, Pb isotopic ratios of metamorphic blue sapphires from different localities overlap (Figure 9). Samples from Andrebabe, MDG have the most unradiogenic Pb isotope ratios (e.g., median $^{207}Pb/^{206}Pb$ ~0.8722), followed by Myanmar with median $^{207}Pb/^{206}Pb$ ~0.8574. Blue sapphires from Sri Lanka and Ilakaka, MDG are the most radiogenic with $^{207}Pb/^{206}Pb$ median values of ~0.8384 and ~0.8436, respectively.

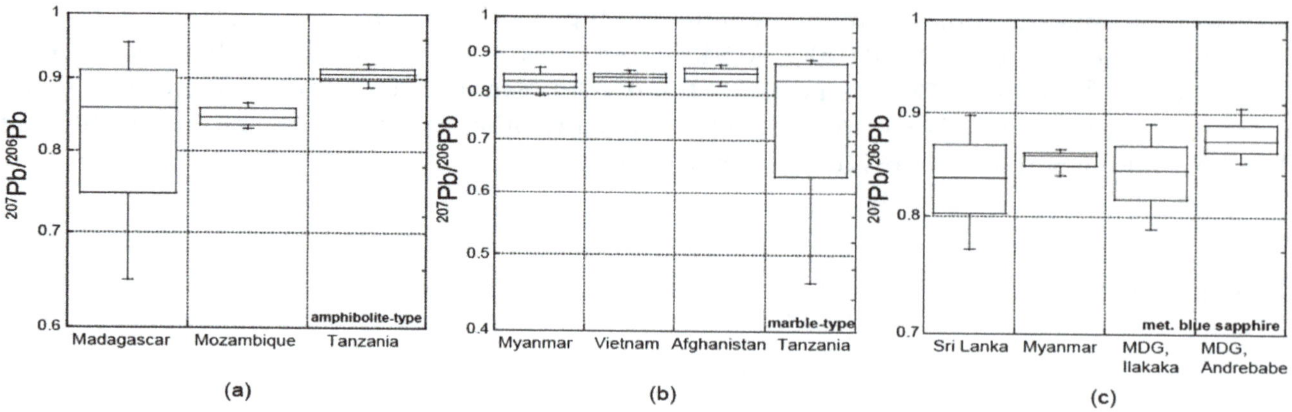

Figure 9. Box and whisker plots displaying $^{207}Pb/^{206}Pb$ isotopic compositions of gem corundum from (**a**) the amphibolite-type ruby localities, (**b**) the marble-type ruby localities, and (**c**) the localities of metamorphic blue sapphire measured in this study.

No correlations between Sr and Pb isotopic compositions are apparent for any of the sample suites analysed in this study (Figure 10).

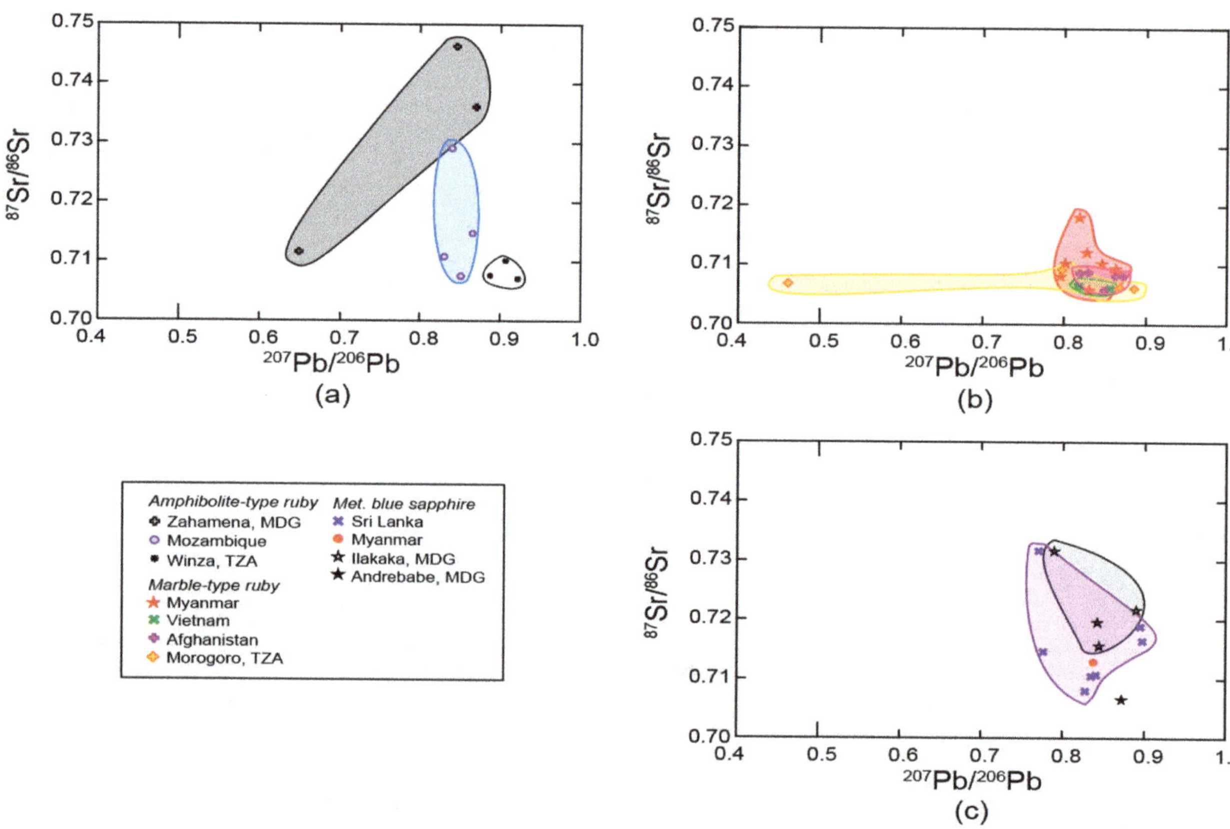

Figure 10. $^{87}Sr/^{86}Sr$ vs. $^{207}Pb/^{206}Pb$ ratios of (**a**) amphibolite-type rubies, (**b**) marble-type rubies, and (**c**) metamorphic blue sapphires measured in this study. Analytical uncertainties are less than the size of the plot symbols. Color of the fields correspond to the color of the symbols.

4. Discussion

4.1. Trace Elements

Quantitative data for the most commonly observed elements in ruby and sapphire—Fe, Cr, Ti, V, and Ga—have been used by many authors (e.g., [8–10,15,33,59,60]) in attempts to determine gem corundum provenance. Figure 11 comprises a number of existing discrimination schemes, showing how our data compare with previously published data from the same localities. A number of classical and non-classical deposits are also plotted to place our data into worldwide context. A commonly used diagram is a logarithmic plot of wt% values Cr_2O_3/Ga_2O_3 versus Fe_2O_3/TiO_2 (e.g., [8,9,61]).

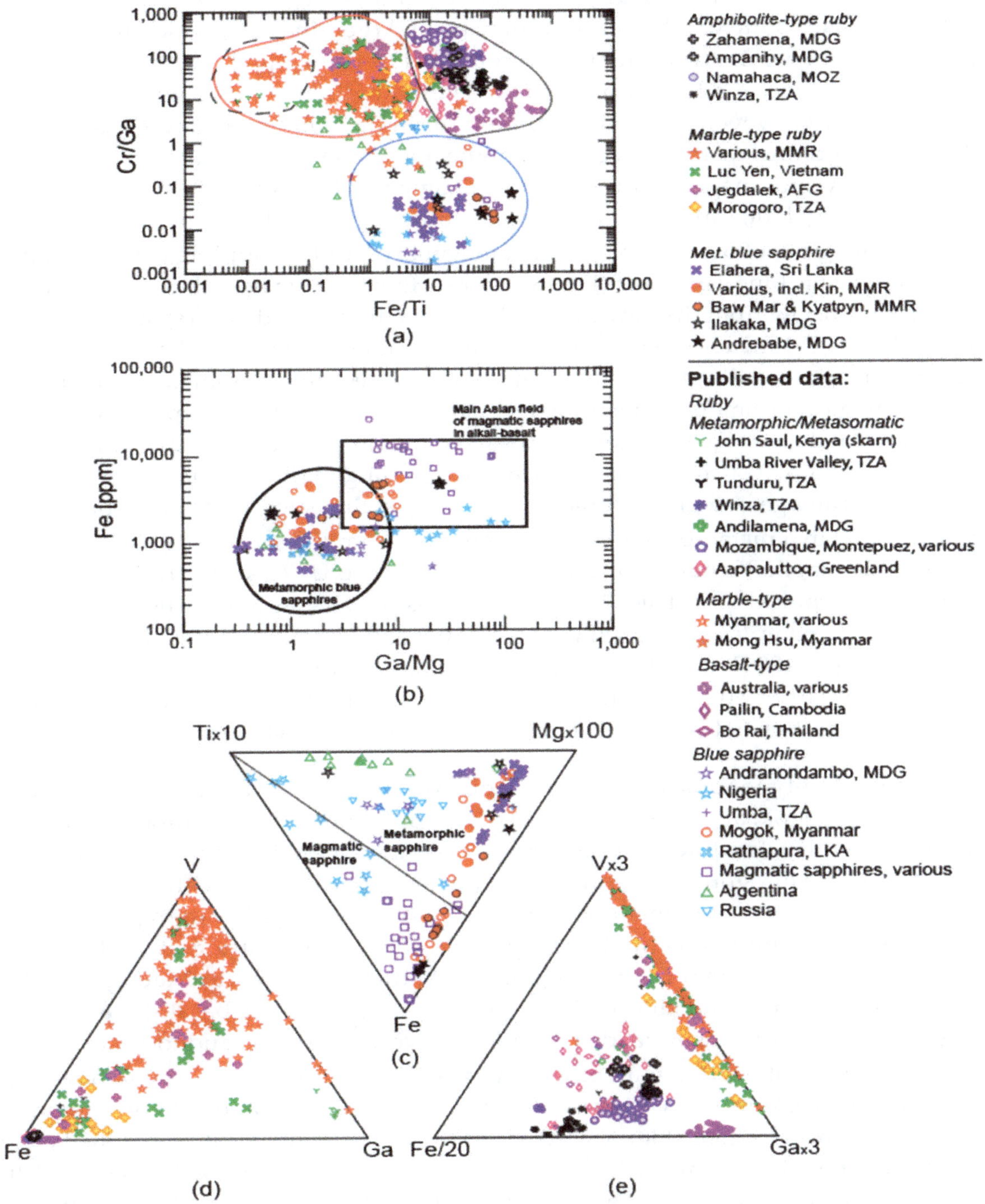

Figure 11. (**a**) Plot of Cr/Ga (ppm) versus Fe/Ti (ppm) (modified from [8]); fields are red = marble-type, black/dashed = Mong Hsu ruby [62], black = high Fe, and blue = blue sapphires. (**b**) Fe (ppm) versus Ga/Mg (ppm) (after [10]); (**c**) Mg × 100-Fe-Ti × 10 ternary plot modified after [10]; and (**d**) Fe-V-Ga ternary plot (after [15]), and (**e**) Fe/20-V × 3-Ga × 3 (after [28]) in gem corundum analyzed in this study. All data plotted are laser ablation inductively coupled plasma mass spectrometry (LA-ICP-MS) values. Additional data from [10,23,25,33,42,43,59,60,62–69].

Figure 11a is a plot of Cr/Ga versus Fe/Ti for the gem corundum suites analyzed in this study. Published datasets used for comparisons are exclusively LA-ICP-MS data, excluding electron microprobe (EMPA) data sets, because it has been shown that data derived from the different methods vary systematically for several elements [33]. In such a plot, marble-type rubies have Cr/Ga > 1 and Fe/Ti < 6, while high-Fe rubies, including amphibolite-type and basalt-type ruby, have Cr/Ga > 1 and Fe/Ti > 6. Blue sapphires, with measurable Cr, have Cr/Ga < 1and Fe/Ti > 1. Exceptions are two non-classical deposits, a suite of alluvial sapphires from Argentina [68], which have Cr/Ga of up to ~4 and Fe/Ti > 0.1, and a suite from the Ural mountains in Russia [67], with Cr/Ga ~1–3. High-Fe rubies, marble-type rubies, and blue sapphires thus fall into discrete fields with only minor overlap along the boundaries. However, different localities within the different deposit type groups overlap significantly. Blue sapphires, with the exception of the above-mentioned suites, overlap regardless of origin (metamorphic versus magmatic) and locality. Diagrams of Fe versus Ga/Mg and Mg × 100-Fe-Ti × 10, first introduced by [10], can be used to separate blue sapphires based on their geological origin. On such plots, the majority of our data of different metamorphic blue sapphire suites fall into the fields defined for metamorphic blue sapphires (Figure 11b,c). However, we note that several of our samples from Myanmar, specifically from the Baw Mar and Kyatpyn mines, and Andrebabe, MDG, fall into the "magmatic sapphire" field. In addition, a suite of blue sapphires from the northern part of the Baw Mar mine also plots as magmatic, despite the fact that these samples were classified as metamorphic based on standard gemological techniques, including inclusion studies and UV/vis spectroscopy [43]. On the basis of the geology at the mine site, these sapphires likely formed either directly in syenitic pegmatites or through the interaction of such pegmatites with surrounding marbles and/or gneisses [43]. In other words, trace element and gemological characteristics suggest that subgroups of sapphires classified as metamorphic formed in a metamorphic environment, but from magmatic rocks. As such, their trace elements reflect the magmatic nature of their host rock, while UV/vis spectroscopy and inclusions speak to their metamorphic origin. This illustrates that existing classification schemes for blue sapphire cannot account for the complex geological processes that form blue sapphires. However, both the Fe versus Ga/Mg and Mg × 100-Fe-Ti × 10 classification diagrams are useful in separating syenite-related metamorphic sapphire from Myanmar and Andrebabe, MDG, from those from Sri Lanka and Ilakaka, MDG. Within the respective fields, however, these populations overlap.

Among the marble-type rubies, only those from Mong Hsu, MMR [62], which is located about 250 km southeast of Mogok, are clearly distinguishable in Figure 11a, with much lower Fe/Ti values than those from the other localities. Rubies from Afghanistan analysed in this study have higher Cr/Ga than those from Morogoro, TZA, however, both localities overlap with rubies from Myanmar and Vietnam. The overlap is somewhat less complete among the amphibolite-type rubies, with those from Winza, TZA, and Ampanihy, MDG, analyzed in this study plotting at slightly lower Cr/Ga and higher Fe/Ti than those from Mozambique, the Andilamena area in Madagascar, Greenland, and basalt-type rubies from Cambodia and Thailand. However, only the basalt-type population from the New England Gem field, Australia, displaying relatively high Fe/Ti and low Cr/Ga indicative of igneous origin [8], does not overlap with other deposits. This deposit is thought to have experienced magmatic–metasomatic inputs in its genesis [66]. When plotting the same data on a ternary diagram depicting Fe–V–Ga concentrations (Figure 11d) after [15], high-Fe rubies all plot at the Fe-apex, while the low-Fe marble-type rubies show a wider distribution. Myanmar rubies plot at higher V; those from Morogoro, TZA, plot close to the Fe-apex; and rubies from Afghanistan and Vietnam overlap both suites. While basalt-type rubies overlap with amphibolite-type ruby here, they are easily separated from other high-Fe rubies using Mg [5]. Smith et al. [28] modified the Fe–V–Ga ternary diagram to Fe/20-V × 3-Ga × 3 to provide more resolution in the Fe-apex area of the plot (Figure 11e). This approach allows the distinction of the basalt-hosted rubies from most rubies of metasomatic and metamorphic origin, but there is still significant overlap for the various localities. Only rubies from Winza, TZA, show minimal overlap.

The overlap of samples from the different localities of both ruby groups examined here, as well as that among metamorphic blue sapphires, in commonly-used discrimination schemes, illustrates that, while they are useful in distinguishing rubies of different geological deposit types, they are not very effective in distinguishing between different localities of the same deposit type.

The broader spectrum of quantitative trace element data we were able to obtain using the ablation and measurement parameters detailed in Section 2.2.1 now makes it possible to evaluate the use of additional elements as provenance discriminators that have not been yet measured consistently. Amphibolite-type rubies in particular show a much broader spectrum of trace elements consistently >LOQ, while such additional elements are fairly limited for the marble-type rubies. The main reason for this is very likely the (on average) higher clarity of the marble-hosted rubies analysed in this study (Figure A2). This suggests that the potential of using such low abundance elements is fairly limited, at least with the current sensitivity of the LA-ICP-MS method, because the main goal of developing a geochemical fingerprint is the determination of provenance for high-quality gemstones. However, the elements Ni, Zn, and Pb are consistently > LOQ for all corundum analysed in this study, regardless of geologic origin or clarity, and were thus included in the discrimination evaluation. Of these elements, Pb has proven to be ineffective as discriminator of geographic locality, at least for the deposit types and localities analysed in this study as contents between the analysed suites overlap significantly. Ni, however, shows good discriminatory power for the separation of different localities of amphibolite-type ruby (Figure 12a), and, in conjunction with Ti and V, defines discrete fields for rubies from Winza, TZA, Mozambique, Zahamena, MDG, and Amphanihy, MDG. Rubies from Greenland, however, have a wide range of Ni, Ti, and V contents, resulting in significant overlap on a Ni × 100-Ti-V × 10 plot (Figure 12a). Some of this overlap can be resolved by further plotting the data on a binary Ga versus V plot (Figure 12b), thus effectively separating the Greenland rubies from those from Madagascar; minor overlap of Mozambique and Greenland persists, however.

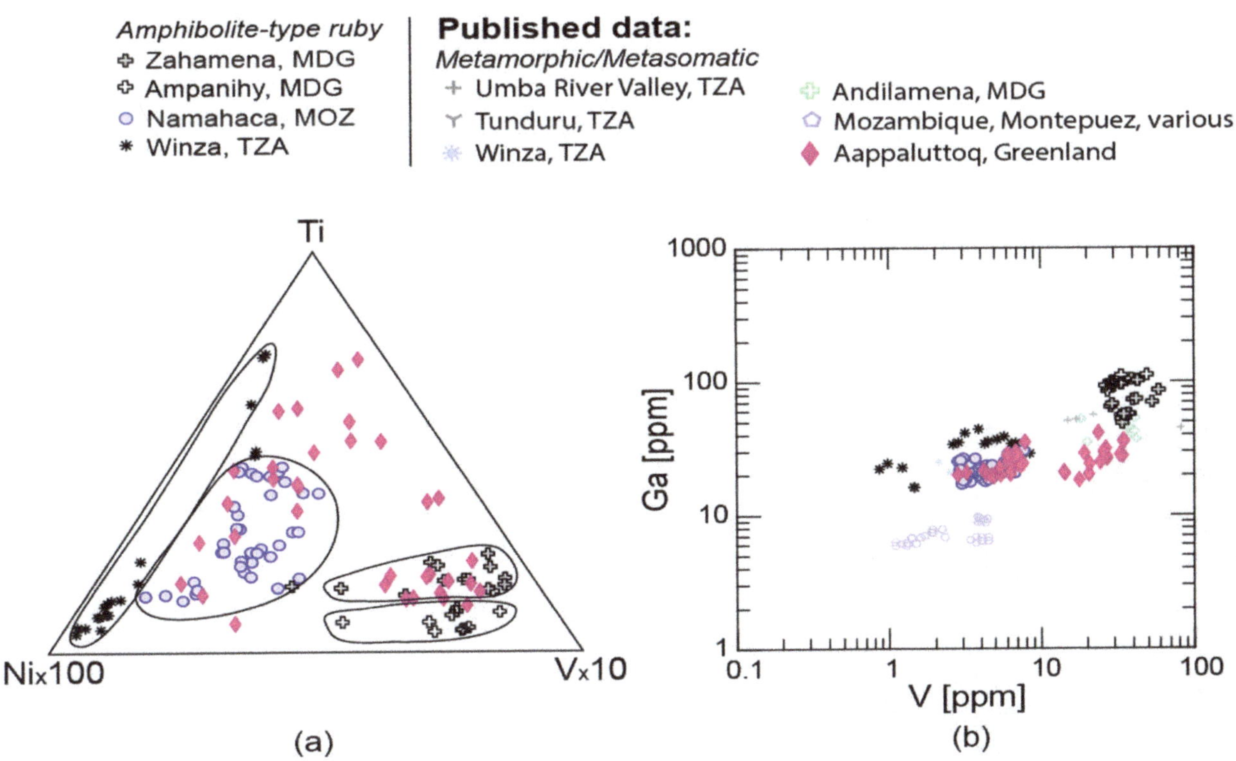

Figure 12. (**a**) Ternary diagram of Ni × 100-Ti-V × 10 for amphibolite-type rubies analyzed using LA-ICP-MS in this study. Greenland data are taken from [12]. (**b**) Binary plot of Ga (ppm) versus V (ppm) abundances in amphibolite-type rubies analyzed in this study. Additional data from [23,25,59].

On the Ga versus V plot, rubies from the Andilamena region in Madagascar [59] plot at slightly lower Ga values than those analyzed in this study. More notably, rubies from Mozambique analyzed by [23] have significantly lower Ga contents than rubies from Mozambique in this study. Both these studies used matrix-matched synthetic corundum standards as calibration materials, while in this study, NIST612 was used, resulting in more accurate analyses for their data. These differences emphasize the importance of using matrix-matched standards for accurate LA-ICP-MS analyses, and also illustrate that, for effective comparative studies, the same LA-ICP-MS system and reference materials should be used.

For the discrimination of different localities among marble-type rubies, none of the additional trace elements obtained in this study add further discriminatory power. Instead, a combination of the most abundant trace elements in corundum delivers the best results (Figure 13). A binary plot of V versus Cr roughly divides rubies from Myanmar; Morogoro, TZA; and Afghanistan into separate fields with only minor overlap (Figure 13a), while a binary plot of Fe versus V clearly separates the Myanmar rubies analyzed in this study from those from Morogoro, TZA (Figure 13b). Ternary diagrams of Fe/Mg × 2-Cr/Ti-V/10 and V-Cr/10-Ti archive similarly passable separations of Myanmar; Morogoro, TZA; and Afghanistan rubies (Figure 13c,d). However, rubies from Vietnam have very variable trace element abundances and overlap with all other marble-type localities analysed in this study. Previously published data of Myanmar rubies [33,62] agree reasonably well with data from this study, however, the data presented here are more variable.

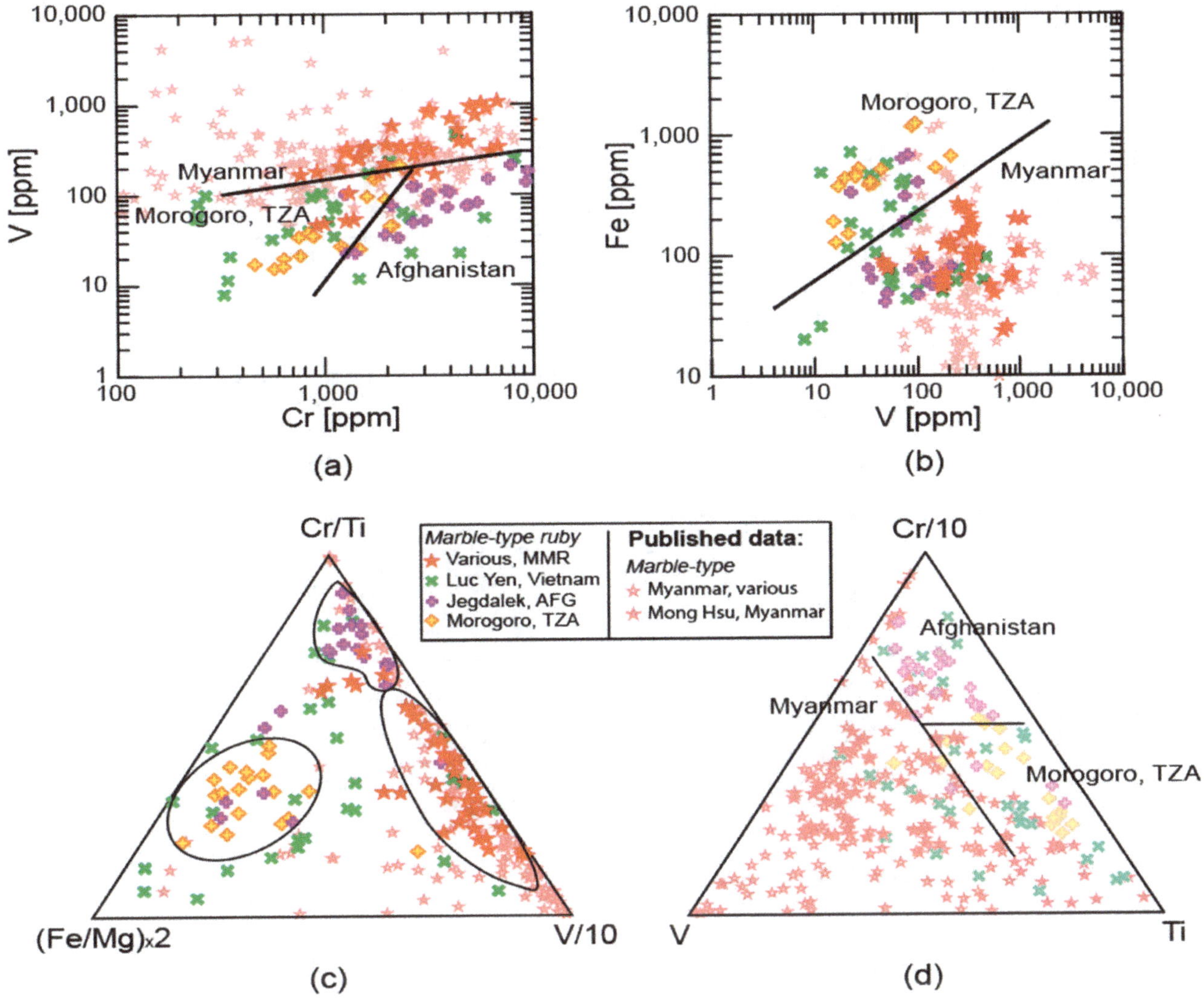

Figure 13. (**a**) Binary plot of V (ppm) versus Cr (ppm), (**b**) binary plot of Fe (ppm) versus V (ppm), (**c**) ternary diagram of Fe/Mg × 2-Cr/Ti-V/10, and (**d**) ternary diagram of V-Cr/10-Ti for marble-type rubies analyzed using LA-ICP-MS in this study. Fields are drawn by eye. Additional data from [33,62].

As for marble-type ruby, none of the additional elements obtained in this study added obvious further discriminatory power for the separation of different localities of metamorphic blue sapphire. However, Cr, which is often reported close to or below the LOD for samples from the localities studied here [10,70], is consistently above LOQ and shows potential as an additional discriminatory element. A binary plot of Mg versus V separates sapphires from Myanmar and Andrebabe, MDG, from sapphires from Sri Lanka and Ilakaka, MDG. In addition, syenite-related metamorphic sapphires have lower Mg (<20 ppm) and V (<4 ppm). A bivariate plot of V versus Cr × Ti separates Sri Lanka and Ilakaka, MDG, sapphires with only minor overlap (Figure 14). Myanmar and Andrebabe, MDG, with magmatic trace element signatures, have low Cr × Ti. Using a bivariate diagram of V versus Ga/Ti, the Andrebabe, MDG sapphires can be separated from both the Myanmar and the overlapping Sri Lanka and Ilakaka, MDG sapphires (Figure 14c). Previously published data for sapphires from Andranondambo, MDG [59] show significant overlap with sapphires from Sri Lanka and Ilakaka, MDG. Trace element compositions previously published for Myanmar sapphires [42,43] are more variable than those analyzed in this study. Metamorphic sapphires from Russia [67] and Argentina [68] are easily distinguished from the other localities by their high Cr × Ti and low and high V contents, respectively (Figure 14b).

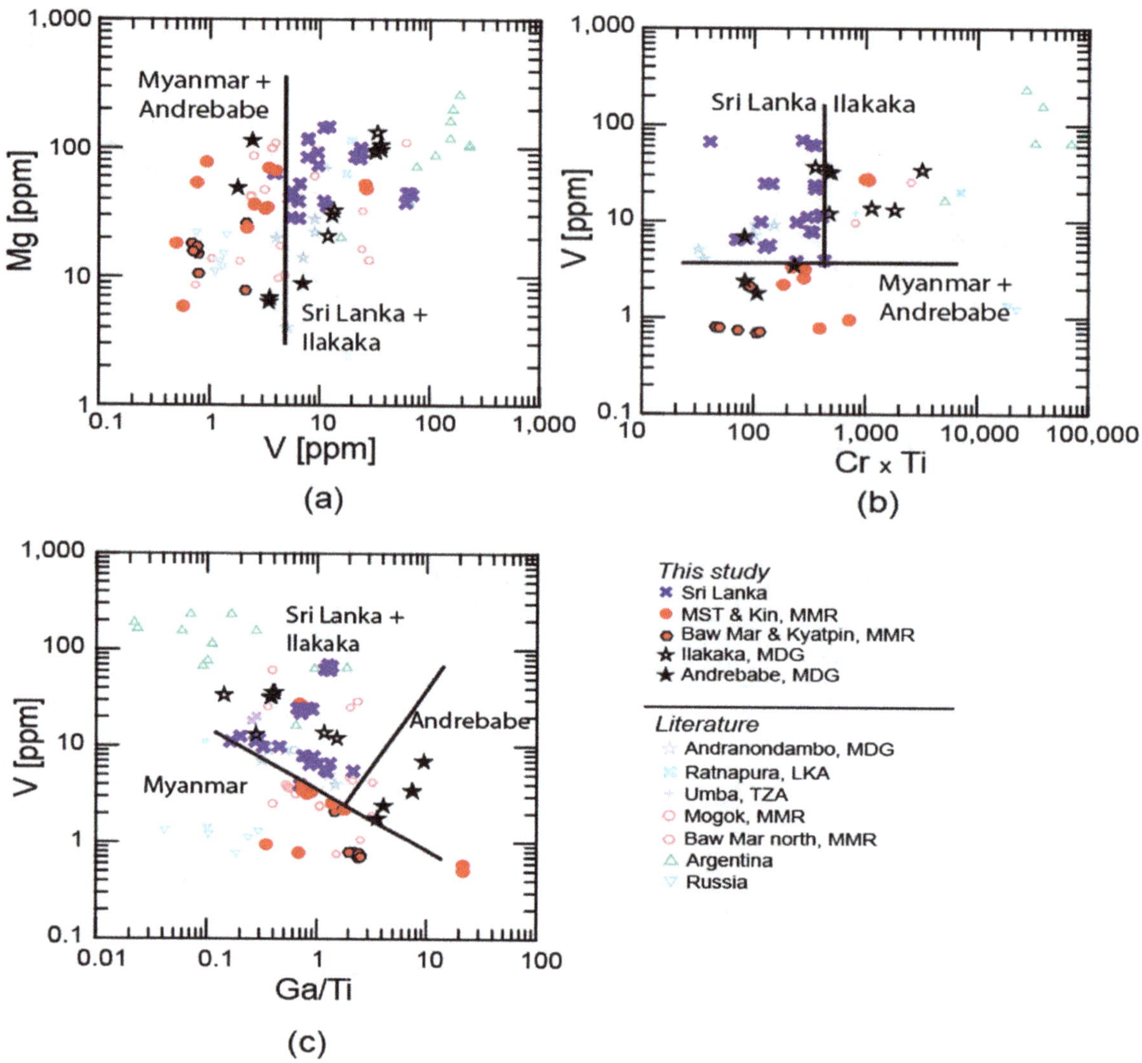

Figure 14. Binary plots of (**a**) Mg (ppm) versus V (ppm), (**b**) V (ppm) versus Cr × Ti (ppm), and (**c**) V versus Ga/Ti for metamorphic blue sapphires analyzed using LA-ICP-MS in this study. Fields are drawn by eye. Additional data from [10,42,43,59].

The results show that distinguishing between blue sapphires from Sri Lanka and Myanmar using trace elements is easily accomplished. Sapphires from Madagascar, however, show extensive chemical overlap with sapphires from other localities, which can be somewhat well-resolved for the small number of samples analyzed here. Madagascar has a vast number of ruby and sapphire deposits that are most often secondary deposits with stones that have formed in a variety of different geological settings [27]. As such, they are not only characterized by very variable trace element characteristics, but also a strong gemological diversity, and are known to overlap (sometimes significantly) with metamorphic sapphires from all other major sources [4].

Multivariate statistical analysis, including principal component analysis (PCA) and linear discriminant analysis (LDA), is a tool to identify patterns and relationships between multiple variables simultaneously and can, potentially, increase the reliability of origin determination as well as detect elements with potential discriminatory power. Both PCA [59,71] and LDA [59] have been successfully applied to rubies and sapphires of different geological deposit types; here, we apply both approaches to visualise the compositional differences for samples of the same geological type from different localities. Calculation of robust quantitative discriminants in this study is restricted by the modest population sizes, so we use PCA and LDA to qualitatively determine which elements might have discriminatory value.

To this end, we applied LDA to our online LA-ICP-MS dataset, using the software R (for details on R runstreams see Supplementary Material S4). LDA is a dimension reduction technique that defines linear discriminants that maximize the "distance" between the means of pre-defined classes. Data for rubies from Greenland [12] were included in the calculations for the amphibolite-type group, because they were collected on the same system using the same ablation and measurement conditions. LDA solutions are derived using the full dataset as calibration data, and the classification error (CE) of the method is determined by testing the calibration data on the method by the following formula:

$$\text{CE} = \text{number of incorrect classifications/total number of samples} \times 100\% \quad (1)$$

As the CE for the calibration data of the method it calibrated will result in an overly optimistic assessment of classification success or failure, solutions are assigned a method uncertainty by K-fold cross validation. This is conducted by subdividing the sample sets into five random folds with 80% of the data classified as "calibration" data, and the remaining 20% as "validation" data. An LDA solution is derived using the calibration set of a single fold, and tested with its respective validation dataset. This is conducted upon all five folds, and the calculated error rates for the folds are then averaged and taken as the method uncertainty.

For the amphibolite-type deposits, LDA was applied iteratively to the datasets using combinations of chemical variables; through this process, we narrowed the number of most useful variables down to four—V, Fe, Mg, and Ni—by excluding those elements that only reproduce existing discrimination power of others. This result confirms the potential of Ni as an element with discriminatory power for the amphibolite-type deposits evaluated in this study. To assess its influence, an LDA solution using only V, Fe, and Mg was applied (Figure 15a).

This solution results in some overlap of rubies from Ampanihy, MDG; Winza, TZA; and Greenland, as well as some overlap of rubies from Greenland with rubies from Mozambique, similar to that observed using a ternary diagram of Ni × 100-Ti-V × 10 (Figure 12a). By adding Ni to the solution, only the overlap of Greenland with Mozambique remains (Figure 15b), and the CE is minimized at ~13.7%. The average K-fold (n = 5 folds) cross-validation error of 17.33 ± 11.93 (2σ) is identical for both solutions (Supplementary Material S5).

For marble-type deposits, an LDA solution using seven elements—V, Fe, Ga, Mg, Ti, Cr, and Zr—gives the best discrimination for the different localities studied here. While Zr shows no obvious discriminatory power in elemental plots, it influenced the LDA calculations positively. Bivariate plots of the linear discriminants (LDs) show a good separation of Morogoro, TZA and Myanmar (Figure 15c,d), similar to the discrimination achieved by plotting elemental abundances against each other (e.g., Figure 13b,c). In addition, the LDA solution results in a better discrimination of Vietnam rubies from Morogoro, TZA (Figure 15c) and from Afghanistan rubies (Figure 15d), which otherwise overlap significantly in the graphical bivariate discrimination schemes using elemental abundances (Figure 13). However, the chemical overlap is still significant with a CE of ~21% and a similar average K-fold (n = 5 folds) cross-validation error of 19.09 ± 11.85 (2σ).

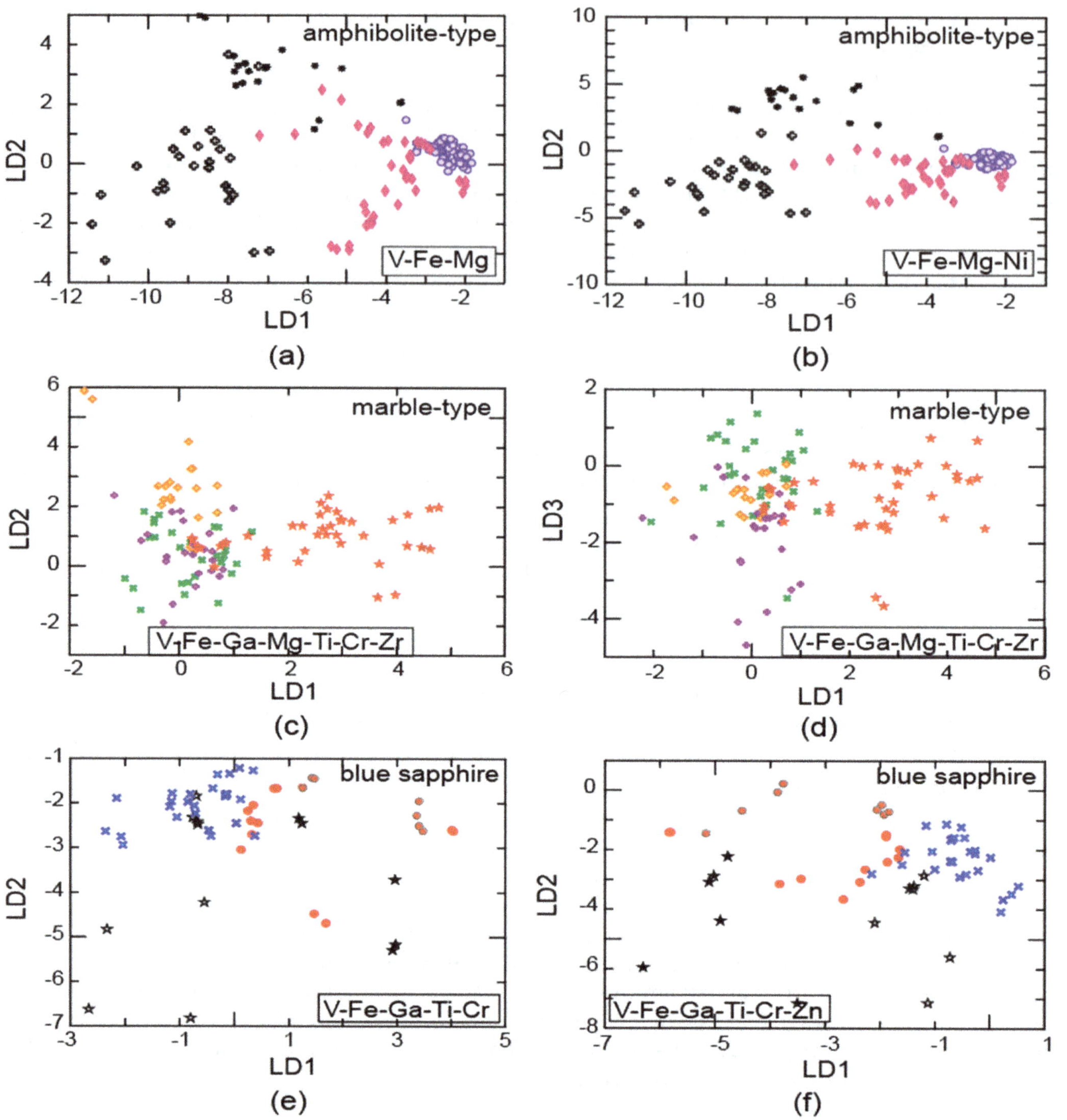

Figure 15. Linear discriminant analysis (LDA) solution for (**a**), (**b**) amphibolite-type rubies; (**c**), (**d**) marble-type rubies; and (**e**), (**f**) metamorphic blue sapphires. Elements used in method calibration are inset on each figure. Symbols as in Figure 11.

For application of LDA to the metamorphic blue sapphires, the Myanmar sapphires were separated into two groups based on the inference that they have discrete geological origins, resulting in five overall groups including the other locations. The best results were obtained using the six elements V, Fe, Ga, Ti, Cr, and Zn. Because Zn shows no obvious discriminatory power in elemental plots, an LDA solution using only V, Fe, Ga, Ti, and Cr was also applied. The results show a clear improvement with the inclusion of Zn, reducing the overlap between sapphires from Sri Lanka and Ilakaka, MDG (Figure 15e,f). The one obvious outlier of the Mogok Stone Tract (MST) & Kin, MMR group behaves similar in the discrimination schemes using elemental abundances (Figure 14), reflecting a syenite-related origin. Whether this indicates that syenite-related samples are also found at the Kin mine or if the sample was mislabeled is unclear. The CE for the LDA solution calibrated on the elements V, Fe, Ga, Ti, and Cr is ~27%, and ~18.6% for the solution calibrated on the elements V, Fe, Ga, Ti, Cr, and Zn, showing a clear improvement of the classification through the addition of Zn. No K-fold cross-validated error is calculated here owing to the small sample sets, which result in erratic calculated error rates, being prone to over- or underestimation of the true success or failures of the LDA methods. This is indicated by the large standard deviation of the average cross-validation error of the amphibolite-type group, which is the largest dataset in this study.

Comparison of Online and Offline Laser Ablation Approaches

Aside from a small number of elements that substitute for Al^{3+} in the crystal structure (e.g., Mg, Cr, Fe, V, Ti, and Ga), trace elements in gem corundum are present as mineral inclusions or as constituents in fractures, and their absolute trace element abundances are primarily a function of impurity density [12]. Therefore, the larger sampling volumes of the offline ablation approach result in more varied trace element characteristics that correspond to the presence of nano- to micrometer-sized mineral and fluid inclusions characteristic of the ruby or sapphire forming environment. Such variations are evident in the differences in Ta/Nb_N, U/Pb_N, Th/U_N, and Zr/Hf_N ratios between online-and offline-ablated material, with the former showing variations for different localities and the latter showing complete overlap for all localities and deposit types (Figure A1a–l). Some of the trace element characteristics of the offline-ablated material show differences that could be utilized for geographic origin determination of different localities of similar deposit types, including, for example, the $Eu/Eu^* > 1$ of Mozambique rubies (Figure 4a), and the $Sr/Sr^* < 1$ for Myanmar rubies. Some suites are also characterized by the absence or certain trace elements, Sr, for example, is below < LOQ in all rubies from Zahamena, MDG, and Cs in sapphires from Myanmar and Andrebabe, MDG. Most of these trace element characteristics, however, are specific to a certain deposit, and as such, cannot be applied universally. In addition, the number of samples ablated offline is small, and some characteristics observed may well be a result of low sample numbers. Owing to the small sample set sizes, we do not apply LDA to explore the data for useful chemical variables, as the derived discriminants may be skewed by the—potentially—non-representative datasets. Given the chemical variation within this modest sample set, outliers weighted more strongly on the solution than they would for a larger dataset. Instead, as a preliminary check, we applied PCA to the offline-ablated trace element datasets to identify possible chemical discriminants for provenance, associated with chemical variance. However, no clear discrimination is apparent in principal component space for the present datasets. Therefore, while some of the trace element characteristics observed may provide additional information valuable for provenance, at this time, they have little potential for the development of a robust geochemical fingerprint of provenance for gem corundum. As more data are acquired, the application of LDA and PCA to variable selection and class discrimination may be more appropriate.

4.2. Radiogenic Isotopes

As demonstrated in Section 4.1, geographic origin determination using trace elements is useful in distinguishing between deposits of different geological deposit types, however, to distinguish between deposits of similar geological type, but different locality is challenging. Similar challenges are faced when using classical gemological techniques [4]. In this study, we obtained Sr and Pb isotopic compositions for ruby and blue sapphire from a variety of different localities of three different geological groups to explore the potential of using isotopic tracers to constrain geographical origin. We evaluated the usefulness of all combinations of isotopic ratios and trace elements, and a selection of the most useful discriminatory plots is shown in Figure 16.

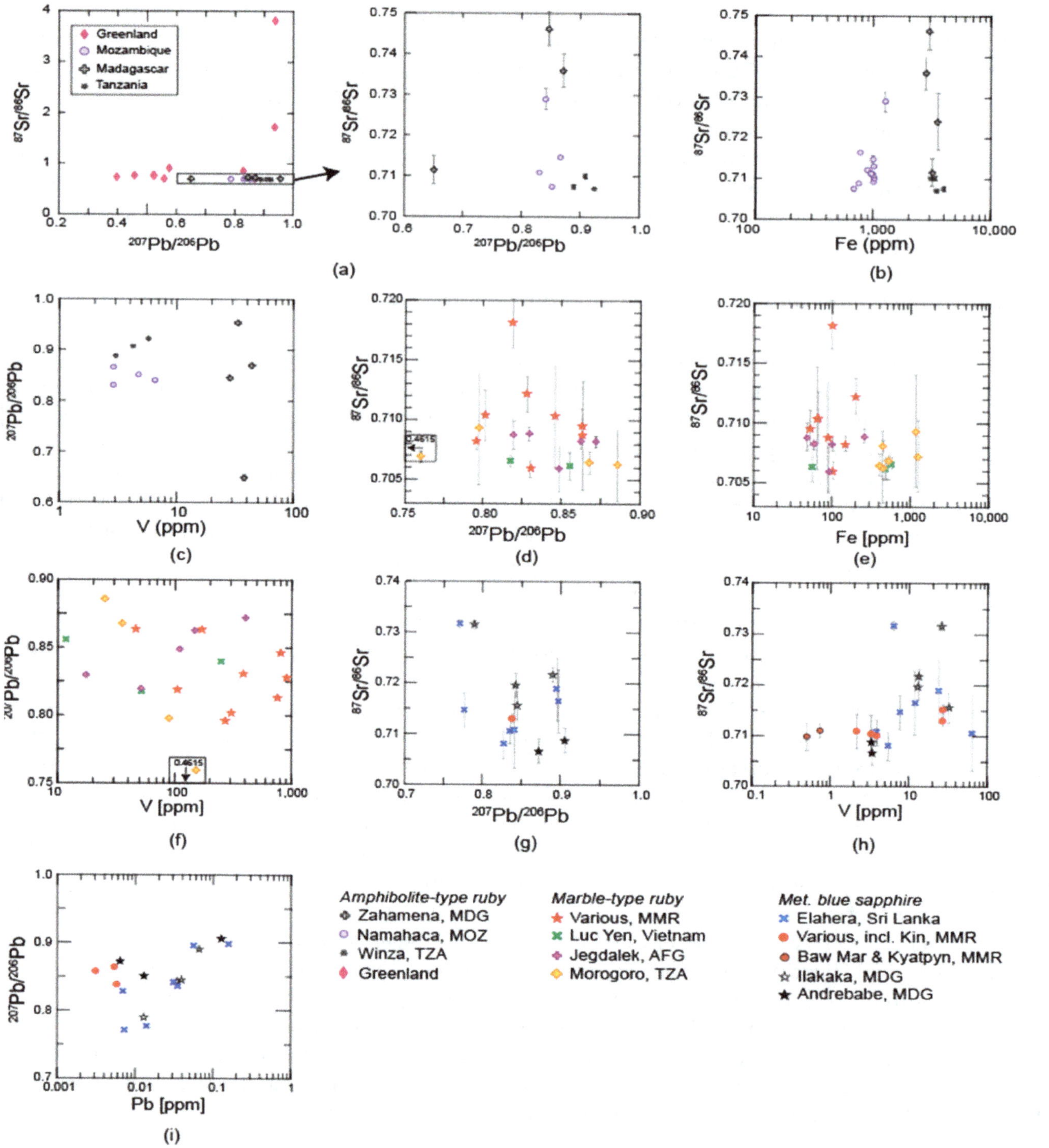

Figure 16. Radiogenic isotope characteristics of rubies and blue sapphires analyzed in this study. (**a**) $^{87}Sr/^{86}Sr$ versus $^{207}Pb/^{206}Pb$, (**b**) $^{87}Sr/^{86}Sr$ versus Fe (ppm), and (**c**) $^{207}Pb/^{206}Pb$ versus V (ppm) of amphibolite-type rubies. (**d**) $^{87}Sr/^{86}Sr$ versus $^{207}Pb/^{206}Pb$, (**e**) $^{87}Sr/^{86}Sr$ versus Fe (ppm), and (**f**) $^{207}Pb/^{206}Pb$ versus V (ppm) of marble-type rubies. (**g**) $^{87}Sr/^{86}Sr$ versus $^{207}Pb/^{206}Pb$, (**h**) $^{87}Sr/^{86}Sr$ versus V (ppm), and (**i**) $^{207}Pb/^{206}Pb$ versus Pb (ppm) of metamorphic blue sapphires. Where error bars are absent, uncertainties are smaller than the symbols.

The discrimination of rubies from different amphibolite-type deposits is fairly straight-forward using trace elements obtained via online LA-ICP-MS, however, some chemical overlap remains between rubies from Greenland with those from Mozambique (Figures 12 and 15a,b). This overlap can be resolved using both Sr and Pb isotopes, because Greenland rubies have been shown to be much more radiogenic than rubies from Mozambique, spanning a wide range of compositions ([12], Figure 16a). It is clear that these very distinct differences in isotopic compositions are largely because of the vast difference in formation age for the Greenland and Mozambique ruby deposits (Pb-Pb = ~2686 Ma [12] and ~550 Ma [72], respectively).

Therefore, it is unsurprising that rubies from Zahamena, MDG and Winza, TZA that formed during the Pan-African orogeny (750–450 Ma) [20] have radiogenic isotope compositions more similar to those of Mozambique, with some minor overlap between the different localities (Figure 7b). Nevertheless, when plotting Sr and Pb isotopic compositions against each other or against trace elements, for example, Fe or V, the different localities fall into discrete fields (Figure 16a–c), demonstrating the usefulness of Sr and Pb isotope ratios for determining the geographic origin of different amphibolite-type deposits.

As shown in Section 3.2.2, the range in Sr isotopes of marble-hosted rubies is much narrower than those of the analysed suites of amphibolite-type ruby, and the overlap among the different localities of marble-type deposits is greater than that of amphibolite-type deposits (Figure 7c). While Myanmar and Afghanistan rubies tend to have more radiogenic Sr compositions than rubies from Vietnam or Morogoro, TZA, the overlap precludes any usefulness such trends might provide, even when combining Sr isotopic ratios with trace elements (e.g., $^{87}Sr/^{86}Sr$ versus Fe, Figure 16e). This overlap of the different localities is even more complete for the Pb isotopic compositions (Figures 9b and 16b).

Similar to the marble-type rubies, metamorphic blue sapphires from different localities show significant overlap. Myanmar and Andrebabe, MDG sapphires generally show less radiogenic isotopic ratios than those from Sri Lanka and Ilakaka, MDG. This could be because of differences in the geological formation of the different deposits and differences in the age and lithology of their source components. In the MST, gem corundum of multiple geological origins was found and is thought to have formed either in connection with intrusions of granitoid rocks (leucogranites and hornblende syenites; between 35 and 23 Ma [62,73]) or regional metamorphism (sillimanite-grade high-temperature metamorphism; from at least 43 to 29 Ma [74]) [20]. The Elahera deposit in Sri Lanka, on the other hand, is an exoskarn deposit formed by the interaction of pegmatitic fluids with dolomitic marble [75]. The Ilakaka deposit in Madagascar is a secondary deposit and carries a wide variety of different sapphires, likely derived from several distinct geological formations. The sources are thought to be multiple and associated with granulite-facies metamorphism [20]. To date, nothing is known about the geology of the Andrebabe deposit in Madagascar, however, the trace element characteristics indicate an association with magmatic rocks. Another explanation could be different formation ages, because sapphires from Myanmar are thought to have formed during the Cenozoic Himalayan orogeny (e.g., zircon U–Pb ~25 Ma for foliated syenite associated with sapphire mineralization [76]), while the sapphires from Sri Lanka and Madagascar formed during the Pan-African orogeny (750–450 Ma) [20,27]. The observed overlap, complete for Sri Lanka and Ilakaka, MDG sapphires, can only partly be resolved when plotting Sr or Pb isotopic ratios against trace elements (Figure 16g–i), and the use of Sr and Pb isotopic compositions for geographic origin determination has proven to be relatively ineffective. No PCA solution yielded positive results.

5. Conclusions

Using an LA-ICP-MS method that combines a large spot size (285 μm) with a high repetition rate (20 Hz), we obtain a broader spectrum of quantitative trace element data than that obtained with more conventional approaches (spot size ~50 μm, repletion rate of 5–10 Hz; for example, [23,62]). This permits the determination of up to 26 elements at levels above the limit of quantification (LOQ) in some rubies and sapphires. Using elemental discrimination schemes and linear discriminant analysis (LDA), we have shown that elements other than the conventional Mg, Ti, V, Cr, Fe, and Ga—which can substitute for Al^{3+} in the corundum crystal lattice—have the potential to be of use in geographic origin determination. The discrimination of the different localities of amphibolite-type ruby analyzed in this study is improved using Ni. For marble-type ruby, LDA results show that adding Zr positively influences the discrimination of the different marble-type localities. For metamorphic blue sapphire, the discrimination of corundum from the different localities analyzed in this study is improved by Zn concentrations. In addition, Cr, which is often below the LOD for blue sapphires from the localities analyzed in this study using conventional methods [10], positively impacts the discrimination. While the discrimination of the different localities of the three different deposit groups is fairly good using LDA, some population overlap persists, particularly among the marble-type rubies and blue sapphires. This is not surprising owing to the very similar geological conditions of formation.

The larger spot size of the LA-ICP-MS method used here precludes it from use in gemological laboratories on precious stones; however, our results indicate directions of focus for future method development. Trace element characteristics of the offline-ablated material show some differences that could be utilized for geographic origin determination of different localities of similar deposit types; however, the majority of these are specific to a certain deposit and, as such, may not be applied universally. Sr and Pb isotope compositions, obtained using offline laser ablation sampling followed by TIMS, have only limited potential for geographic origin determination at this time. Measured $^{87}Sr/^{86}Sr$ and Pb isotope ratios of rubies from different amphibolite-type deposits show distinct ranges and separate fields on bivariate plots that are positive indicators of discrimination. For marble-type rubies and metamorphic blue sapphires, however, different geographic origins overlap.

Overall, our results show that the discrimination of different geographic regions of similar geological deposit types using trace elements is fairly simple for amphibolite-type rubies, while it is much more challenging for marble-type rubies and metamorphic sapphires. As more data from these localities are acquired and analyzed, discriminants can be refined and developed further. However, adding additional localities might lead to more significant overlap. Similarly, radiogenic isotopes show discriminatory potential for amphibolite-type rubies, but not marble-type rubies or metamorphic blue sapphires, likely owing to significant isotopic variability in the different components that combine to form these gems. Our results re-emphasize both the promise of using combined isotopic and trace element fingerprints and the challenge of geographic origin determination, emphasizing the need for a more powerful discriminatory tool than can be formulated using current technology.

Supplementary Materials
Table S1: Elemental concentrations (ppm) of spot analyses of rubies and blue sapphires analyzed by LA-ICP-MS; Table S2: Trace element concentrations (ppm) in NIST SRM 614, NIST SRM 616, and corundum standard 02-1032-B analyzed by LA-ICP-MS; Table S3: Trace element abundances (ppm) and Sr and Pb isotopic ratios determined using offline LA followed by solution ICP-MS and TIMS for amphibolite- and marble-type ruby and metamorphic blue sapphires analyzed in this study. Quantification is achieved via normalization to the weight loss of the corundum crystals during ablation; Supplementary Material S4: LDA and PCA methodology and R runstreams, Supplementary Material S5: Summary of LDA results on LA-ICP-MS data, including error calculations, Supplementary Material S6: Sample classification codes.

Author Contributions: M.Y.K. performed the analyses, interpreted the results of the analyses, and wrote the manuscript. D.G.P. and M.Y.K. were responsible for the conception of this project. M.F.H. performed the statistical analyses and assisted with writing parts of the manuscript. Y.L. and C.S. provided technical input and assisted with the analyses. A.J.F. provided geology background and additional samples. All authors have read and agreed to the published version of the manuscript.

Acknowledgments: The authors would like to thank Aaron Palke, of the Gemological Institute of America, for providing insight on sample gemology and fruitful discussions, and Sarah Woodland for assisting with measurements. In addition, the authors thank George Harlow, of the American Museum of Natural History, for providing samples from Myanmar, Vietnam, and Madagascar. Thank you to the three anonymous reviewers and the editors, Lin Sutherland and Khin Zaw, for providing constructive comments that improved this paper.

Appendix A

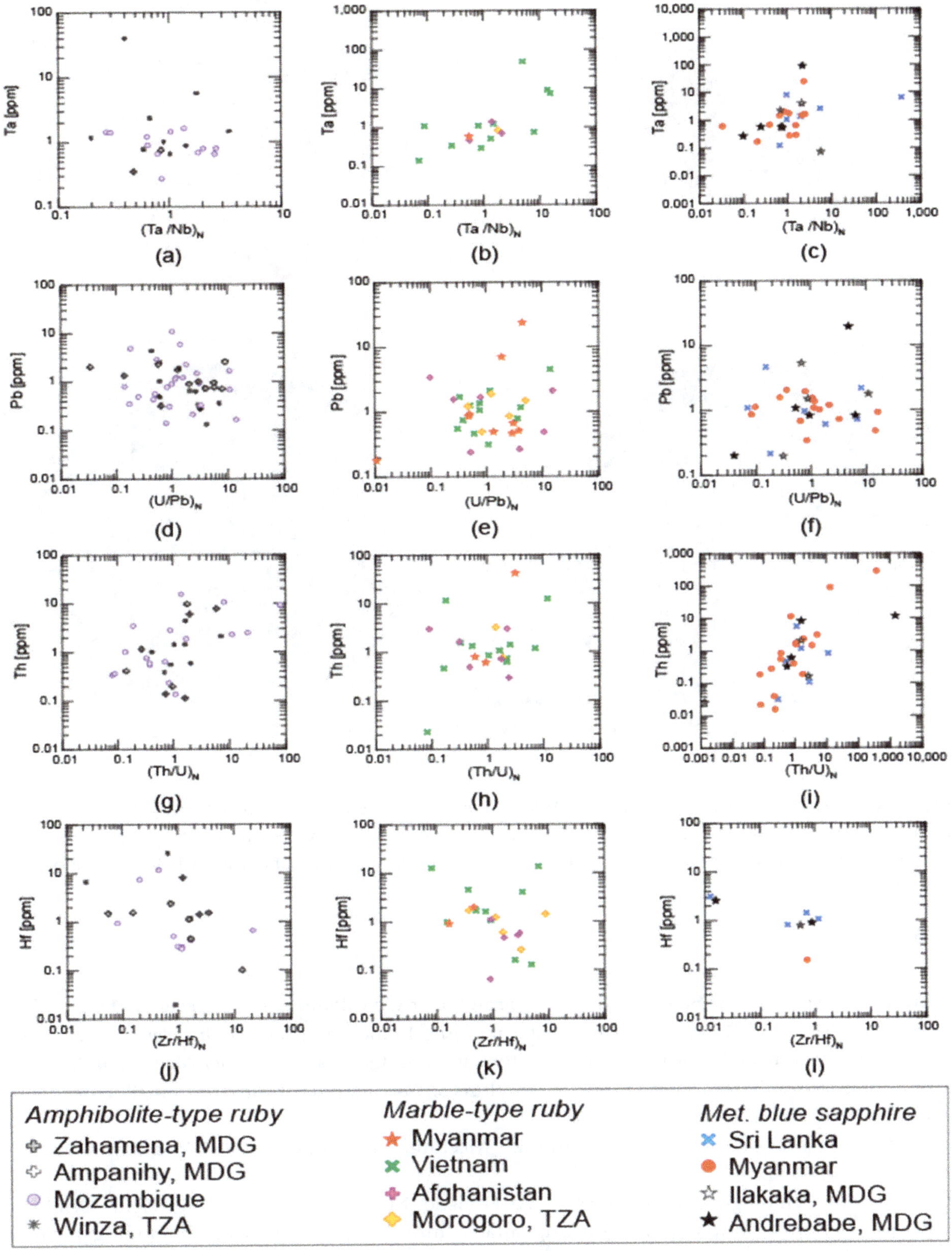

Figure A1. Primitive mantle-normalized [58] LA-ICP-MS data of gem corundum analyzed in this study. Points represent individual analyses. **(a–c)** Ta versus $(Ta/Nb)_N$; **(d–f)** Pb versus $(U/Pb)_N$, **(g–i)** Th versus $(Th/U)_N$, and **(j–l)** Hf versus $(Zr/Hf)_N$ for amphibole-type ruby, marble-type ruby, and met. blue sapphire, respectively.

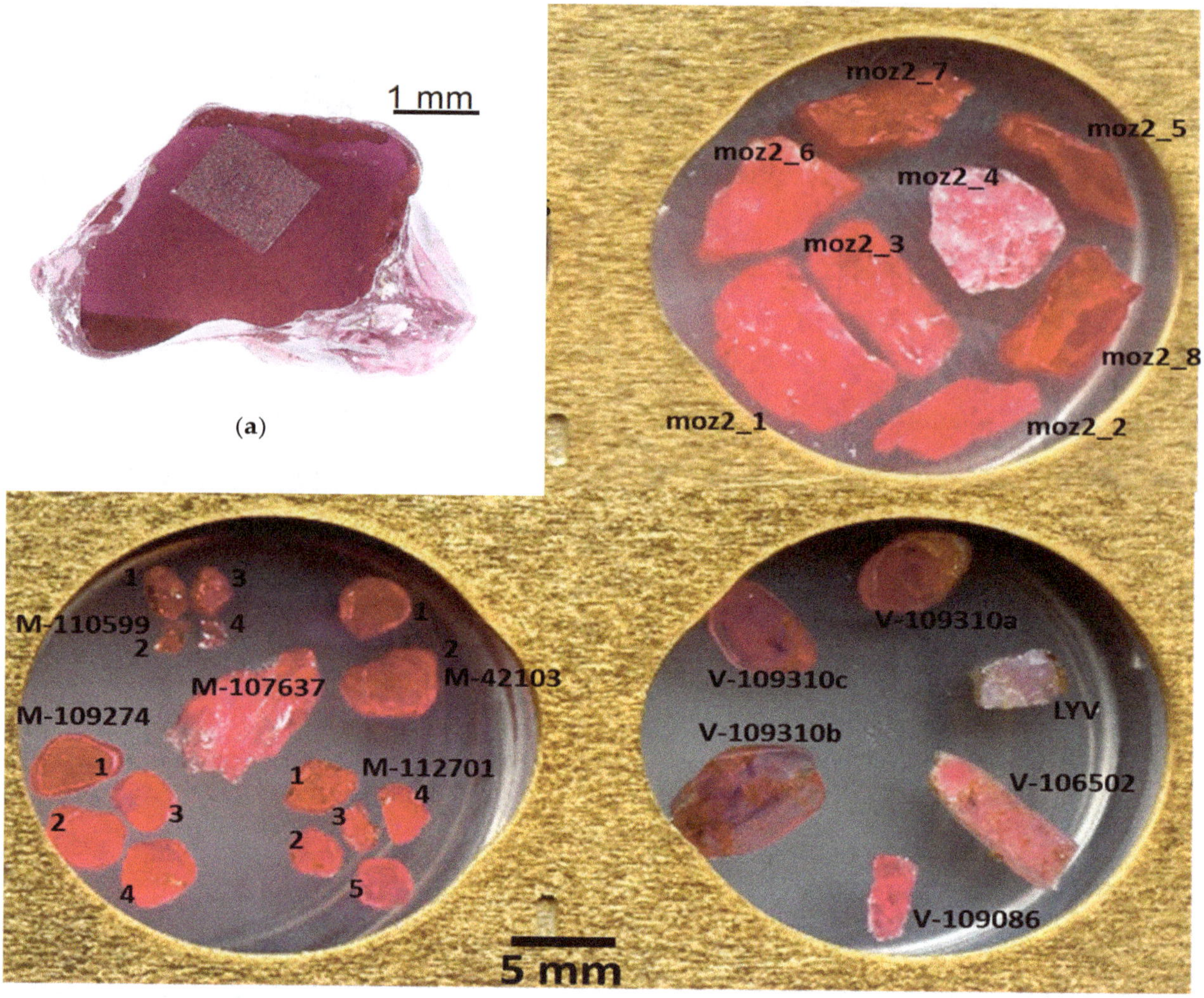

(b)

Figure A2. (**a**) Ruby from Zahamena, MDG after ablation clearly demonstrating the size of the offline ablation pit. (**b**) One inch resin sample mounts for LA-ICP-MS analysis showing rubies from Mozambique (moz), Myanmar (M-xxx), and Vietnam (V-xxx). A range of color and clarity is shown by these rubies.

References

1. McClure, S.F.; Moses, T.M.; Shigley, J.E. The geographic origin dilemma. *Gems Gemol.* **2019**, *55*, 457–462.
2. Bui, H.A.N.; Fritsch, E.; Rondeau, B. Geographical origin: Branding or science. *InColor Mag.* **2012**, *19*, 30–39.
3. Cartier, L. Environmental stewardship in gemstone mining: Quo vadis? *InColor Mag.* **2010**, *15*, 9.
4. Palke, A.C.; Saeseaw, S.; Renfro, N.D.; Sun, Z.; Mcclure, S.F. Geographic origin determination of blue sapphire. *Gems Gemol.* **2019**, *55*, 536–579. [CrossRef]
5. Palke, A.C.; Saeseaw, S.; Renfro, N.D.; Sun, Z.; Mcclure, S.F. Geographic origin determination of ruby. *Gems Gemol.* **2019**, *55*, 580–613. [CrossRef]
6. The Gübelin Gemological Laboratory. A Holistic Method to Determing Gem Origin. 2006. Available online: https://www.gubelingemlab.com/tl_files/content/03%20Gemmology/Gem%20Lab/Documents/Origin_Article_Sep2006.pdf (accessed on 15 May 2020).
7. Giuliani, G.; Fallick, A.E.; Garnier, V.; France-Lanord, C.; Ohnenstetter, D.; Schwarz, D. Oxygen isotope composition as a tracer for the origins of rubies and sapphires. *Geology* **2005**, *33*, 249–252. [CrossRef]
8. Saminpaya, S.; Manning, D.A.C.; Droop, G.T.R.; Henderson, C.N.B. Trace elements in Thai gem corundums. *J. Gemmol.* **2003**, *28*, 399–416. [CrossRef]

9. Abduriyim, A.; Kitawaki, H. Applications of Laser Ablation—Inductively Coupled Plasma—Mass Spectrometry (LA-ICP-MS) to gemology. *Gems Gemol.* **2006**, *42*, 98–118. [CrossRef]
10. Peucat, J.J.; Ruffault, P.; Fritsch, E.; Bouhnik-Le Coz, M.; Simonet, C.; Lasnier, B. Ga/Mg ratio as a new geochemical tool to differentiate magmatic from metamorphic blue sapphires. *Lithos* **2007**, *98*, 261–274. [CrossRef]
11. Graham, I.T.; Sutherland, L.; Zaw, K.; Nechaev, V.; Khanchuk, A. Advances in our understanding of the gem corundum deposits of the West Pacific continental margins intraplate basaltic fields. *Ore Geol. Rev.* **2008**, *34*, 200–215. [CrossRef]
12. Krebs, M.Y.; Pearson, D.G.; Fagan, A.J.; Bussweiler, Y.; Sarkar, C. The application of trace elements and Sr–Pb isotopes to dating and tracing ruby formation: The Aappaluttoq deposit, SW Greenland. *Chem. Geol.* **2019**, *523*, 42–58. [CrossRef]
13. Dickin, A.P. *Radiogenic Isotope Geology*, 2nd ed.; Cambridge University Press: New York, NY, USA, 2005; ISBN 9780521823166.
14. Garnier, V.; Giuliani, G.; Ohnenstetter, D.; Schwarz, D. Saphirs et rubis. Classification des gisements de corindon. *Règne Minér.* **2004**, *55*, 4–47.
15. Muhlmeister, S.; Fritsch, E.; Shigley, J.E.; Devouard, B.; Laurs, B.M. Separating natural and synthetic rubies on the basis of trace-element chemistry. *Gems Gemol.* **1998**, *34*, 80–101. [CrossRef]
16. Kievlenko, E.Y. *Geology of Gems*; Ocean Pictures Ltd.: Littleton, CO, USA, 2003.
17. Simonet, C.; Fritsch, E.; Lasnier, B. A classification of gem corundum deposits aimed towards gem exploration. *Ore Geol. Rev.* **2008**, *34*, 127–133. [CrossRef]
18. Giuliani, G.; Fallick, A.E.; Rakotondrazafy, M.; Ohnenstetter, D.; Andriamamonjy, A.; Ralantoarison, T.; Rakotosamizanany, S.; Razanatseheno, M.; Offant, Y.; Garnier, V.; et al. Oxygen isotope systematics of gem corundum deposits in Madagascar: Relevance for their geological origin. *Miner. Depos.* **2007**, *42*, 251–270. [CrossRef]
19. Stern, R.J.; Tsujimori, T.; Harlow, G.; Groat, L. Plate tectonic gemstones. *Geology* **2013**, *41*, 723–726. [CrossRef]
20. Giuliani, G.; Ohnenstetter, D.; Fallick, A.E.; Groat, L.A.; Fagan, A.J. The geology and genesis of gem corundum deposits. In *Mineralogical Association of Canada Short Course*; Mineralogical Association of Canada: Québec, QC, Canada, 2014; Volume 44, pp. 29–112.
21. Harlan, S.S. Timing of emplacement of the sapphire-bearing Yogo Dike, Little Belt Mountains, Montana. *Econ. Geol.* **1996**, *91*, 1159–1162. [CrossRef]
22. Vertriest, W.; Palke, A.C.; Renfro, N.D. Field gemology: Building a research collection and understanding the development of gem deposits. *Gems Gemol.* **2019**, *55*, 490–511. [CrossRef]
23. Pardieu, V.; Sangsawong, S.; Muyal, J.; Chauviré, B.; Massi, L.; Sturman, N. Rubies from the Montepuez area (Mozambique). *GIA News*; Bangkok, Thailand, 2013. Available online: https://www.gia.edu/doc/GIA_Ruby_Montepuez_Mozambique.pdf (accessed on 10 May 2020).
24. Peretti, A.; Peretti, F.; Kanpraphai, A.; Bieri, W.P.; Mametner, K.; Guenther, D. *Winza Rubies Identified*, 2nd ed.; Peretti, A., Ed.; GRS Gem Research Swisslab: Lucerne, Switzerland, 2008; ISBN 9783952335970.
25. Schwarz, D.; Pardieu, V.; Saul, J.M.; Schmetzer, K.; Laurs, B.M.; Giuliani, G.; Klemm, L.; Malsy, A.; Erel, E.; Hauzenberger, C.; et al. Rubies and sapphires from Winza, Central Tanzania. *Gems Gemol.* **2008**, *44*, 322–347. [CrossRef]
26. Pardieu, V.; Sangsawong, S.; Detroyat, S. Rubies from a new deposit in zahamena national park, Madagascar. *Gems Gemol.* **2015**, *51*, 454–456.
27. Rakotondrazafy, A.F.M.; Giuliani, G.; Ohnenstetter, D.; Fallick, A.E.; Rakotosamizanany, S.; Andriamamonjy, A.; Ralantoarison, T.; Razanatseheno, M.; Offant, Y.; Garnier, V.; et al. Gem corundum deposits of Madagascar: A review. *Ore Geol. Rev.* **2008**, *34*, 134–154. [CrossRef]
28. Smith, C.P. Ruby and pink sapphire from Aappaluttoq, Greenland. *J. Gemmol.* **2016**, *35*, 294–306. [CrossRef]
29. Keulen, N.; Kalvig, P. Fingerprinting of corundum (ruby) from Fiskenaesset, West Greenland. *Geol. Surv. Den. Greenl. Bull.* **2013**, *28*, 53–56. [CrossRef]
30. Fagan, A.J. The Ruby and Pink Sapphire Deposits of SW Greenland: Geological Setting, Genesis, and Exploration Techniques. Ph.D. Thesis, The University of British Columbia, Vancouver, BC, Canada, 2018.
31. Fanka, A.; Sutthirat, C. Petrochemistry, mineral chemistry, and pressure—Temperature model of corundum-bearing amphibolite from Montepuez, Mozambique. *Arab. J. Sci. Eng.* **2018**, *43*, 3751–3767. [CrossRef]

32. Bowersox, G.W.; Foord, E.E.; Laurs, B.M.; Shigley, J.E.; Smith, C.P. Ruby and sapphire from Jegdalek, Afghanistan. *Gems Gemol.* **2000**, *36*, 110–126. [CrossRef]
33. Harlow, G.E.; Bender, W. A study of ruby (corundum) compositions from the Mogok Belt, Myanmar: Searching for chemical fingerprints. *Am. Miner.* **2013**, *98*, 1120–1132. [CrossRef]
34. Van Long, P.; Giuliani, G.; Fallick, A.E.; Boyce, A.J.; Pardieu, V. Trace elements and oxygen isotopes of gem spinels in marble from the Luc Yen—An Phu areas, Yen Bai province, North Vietnam. *Vietnam J. Earth Sci.* **2018**, *40*, 166–179.
35. Van Long, P.; Vinh, H.Q.; Garnier, V.; Giuliani, G.; Ohnenstetter, D.; Lhomme, T.; Schwarz, D.; Fallick, A.E.; Dubessy, J.; Trinh, P.T.; et al. Gem corundum deposits in Vietnam. *J. Gemmol.* **2004**, *29*, 129–147. [CrossRef]
36. Garnier, V.; Giuliani, G.; Ohnenstetter, D.; Fallick, A.E.; Dubessy, J.; Banks, D.; Vinh, H.Q.; Lhomme, T.; Maluski, H.; Pêcher, A.; et al. Marble-hosted ruby deposits from Central and Southeast Asia: Towards a new genetic model. *Ore Geol. Rev.* **2008**, *34*, 169–191. [CrossRef]
37. Haenni, H.A.; Schmetzer, K. New rubies from the Morogoro area, Tanzania. *Gems Gemol.* **1991**, *27*, 156–167. [CrossRef]
38. Balmer, W.A.; Hauzenberger, C.A.; Fritz, H.; Sutthirat, C. Marble-hosted ruby deposits of the Morogoro Region, Tanzania. *J. Afr. Earth Sci.* **2017**, *134*, 626–643. [CrossRef]
39. Kan-Nyunt, H.P.; Karampelas, S.; Link, K.; Thu, K.; Kiefert, L.; Hardy, P. Blue sapphires from the Baw Mar mine in Mogok. *Gems Gemol.* **2013**, *49*, 223–232. [CrossRef]
40. Hughes, E.B.; Perkins, R. Madagascar sapphire: Low-temperature heat treatment experiments. *Gems Gemol.* **2019**, *55*, 184–196. [CrossRef]
41. Guznawardenea, M.; Rupasinghe, M.S. The elahera gem field in Central Sri Lanka. *Gems Gemol.* **1986**, *22*, 80–95. [CrossRef]
42. Atikarnsakul, U.; Vertriest, W.; Soonthorntantikul, W. Characterization of Blue Sapphires from the Mogok Stone Tract, Mandalay Region, Burma (Myanmar). *GIA News*. 19 June 2018, pp. 1–56. Available online: https://www.gia.edu/doc/Characterization-of-blue-sapphires-from-the-Mogok-Stone-Tract-Mandalay-region-Burma-Myanmar.pdf (accessed on 10 May 2020).
43. Soonthorntantikul, W.; Vertriest, W.; Raynaud-flattot, V.L.; Sangsawong, S. *An in-Depth Gemological Study of Blue Sapphires From the Baw Mar Mine (Mogok, Myanmar)*; GIA: Bangkok, Thailand, 2017.
44. Milisenda, C.C.; Henn, U.; Henn, J. New gemstone occurrences in the south-west of Madagascar. *J. Gemmol.* **2001**, *27*, 385–394. [CrossRef]
45. Pardieu, V.; Rakotosaona, N. *Ruby and Sapphire Rush Near Didy, Madagascar (April–June 2012)*; GIA: Bangkok, Thailand, 2012.
46. Dissanayake, C.B.; Chandrajith, R.; Tobschall, H.J. The geology, mineralogy and rare element geochemistry of the gem deposits of Sri Lanka. *Bull. Geol. Soc. Finl.* **2000**, *72*, 5–20. [CrossRef]
47. Pardieu, V.; Sangsawong, S.; Vertriest, W.; Detroyat, S.; Raynaud, V.; Engniwat, S. Blue sapphires from a new deposit near Andranondambo, Madagascar. *Gems Gemmol.* **2016**, *52*, 96–97.
48. Schwarz, D.; Petsch, E.J.; Kanis, J. Sapphires from the Andranondambo Region, Madagascar. *Gems Gemol.* **1996**, *32*, 80–99. [CrossRef]
49. Stone-Sundberg, J.; Thomas, T.; Sun, Z.; Guan, Y.; Cole, Z.; Equall, R.; Emmett, J.L. Accurate reporting of key trace elements in ruby and sapphire using matrix-matched standards. *Gems Gemol.* **2017**, *53*, 438–451. [CrossRef]
50. Paton, C.; Hellstrom, J.; Paul, B.; Woodhead, J.; Hergt, J. Iolite: Freeware for the visualisation and processing of mass spectrometric data. *J. Anal. At. Spectrom.* **2011**, *26*, 2508. [CrossRef]
51. McNeill, J.; Pearson, D.G.; Klein-Bendavid, O.; Nowell, G.M.; Ottley, C.J.; Chinn, I. Quantitative analysis of trace element concentrations in some gem-quality diamonds. *J. Phys. Condens. Matter* **2009**, *21*, 364207. [CrossRef]
52. Klein-BenDavid, O.; Pearson, D.G.; Nowell, G.M.; Ottley, C.; McNeill, J.C.R.; Cartigny, P. Mixed fluid sources involved in diamond growth constrained by Sr-Nd-Pb-C-N isotopes and trace elements. *Earth Planet. Sci. Lett.* **2010**, *289*, 123–133. [CrossRef]
53. Krebs, M.Y.; Pearson, D.G.; Stachel, T.; Laiginhas, F.; Woodland, S.; Chinn, I.; Kong, J. A common parentage-low abundance trace element data of gem diamonds reveals similar fluids to fibrous diamonds. *Lithos* **2019**, *324–325*. [CrossRef]

54. Charlier, B.L.A.; Ginibre, C.; Morgan, D.; Nowell, G.M.; Pearson, D.G.; Davidson, J.P.; Ottley, C.J. Methods for the microsampling and high-precision analysis of strontium and rubidium isotopes at single crystal scale for petrological and geochronological applications. *Chem. Geol.* **2006**, *232*, 114–133. [CrossRef]
55. Harlou, R.; Pearson, D.G.; Nowell, G.M.; Ottley, C.J.; Davidson, J.P. Combined Sr isotope and trace element analysis of melt inclusions at sub-ng levels using micro-milling, TIMS and ICPMS. *Chem. Geol.* **2009**, *260*, 254–268. [CrossRef]
56. Sarkar, C.; Pearson, D.G.; Heaman, L.M.; Woodland, S.J. Precise Pb isotope ratio determination of picogram-size samples: A comparison between multiple Faraday collectors equipped with $10^{12}\Omega$ amplifiers and multiple ion counters. *Chem. Geol.* **2015**, *395*, 27–40. [CrossRef]
57. Todt, W.; Cliff, R.A.; Hanser, A.; Hofmann, A.W. Evaluation of a 202Pb–205Pb double spike for high-precision lead isotope analysis. In *Earth Processes: Reading the Isotopic Code*; Basu, A., Hart, S.R., Eds.; American Geophysical Union Geophysical Monograph: Washington, DC, USA, 1996; pp. 429–437.
58. McDonough, W.F.; Sun, S.-S. The composition of the earth. *Chem. Geol.* **1995**, *120*, 223–253. [CrossRef]
59. Pornwilard, M.M.; Hansawek, R.; Shiowatana, J.; Siripinyanond, A. Geographical origin classification of gem corundum using elemental fingerprint analysis by laser ablation inductively coupled plasma mass spectrometry. *Int. J. Mass Spectrom.* **2011**, *306*, 57–62.
60. Sutherland, F.; Zaw, K.; Meffre, S.; Yui, T.-F.; Thu, K. Advances in trace element "Fingerprinting" of gem corundum, ruby and sapphire, Mogok Area, Myanmar. *Minerals* **2014**, *5*, 61–79. [CrossRef]
61. Sutherland, F.L.; Abduriyim, A. Geographic typing of gem corundum: A test case from Australia. *J. Gemmol.* **2009**, *31*, 203–210. [CrossRef]
62. Sutherland, F.; Zaw, K.; Meffre, S.; Thompson, J.; Goemann, K.; Thu, K.; Nu, T.; Zin, M.; Harris, S. Diversity in ruby geochemistry and its inclusions: Intra- and inter-continental comparisons from Myanmar and Eastern Australia. *Minerals* **2019**, *9*, 28. [CrossRef]
63. Sutherland, F.L.; Zaw, K.; Meffre, S.; Giuliani, G.; Fallick, A.E.; Graham, I.T.; Webb, G.B. Gem-corundum megacrysts from east Australian basalt fields: Trace elements, oxygen isotopes and origins. *Aust. J. Earth Sci.* **2009**, *56*, 1003–1022. [CrossRef]
64. Zaw, K.; Sutherland, L.; Yui, T.F.; Meffre, S.; Thu, K. Vanadium-rich ruby and sapphire within Mogok Gemfield, Myanmar: Implications for gem color and genesis. *Miner. Depos.* **2014**, *50*, 25–39. [CrossRef]
65. Sutherland, F.L.; Piilonen, P.C.; Zaw, K.; Meffre, S.; Thompson, J. Sapphire within zircon-rich gem deposits, Bo Loei, Ratanakiri Province, Cambodia: Trace elements, inclusions, U–Pb dating and genesis. *Aust. J. Earth Sci.* **2015**, *62*, 761–773.
66. Sutherland, F.L.; Graham, I.T.; Harris, S.J.; Coldham, T.; Powell, W.; Belousova, E.A.; Martin, L. Unusual ruby–sapphire transition in alluvial megacrysts, Cenozoic basaltic gem field, New England, New South Wales, Australia. *Lithos* **2017**, *278–281*, 347–360. [CrossRef]
67. Sorokina, E.S.; Rassomakhin, M.A.; Nikandrov, S.N.; Karampelas, S.; Kononkova, N.N.; Nikolaev, A.G.; Anosova, M.O.; Somsikova, A.V.; Kostitsyn, Y.A.; Kotlyarov, V.A.; et al. Origin of blue sapphire in newly discovered spinel–chlorite–muscovite rocks within meta-ultramafites of ilmen mountains, South Urals of Russia: Evidence from mineralogy, geochemistry, RB-SR and Sm-Nd isotopic data. *Minerals* **2019**, *9*, 36. [CrossRef]
68. Graham, I.T.; Harris, S.J.; Martin, L.; Lay, A.; Zappettini, E. Enigmatic alluvial sapphires from the Orosmayo Region, Jujuy Province, Northwest Argentina: Insights into their origin from in situ oxygen isotopes. *Minerals* **2019**, *9*, 390. [CrossRef]
69. Cartier, L.E. Ruby and sapphire from Marosely, Madagascar. *J. Gemmol.* **2009**, *31*, 171–180. [CrossRef]
70. Baldwin, L.C.; Ballhaus, F.T.C.; Fonseca, A.G.R.O.C. Petrogenesis of alkaline basalt-hosted sapphire megacrysts. Petrological and geochemical investigations of in situ sapphire occurrences from the Siebengebirge Volcanic Field, Germany. *Contrib. Miner. Pet.* **2017**, *172*, 1–27. [CrossRef]
71. Guillong, M.; Günther, D. Quasi "non-destructive" laser ablation-inductively coupled plasma-mass spectrometry fingerprinting of sapphires. *Spectrochim. Acta-Part B At. Spectrosc.* **2001**, *56*, 1219–1231. [CrossRef]
72. Boyd, R.; Nordgulen, O.; Thomas, R.J.; Bingen, B.; Bjerkgård, T.; Grenne, T.; Henderson, I.; Melezhik, V.A.; Often, M.; Sandstad, J.S.; et al. The geology and geochemistry of the East African orogen in northeastern Mozambique. *S. Afr. J. Geol.* **2010**, *113*, 87–129. [CrossRef]

73. Barley, M.E.; Pickard, A.L.; Zaw, K.; Rak, P.; Doyle, M.G. Jurassic to Miocene magmatism and metamorphism in the Mogok metamorphic belt and the India-Eurasia collision in Myanmar. *Tectonics* **2003**, *22*, 22. [CrossRef]
74. Searle, M.P.; Noble, S.R.; Cottle, J.M.; Waters, D.J.; Mitchell, A.H.G.; Hlaing, T.; Horstwood, M.S.A. Tectonic evolution of the Mogok metamorphic belt, Burma (Myanmar) constrained by U-Th-Pb dating of metamorphic and magmatic rocks. *Tectonics* **2007**, *26*, 26. [CrossRef]
75. Silva, K.K.M.W.; Siriwardena, C.H.E.R. Geology and the origin of the corundum-bearing skarn at Bakamuna, Sri Lanka. *Miner. Depos.* **1988**, *23*, 186–190. [CrossRef]
76. Thu, K. The Igneous Rocks of the Mogok Stone Tract. Ph.D. Thesis, University of Yangon, Yangon, Myanmar, 2007.

4

Petrogenesis of the Snezhnoe Ruby Deposit, Central Pamir

Andrey K. Litvinenko [1], Elena S. Sorokina [2,3], Tobias Häger [2,*], Yuri A. Kostitsyn [3], Roman E. Botcharnikov [2], Alina V. Somsikova [3], Thomas Ludwig [4], Tatiana V. Romashova [3] and Wolfgang Hofmeister [2]

1 Department of Mineralogy and Gemology, Russian State Geological Prospecting University (MGRI), Miklukho-Maklai str. 23, 117485 Moscow, Russia; litvinenkoak@mgri.ru

2 Institut für Geowissenschaften, Johannes Gutenberg-Universität Mainz (JGU), J.-J.-Becher-Weg 21, 55128 Mainz, Germany; esorokin@uni-mainz.de (E.S.S.); rbotchar@uni-mainz.de (R.E.B.); hofmeister@uni-mainz.de (W.H.)

3 Vernadsky Institute of Geochemistry and Analytical Chemistry, Russian Academy of Sciences (GEOKHI RAS), Kosygin str. 19, 119991 Moscow, Russia; kostitsyn@geokhi.ru (Y.A.K.); somsikova@geokhi.ru (A.V.S.); romashova@geokhi.ru (T.V.R.)

4 Institut für Geowissenschaften, Universität Heidelberg, Im Neuenheimer Feld 234-236, 69120 Heidelberg, Germany; thomas.ludwig@geow.uni-heidelberg.de

* Correspondence: haeger@uni-mainz.de

Abstract: The Snezhnoe ruby deposit is located in the Muzkol–Rangkul anticlinorium within the Cimmerian zone of the Central Pamir. On the local scale, the deposit occurs on discrete relict bedding planes of calcitic marbles belonging to the Sarydzhilgin suite. Four ruby-bearing mineral assemblages are present within the main parts of the deposit: (1) scapolite + phlogopite + muscovite + margarite; (2) plagioclase + muscovite + margarite; (3) muscovite + phlogopite + margarite; (4) calcite. The ruby + calcite association is the most economically important, whereas the association of plagioclase + scapolite + phlogopite + muscovite is typical for the ruby-free parts of the deposit. Mica group minerals with a distinctive green color due to enhanced Cr and V concentrations are the main prospecting indicators for the ruby mineralization. The oxygen isotopic composition of the rubies is +15.3‰, a common value for crustal metamorphic and sedimentary rocks. The ratios of indicative trace elements in the rubies are Ga/Mg < 8.2, Fe/Mg < 51.2, Cr/Ga > 6.9 and Fe/Ti < 31.6. These values are characteristic for metamorphic corundum. The bulk ruby-bearing rocks have an initial $^{87}Sr/^{86}Sr$ ratio of ~0.70791 and εNd of ~−9.6, also pointing to the crustal origin of the deposit in agreement with the geological data. Ancient Al-enriched sediments are suggested to be a possible protolith for the ruby-bearing rocks. The temperature of the metamorphic processes was estimated at 760 ± 30 °C using Zr-in-rutile geothermometry. Raman mapping of rutile inclusions trapped within the ruby crystal indicates that the minimum pressure of mineralization was about one kilobar. The age determined by the Rb–Sr thermal ionization mass spectrometry of phlogopite, plagioclase and bulk rock is 23 ± 1.6 Ma, corresponding to the timing of relaxation after peak metamorphism during the Alpine–Himalayan Orogeny.

Keywords: ruby; mineralogical association; geochemistry; Snezhnoe deposit; Tajikistan; Central Pamir; Muzkol–Rangkul anticlinorium; in situ U–Pb LA–ICP–MS rutile dating; oxygen isotopes; Rb–Sr and Sm–Nd isotopes

---

1. Introduction

Corundum (α-Al_2O_3) is the aluminum end-member of the hematite group of minerals and contains a limited number of transitional elements acting as its chromophores. Thus, ruby is assigned to the red-colored corundum variety that is due to the isomorphic substitution of Al^{3+} by Cr^{3+}. Gem-quality ruby and sapphire (the blue–green–yellow-colored gem varieties of corundum) deposits are rare resulting in their high economic value with annual worldwide production exceeding several billion dollars [1]. The occasional discovery of large-scale commercial ruby deposits is related to the rarity of geological processes that produce them. Ruby mineralization commonly occurs as a result of moderate to high temperature metamorphic and magmatic processes in continent–continent collision zones ([2–6], etc.), less frequently in subduction areas [7]. Hence, their formation can be caused by several different processes over various timescales. This makes the reconstruction of the genesis and evolution of ruby mineralization challenging.

The Snezhnoe marble-hosted ruby deposit is located in the Tajik eastern part of the Central Pamir region within the Alpine–Himalayan Belt and is linked to Cenozoic Indo-Asia collision [4]. Due to being part of this collision, it shares many mineralogical and geochemical features common for metamorphic ruby sites formed along this mobile belt and related syn-tectonic structures including Mogok in Myanmar, Jegdalek in Afghanistan, the Hunza Valley in Pakistan, Luc Yen in Vietnam, etc. [4]. However, despite numerous studies ([8–12], etc.), the thermodynamic conditions during the mineralization process and the formation age of the ruby-bearing rocks of the Snezhnoe deposit remain poorly understood. New mineralogical and petrochemical features of the ruby-bearing rocks, along with data on the oxygen and radiogenic isotope signatures are presented in this study. This new dataset is then applied for a better understanding the geological processes responsible for the formation of the Snezhnoe ruby deposit as well as for the establishment of exploration criteria for discovering future gem corundum mineralization and deposits.

2. Geological Setting

On the regional scale, the Snezhnoe deposit is located within the Muzkol–Rangkul anticlinorium belonging to the eastern part of the Tajik Central Pamir region (Figure 1). The western part of the area is adjacent to the Vanch–Yazguliem anticlinorium. Both anticlinoria are linked to the exposed Precambrian crystalline basement which was tectonically re-worked during the Cimmerian Orogeny [13].

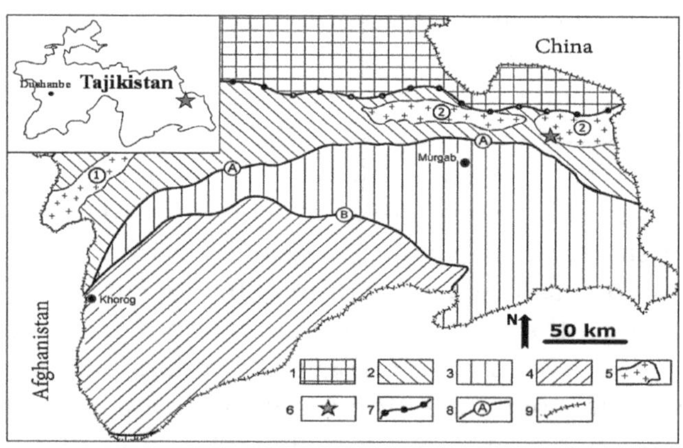

Figure 1. A location of the Snezhnoe deposit in relation to the main regional structures of the Pamir Mountains. Numbers 1–4 are folded regions: 1—northern Pamir (Hercynides); 2 to 4—southern Pamir (Cimmerian Plate). The southern Pamir is subdivided into zone 2–central Pamir, 3–south-eastern Pamir and 4—southwestern Pamir; 5—Early Proterozoic blocks, 6– Snezhnoe deposit; 7 and 8—major faults: 7—between folded regions (Vanch–Akbaytal) and 8—between tectonic zones: A—Rushan–Pshart and B—Gunt–Alichur; 9—border of Tajikistan; 1 (in circle) —Vanch–Yazgulem anticlinorium, 2 (in circle) –Muzkol–Rangkul anticlinorium. The map is modified after [13].

The Muskol–Rangkul anticlinorium consists of two tectonic blocks, the Sarymulin anticline in the west and the Shatput anticline in the east. In the central part, it is covered by Paleozoic–Mesozoic volcanic-sedimentary rocks. The Snezhnoe deposit is located on the southwestern flank of the Shatput block [14,15]. This block represents a steeply dipping monocline with the main foliation trending from 200° to 230° and a dip angle varying from 45° to 70°. Its length is ~15 km and the thickness is ~1.5 km. The monocline structure is adjacent to the Permian—Triassic system of the southeastern Pamir and the steeply dipping Muzkol fault (Figure 2).

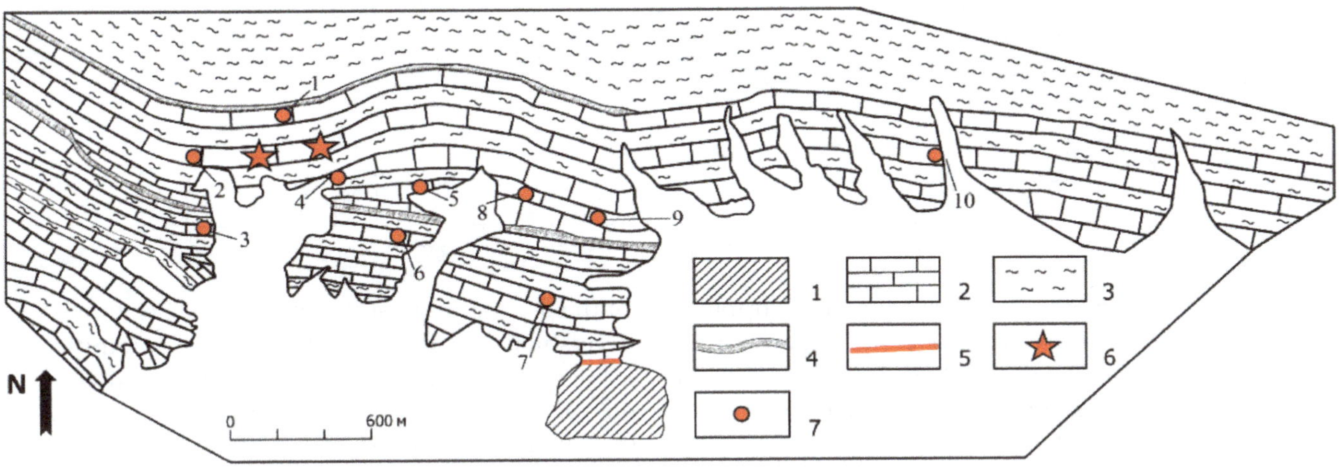

Figure 2. Geological map of the southwestern limb of the Shatput anticline modified after [8]. Geomorphologically, this is the southern slope of the Tura–Kuloma ridge. 1—Triassic system (lines indicate the strike of rocks; azimuth of dip is to the southeast). 2–4—Sarydzhilgin suite (sub-latitudinal strike): 2—marbles, 3—crystalline schists, 4—diopside–scapolite rocks, 5—deep fault, 6—ruby deposits: in the west—Nadezhda, in the east—Snezhnoye. 7—ruby occurrences close to red circles: 1—Timosha, 2—Corundum-4, 3—Corundum-7, 4—Lagernoye, 5—Alyonushka, 6—Corundum-2, 7—Corundum-1, 8—Corundum-8, 9—Corundum-6, 10—Trika. White area is the Quaternary sediments.

The thickness of the Muzkol metamorphic series within the Muzkol–Rangkul anticlinorium is about 6 km. It consists of four suites [16], with the calcitic marbles of the Sarydzhilgin suite present as layers intercalated with amphibole–pyroxene and scapolite calciphyres, gneisses, crystalline schists and quartzites (Figure 3). These calcitic marbles host the Snezhnoe deposit, with the magnesium content of the marbles varying from 1.2% MgO at the top to 2.7 wt% MgO at the base. In the Snezhnoe deposit, micaceous ruby-bearing lenses are arranged as discrete en echelon lenses along the marble bedding surface and can be traced along strike for more than 200 m with the maximum thickness of individual lenses being 1 m [15]. These ruby-bearing lenses are cross-cut by numerous much smaller veins up to 5 cm thick and filled by later scapolite and mica group minerals, most likely linked to later hydrothermal processes [14]. The veins are zonal with the rim (up to 0.5 cm thick) composed of Cr-bearing muscovite and margarite and the core mainly comprising scapolite, which forms distinctive concentric aggregates.

The Muzkol series underwent multi-cyclic metamorphism from high-temperature amphibolite to epidote–amphibolite facies and then to retrogressive greenschist facies [16]. The first metamorphic cycle occurred at 1.9–1.6 Ga under the high-temperature conditions of amphibolite facies at T = 700–750 °C and P = 8–9 kbar [16,17]. The second cycle occurred between 100 to 20 Ma and had a peak metamorphic temperature of about 800 °C at a pressure close to 9 kbar (with partial melting of protolith rocks) in the central part of the anticlinorium. The metamorphic facies evolved from high-temperature and high-pressure ones in the central part through medium *T*-P epidote–amphibolite to greenschist facies at 350 °C and 4 kbar towards the periphery [18,19].

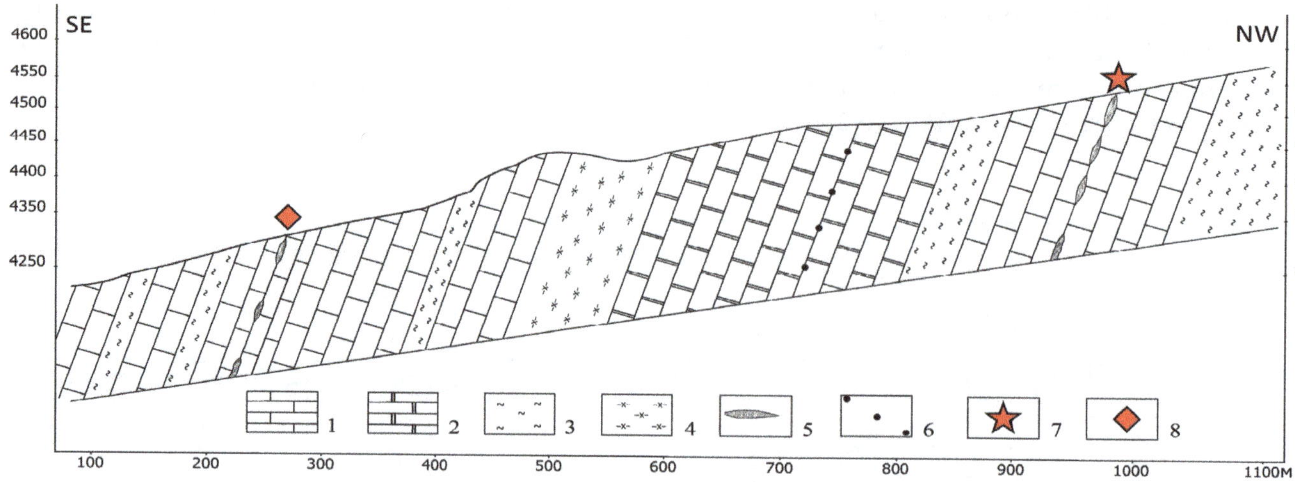

Figure 3. A geological cross-section of the upper part of the Sarydzhilgin suite through the Snezhnoe deposit and the ruby occurrences of Lagernoye and Alenushka: 1—calcitic marble and 2—dolomitic marble (dip angle is 60–70°), 3—high alumina crystalline schists, 4—diopside–scapolite rocks, 5—ruby deposits–micaceous lenses, 6—monomineralic ruby mineralization. 7—Snezhnoe deposit. 8—Alyonushka ruby occurrence. The cross-section is modified after [20]. The area above of the upper curve represents the modern topographic surface, whereas the white area below is the cutoff of the geological cross-section.

Since the last century, different types of prospecting, evaluation and exploration activity was carried out at the Snezhnoe deposit. Currently, mining activity is conducted from the surface and by underground methods.

3. Materials and Methods

During field trips in 2010–2017, the Snezhnoe deposit was traced along its strike in order to examine a complete profile along the ruby-bearing rocks (Table S1). The concentrations of major and trace elements in the ruby-bearing rocks were analyzed using X-ray fluorescence spectrometry. The paragenetic mineral associations with ruby were determined by petrographic studies. Minerals were studied in petrographic thin-sections and by X-ray diffraction (XRD), while their composition was measured using electron microprobe analysis (EMPA) and laser-ablation inductively coupled plasma—mass spectrometry (LA–ICP–MS). In situ U–Pb LA–ICP–MS analyses of rutile grains, syn-genetically associated with the ruby, were applied for geochronological reconstructions. Age determinations of bulk corundum-bearing rocks and mineral separates were conducted using thermo-ionization mass-spectrometry (TIMS). Oxygen isotope analyses of rubies using secondary ionization mass-spectrometry (SIMS) were done for identification of the possible protolith of the corundum-bearing rocks.

The whole-rock chemistry of the studied samples was determined using an AXIOS Advanced X-ray fluorescence spectrometer at the Vernadsky Institute of Geochemistry and Analytical Chemistry, Russian Academy of Sciences (GEOKHI RAS) with an X-ray tube equipped with a Rh anode, 3 kW power, a Soller scanning channel with the analyzing crystals and a detonating device.

The selected minerals extracted from the ruby-bearing rocks were identified by X-ray diffraction analysis using the DRON-3 M facility at the Russian State Geological Prospecting University. The X-ray source is a Co–anode using a Fe-filter with 30 kV voltage, current of 30 mA and 2θ scanning angle from 5° to 60°.

The mineral composition was determined by EMPA using a Cameca SX 100 in the wavelength-dispersive detection mode at the GEOKHI RAS with an electron beam of 15 kV, 30 nA and a beam size of 5 μm. A set of natural and synthetic reference materials was used for calibration.

The detection limit for almost all elements was less than 0.01 wt%. Elemental concentrations were calculated using the PAP correction (atomic number, fluorescence and absorption correction).

The LA–ICP–MS analysis was performed to determine the trace element concentrations within discrete zones of ruby grains from the Snezhnoe deposit using an ESI NWR193 ArF Excimer Laser combined with an Agilent 7500 ce quadrupole-ICP-MS at the Institut für Geowissenschaften, Johannes Gutenberg-Universität Mainz (JGU), Germany. The samples were ablated using a spot size of 70 μm, a repetition rate of 10 Hz and an energy density of approximately 3.0 J/cm^2. Warmup/background time was 15 s, dwell time 30 s and wash out time 20 s. Trace elements of interest within the ruby structure were analyzed using the following isotopes: ^{24}Mg, ^{47}Ti, ^{51}V, ^{53}Cr, ^{56}Fe and ^{69}Ga. The isotopes of additional elements ^{6}Li, ^{9}Be, ^{23}Na, ^{24}Mg, ^{29}Si, ^{39}K, ^{43}Ca, ^{47}Ti, ^{55}Mn, ^{86}Sr, ^{90}Zr, ^{93}Nb, ^{137}Ba, ^{179}Hf, ^{181}Ta and ^{208}Pb were analyzed as well to exclude the areas with possible contamination from solid inclusions in time-resolved spectra. NIST (National Institute of Standards and Technology) SRM (standard reference material) 612 was used as primary reference material. NIST SRM 610 and basaltic USGS (United States Geological Survey) BCR-2G basalt glass were used as quality control materials. Reference and quality control materials were measured after every 30 corundum unknowns to monitor the accuracy and precision of the analyses and calibration. The time-resolved signal spectra were processed in GLITTER 4.4.1 software (www.glitter-gemoc.com, Macquarie University, Sydney, Australia) using ^{27}Al as the internal standard applying a theoretical value of Al_2O_3 as 100 wt% for the corundum samples and the values given in GeoReM database for the reference material [21,22]. The measured concentrations for most elements in both QCM agree within 18% with the preferred values provided in the GeoReM database.

Trace-elements and U–Pb isotopic composition of rutile grains and rutile inclusions within ruby were determined using the Element2 ICP-MS (ThermoFisher) coupled with an Analyte G2 193 nm excimer laser (PhotonMachines) at the Westfälische Wilhelms-Universität Münster. Twenty-three rutile grains from the Snezhnoe deposit were chosen for in situ LA–ICP–MS trace element measurements and U–Pb dating. Prior to analysis, backscattered electron images (BSE) were obtained at the Institut für Geowissenschaften, Johannes Gutenberg-Universität Mainz (JGU), Germany using a Zeiss DSM 962 SEM coupled with an Oxford energy-dispersive spectrometer. For LA–ICP–MS measurements, gas flow rates were about 1.1 L/min for He, 0.9 L/min and 1.1 L/min for the Ar-auxiliary and sample gas. The cooling gas flow rate was set to 16 l/min. Trace element concentrations within the rutile grains were determined by measuring the following isotopes: ^{51}V, ^{53}Cr, ^{90}Zr ^{93}Nb, ^{95}Mo, ^{118}Sn, ^{121}Sb, ^{178}Hf, ^{181}Ta, ^{182}W, ^{208}Pb, ^{232}Th and ^{238}U. The analysis of ^{29}Si, ^{43}Ca, ^{49}Ti, ^{89}Y, ^{139}La and ^{140}Ce was applied to control any possible contamination with intergrown minerals. The spot-size during the analyses was 70 μm at a repetition rate of 10 Hz and an energy density of 3–4 J/cm^2. The background time was 15 s, dwell time was 40 s and wash out time was 20 s. The NIST SRM 612 was used as the reference material, while R10 [23] and R632 [24] were applied for quality control (QCM). The time-resolved spectra were processed in GLITTER 4.4.1 software using Ti as the internal standard applying the stoichiometric value of TiO_2 as 100 wt.% in pure crystalline rutile for analyzing the rutile unknowns.

For U–Pb rutile dating, ^{206}Pb, ^{207}Pb, ^{208}Pb, ^{232}Th and ^{238}U isotopes were measured. The repetition rate was 10 Hz using an energy of about 3 J/cm^2. The spots size of 70 μm were ablated in close vicinity to the spots used for trace element measurements. Five unknowns were bracketed with three measurements on the R10 calibration material to correct for instrumental mass bias. The data reduction was performed using an in-house Excel spreadsheet [25]. The rutile reference material SP-1 (Adirondack terrane, New York) and R632 yielded a Concordia age (2σ) of 911 ± 12 Ma and 504 ± 10 Ma, which is in agreement with the ID-TIMS age of 911 ± 2 Ma and 496 ± 2 Ma, respectively [24,26]. The Concordia diagrams and the age calculations were made using Isoplot v. 4.13 (Berkeley Geochronology Center, USA) [27].

Minimum entrapment pressures of rutile inclusions in the ruby host was estimated from the luminescence shift of the Cr^{3+} R-lines in the ruby host in close vicinity to the inclusions [28]. A Confocal Raman Spectrometer Horiba Jobin Yvon HR800 (HORIBA France SAS, Longjumeau, France) coupled with an Olympus BX41 microscope (Olympus Optical Co., Ltd., Tokyo, Japan) and automatic XYZ-stage were used for this technique. A red helium-neon laser with $\lambda = 632.816$ nm (polarized during the measurements) was used with a grating of 1800 grooves/mm. The confocal hole of 400 mm and the entrance slit of 100 mm produced a resolution of 0.7 (blue) to 0.5 (red) cm^{-1} for an exposition time of 0.5 s with measured steps of 100 mm. Magnification of ×50 was used for a measured range of 690.0–700.0 nm. The spectrometer was calibrated at 520.7 cm^{-1} using Si as a reference.

The oxygen isotope composition of the ruby grains from Snezhnoe was determined by using the Cameca IMS1280-HR ion microprobe at Heidelberg University (HIP), Germany. Before the SIMS measurements, BSE images were taken using a LEO 440 scanning electron microscope coupled with an Oxford Instruments X-Max energy dispersive spectrometer in order to avoid the areas with inclusions. A 2 nA and 20 keV Cs^+ primary ion beam with a raster size of 10 μm (12 μm during pre-sputtering) was used for measurements targeting rim-to-rim areas of the ruby samples. Negative secondary ions were accelerated to 10 keV. The secondary ion image was limited to ~23 μm, the dynamic transfer optical system (DTOS) was activated and sample charging was compensated with the electron gun (NEG). The ^{16}O and ^{18}O isotopes were detected simultaneously in two Faraday Cups. The mass resolving power was ~2300. Prior to each analysis, the secondary beam was centered automatically in the field aperture (X and Y) and the entrance slit (X only). Including the time for beam centering, the analyses started after a total pre-sputtering time of 90 s and each analysis had 20 cycles with 4 s integration time per cycle. The internal precision reported is the standard deviation (SD) of the mean value of the isotope ratios. The baseline of the FC amplifiers was determined with an integration time of 200 s. The instrumental mass fractionation was determined using a laser ruby reference material ($\delta^{18}O$ = 15.7‰ ± 0.2‰, [29]) embedded within the same amount as the samples. The reproducibility of the IMF was 0.09‰ (1 SD) for the session with this sample.

Powders of bulk ruby-bearing rocks were produced for Rb–Sr and Sm–Nd isotope analyses. Additionally, single mineral fractions of plagioclase and phlogopite were selected for Rb–Sr geochronology. Some 0.02–0.03 g of powder from a larger portion of 200–300 g was dissolved in a mixture of hydrofluoric and nitric acids with a 5:1 ratio on a shaker under incandescent lamps within three days. The obtained solutions were evaporated on a heating plate and, after evaporation, 1 mL of concentrated HCl acid was added three times to the dry residue. Rubidium, Sr and Sm + Nd mixtures were extracted from the solutions using fluoroplastic chromatographic columns with a DowexW 50 × 8 synthetic ion-exchange resin. The extraction was done by stepwise elution with 2.2 acid normality (N) of HCl (for Rb and Sr) and 4.0-N HCl (for the Sm + Nd mixture). From the Sm + Nd mixture, Sm and Nd were extracted by stepwise elution with 0.15-N HCl, 0.3-N HCl and 0.7-N HCl using polyethylene columns with synthetic ion-exchange Ln-spec resin. Measurements were done using a Finnigan™ Triton multi-collector thermal ionization mass spectrometer (TIMS) by Thermo Scientific at GEOKHI RAS using a two-tape (Re-Re) ion source for Rb, Sm and Nd and a single-tape (Re) ion source for Sr. Measurements were performed in a static mode with simultaneous recordings of ion currents for different isotopes. Normalization was carried-out according to the exponential law for $^{86}Sr/^{88}Sr = 0.1194$ and $^{148}Nd/^{144}Nd = 0.241572$ to eliminate mass-discrimination. The analyses of the international standards Sr–SRM-987 and JNdi–1 monitored the reproducibility and accuracy of isotope measurements for Sr and Nd. The measured ratios in reference material are $^{87}Sr/^{86}Sr = 0.710233 \pm 0.000010$ and $^{143}Nd/^{144}Nd = 0.512093 \pm 0.000009$. Concentrations of Rb, Sr, Sm and Nd were determined by isotopic dilution using ^{85}Rb-^{84}Sr and ^{149}Sm-^{150}Nd tracers. The data-reduction was performed using an in-house Excel spreadsheet.

4. Results

4.1. Petrology of the ruby-bearing rocks

The ruby-bearing rocks vary significantly in terms of their mineral composition changing along both strike and depth. Their structure is characterized by a wide range of crystal sizes and shapes.

There are four groups of mineral associations with ruby forming isolated areas in the deposit with one, three and four-mineral parageneses (Table 1): (1) scapolite (marialite) + phlogopite + muscovite + margarite; (2) plagioclase (albite–andesine) + muscovite + margarite (Figure 4A); (3) muscovite + phlogopite + margarite; (4) calcite (Figure 4B). Ruby-free areas of the deposit are characterized by plagioclase + scapolite + phlogopite + muscovite mineral paragenesis.

Table 1. List of major, minor and accessory minerals from the Snezhnoe deposit.

Minerals/Mineral Class	Major and Minor Minerals (Methods Applied *)	Accessory Minerals (Methods Applied)
Oxides	Corundum (LA–ICP–MS geochemistry, SIMS oxygen isotopy)	Rutile (U–Pb LA–ICP–MS dating, Raman mapping) Ilmenite + Nb-rich rutile Magnetite
Alumino-silicates	Plagioclase (TIMS Rb–Sr/Sm–Nd isotopy) Phlogopite (TIMS Rb–Sr/Sm–Nd isotopy) Muscovite Margarite	Corundophillite
Silicates	Scapolite	Zircon Titanite Tourmaline Thorite Dissakisite-(Ce)
Carbonates	Calcite	REE carbonates Dolomite
Sulfides		Pyrite Pyrrhotite Chalcopyrite
Phosphates		Xenotime-(Y) Monazite-(Ce) Fluorapatite
Native element		Graphite

* EMPA was applied to all minerals listed.

Figure 4. Images of hand specimens of Snezhnoe ruby-bearing rocks with: (**A**) ruby + plagioclase (albite–andesine) + muscovite (variety fuchsite) mineral paragenesis; (**B**) ruby + calcite mineral paragenesis. The images were modified from [11].

Gem-quality ruby is restricted in occurrence to Cr-bearing mica lenses or graphite-free areas within the host marbles. The gem quality of ruby (i.e., degrees of transparency and color intensity) increases with decreasing number of associated paragenetic minerals. Therefore, the calcite mono-mineralic assemblage is the most economically valuable.

4.2. *Mineralogy of Ruby-Bearing Rocks*

4.2.1. Major and Minor Minerals

Mica group minerals were found in most ruby associations. Micas often contain traces of Cr and V (Table 1, Tables S2–S4) resulting in their characteristic green color (Figure 4A). These green-colored mica group minerals serve as the main prospecting indicator for the ruby mineralization.

Muscovite forms small- to medium-sized flaky aggregates ranging from ~1 to 4 mm. It is also found as inclusions in plagioclase and scapolite group minerals and in ruby (Figure 5, Table S2). It is rarely replaced by the chlorite group minerals, likely clinochlore containing up to 25.35 wt% SiO_2, 22.52 wt% Al_2O_3, 22.45 wt% FeO, 13.98 wt% MgO, 1.35 wt% Cr_2O_3 [11] and forming thin films on the surface of muscovite or infilling fractures in corundum.

Fuchsite (the chromium-enriched variety of muscovite) is found in the form of a fine-medium flaky aggregates or individual flakes with size to 1–3 mm. These aggregates form micaceous ruby-bearing lenses as well as rare impregnations in calcitic marbles. The fuchsites are characterized by high concentrations of TiO_2 up to 4.78 wt%, MgO up to 5.5 wt%, Na_2O up to 2.07 wt%, Cr_2O_3 up to 2.5 wt%, V_2O_3 up to 0.89 wt% and F from 0.03 wt% to 0.89 wt% (Table S2). Fuchsite crystals were also detected in cross-cutting veins, which were considered as hydrothermal in origin [14]. These minerals form concentrically zoned crystals with a diameter up to 6 mm and they are found in paragenetic association with later margarite flakes, concentric scapolite crystals and likely clinochlore. Rarely, the hydrothermal veins contain later tourmaline and quartz (never observed in the original ruby-bearing association).

Margarite (the Ca-end member of the brittle mica group) was detected as a syngenetic rock-forming mineral within the Snezhnoe deposit [11,14]. Margarite forms coarse- and finely flaked mineral grains. It is localized in occurrence mainly among calcite porphyroblasts and included within host corundum. The margarite contains Na_2O from 1.24 wt% to 4.37 wt%, Cr_2O_3 up to 1.27 wt%, V_2O_3 up to 0.1 wt% and F up to 0.32 wt% (Figure 5; Table S3).

Phlogopite forms lustrous brown-colored well-formed crystals some 2–5 to 10 mm in size. It is enriched in Cr_2O_3 up to 3.20 wt% and F up to 1.87 wt% (Table S4). As in the case of muscovite, it is only rarely found to be replaced by chlorite group minerals.

Plagioclase forms white to black (due to the abundance of mineral inclusions including graphite, rutile, mica, zircon and rarely, ruby) prismatic crystals up to 10 mm in size. The composition is albite–andesine (An_{6-41}) (Table S5). Dufour et al. [9] noted anorthite in the paragenesis with phlogopite and muscovite, however, albite occurs within rims around the ruby grains.

Scapolite forms two generations: the early one is metamorphic in origin and syngenetic with ruby and associated minerals. For this generation, the meionite component ranges from 23 wt% to 74 wt%, while chlorine varies from 0.01 wt% to 2.29 wt%, sulfur content was up to 0.5 wt% and fluorine values were up to 0.25 wt% (Table S6). The later generation is hydrothermal in origin and occurs in veins cross-cutting the ruby-bearing rocks.

Calcite occurs within the micaceous lenses as irregular-shaped grains up to 0.4 mm and locally replaced by scapolite [11]. In the host marbles, it comprises the porphyroblastic rock matrix with crystal size varying from 1 to 6 mm.

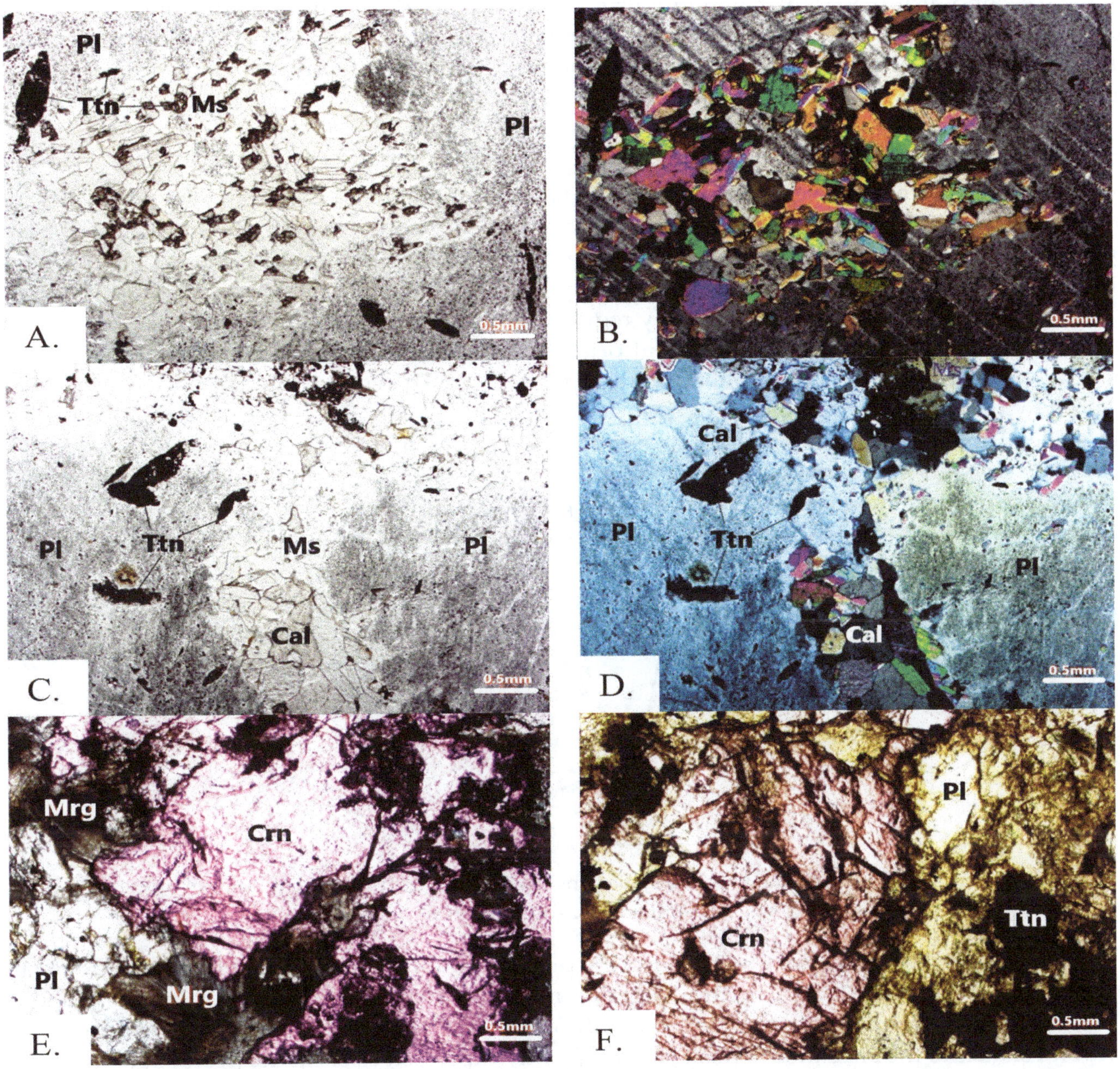

Figure 5. Petrographic thin-sections of titanite (Ttn), muscovite (Ms) and calcite (Cal) grains intergrown with larger plagioclase (Pl) crystals in plane-polarized light (**A**,**C**) and under crossed-polarized light (**B**,**D**); corundum (Crn) intergrowths with margarite (Mrg) and plagioclase crystals (**E**,**F**), plane-polarized light.

4.2.2. Accessory Minerals

The accessory minerals identified in the studied ruby-bearing rocks are divided into two groups based on their typical size: (1) the large, up to millimeter size minerals, relatively easily distinguishable by eye and (2) those reaching only a few hundred micrometers in size and visible only under high magnification (Table 1). Currently, 20 accessory minerals were detected within the Snezhnoe deposit.

The first group includes graphite forming thin films around ruby grains, corundophyllite replacing muscovite, calcite with a minor Mg content, titanite (in 2 generations, with one replacing rutile and the other forming moderately rounded discrete crystals, Table S7), rutile (in 2 types with the earlier grains forming prisms and rounded grains linked to ruby and mica group minerals, while the later crystals replace titanite), tourmaline of the dravite–elbaite series (Table S9) and rare pyrite.

The second group of accessory minerals was only detected at high magnification and includes ilmenite and Nb-rich rutile, fluorapatite, pyrite, pyrrhotite, magnetite, chalcopyrite, thorite, dissakisite-(Ce) (allanite group), REE-rich carbonates, dolomite, xenotime-(Y), monazite-(Ce) and zircon containing up to 2 wt% HfO_2.

4.2.3. Mineralogy, Geochemistry and Oxygen Isotopes of Ruby

For the ruby-bearing assemblage, ruby is the largest mineral in these rocks with crystals reaching some 10 × 20 cm. Different ruby generations were observed during petrographic studies. The earlier crystals were found in the form of poikiloblastic grains in association with scapolite, plagioclase and micas. While the later generation of ruby was only observed in the calcitic marbles [30].

The ruby crystals are characterized by regular, distorted and combined forms. Regular crystal shapes are found in the form of hexagonal prisms combined with hexagonal pyramids, however, flattened by a pinacoid and rhombohedral faces (Figure 6). The crystal faces are often smooth and covered by fine-grained graphite, mica group minerals and calcite. Distorted crystals form sinuous and skeletal forms or nodules with a large number of faces poorly visible under the microscope. Combined crystals are formed by prisms complicated by skeletal surfaces. The surfaces of these crystals along with smooth ones, are tuberous, perforated (embossed) with noticeable resorption features. Margarite flakes of sometimes spherical shape, wedge-shaped titanite and calcite crystals are intergrown with the rubies.

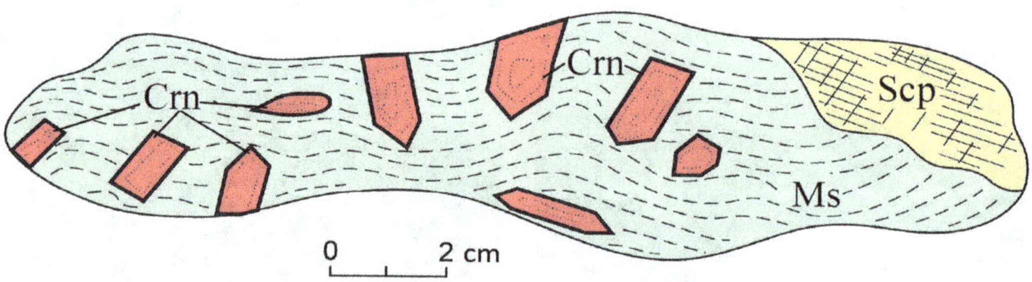

Figure 6. The sketch of a fragment of a ruby-bearing rock showing the apparent random orientation of ruby crystals. Crn—ruby crystals, Ms—muscovite (dashed lines), Scp—scapolite.

Using EMPA, the content of Cr_2O_3 within the pink corundum crystals varies from below detection limit (bdl; <0.01 wt%) to 0.24 wt%. In the red grains, it reaches 1.5 wt% with a mean value of 0.3–0.6 wt%. In fact, the variations in Cr^{3+} concentration within the ruby lattice are linked to the variety of ruby tones and shades at the Snezhnoe deposit [15], frequently expressed as oscillatory color zonation within some crystals. Moreover, the rubies also contain significant concentrations of other trace elements including FeO_t up to 0.41 wt% and V_2O_3 up to 0.13 wt%. The dark pink to red ruby crystals are characterized by Cr_2O_3 ranging from 0.01 wt% to 0.55 wt%, FeO_t ranging from 0.03 wt% to 0.19 wt%, TiO_2 values from 0.01 wt% to 0.17 wt%, MgO from bdl (<0.01 wt%) to 0.04 wt% and V_2O_3 from 0.01 wt% to 0.07 wt% (Table S10).

Three gem-quality rubies with bright red color and purple hue from the Snezhnoe deposit were analyzed using LA–ICP–MS to determine their concentrations of Cr, Fe, V, Mg, Ti and Ga (Table S11). The obtained values are in good agreement with the EMPA analyses for similar colored ruby crystals. Chromium (512 to 4720 μg/g), Fe (371–623 μg/g) and Ga (65.9–98.1 μg/g) all showed a heterogeneous distribution. However, these trace elements demonstrated a positive correlation between each other on different binary plots (Cr + V vs. Fe+ Ga, Fe vs. Ga, Cr vs. Ga, Figure 7). Such a correlation is linked presumably to the mechanism of Al^{3+} ion substitution within the corundum structure along the pinacoid faces in the direction perpendicular to the *c*-axis. It is frequently observed in other ruby locations [31]. The titanium content also shows a heterogeneous distribution ranging from 13.9 to 177 μg/g due to minor fluctuations, however, without any correlation with the Cr, Ga or Fe values.

Magnesium shows a mostly homogeneous distribution within the studied ruby grains (from 9.07 to 26.9 μg/g).

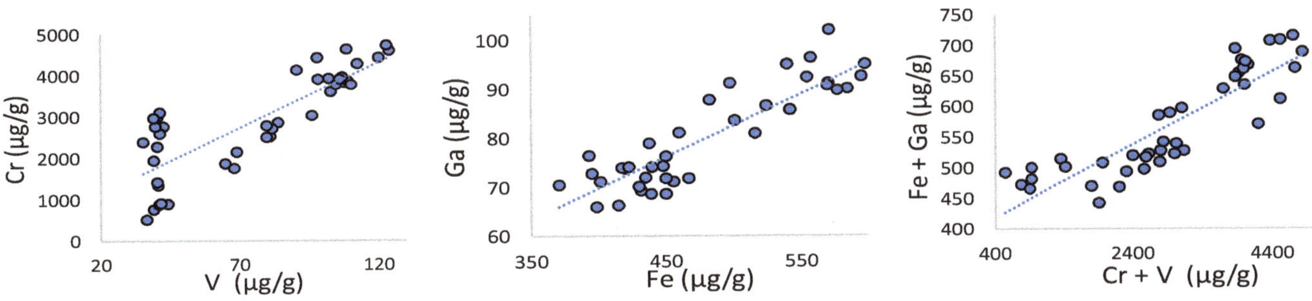

Figure 7. Trace element distribution within ruby grains oriented perpendicular to the *c*-axis on Cr vs. V (left panel), Ga vs. Fe (middle panel) and Fe+Ga vs. Cr + V (right panel) binary plots.

The trace element ratios of Ga/Mg < 8.2, Fe/Mg < 51.2, Cr/Ga > 6.9 and Fe/Ti < 31.6 are all in the range common for metamorphic corundum [32,33]. On the FeO–Cr_2O_3–MgO–V_2O_3 vs. FeO + TiO_2 + Ga_2O_3 plot [34], Snezhnoe rubies overlap the "John Saul ruby mine" area in the "ruby in marble" field in close proximity to the "metasomatic" corundum field (Figure 8).

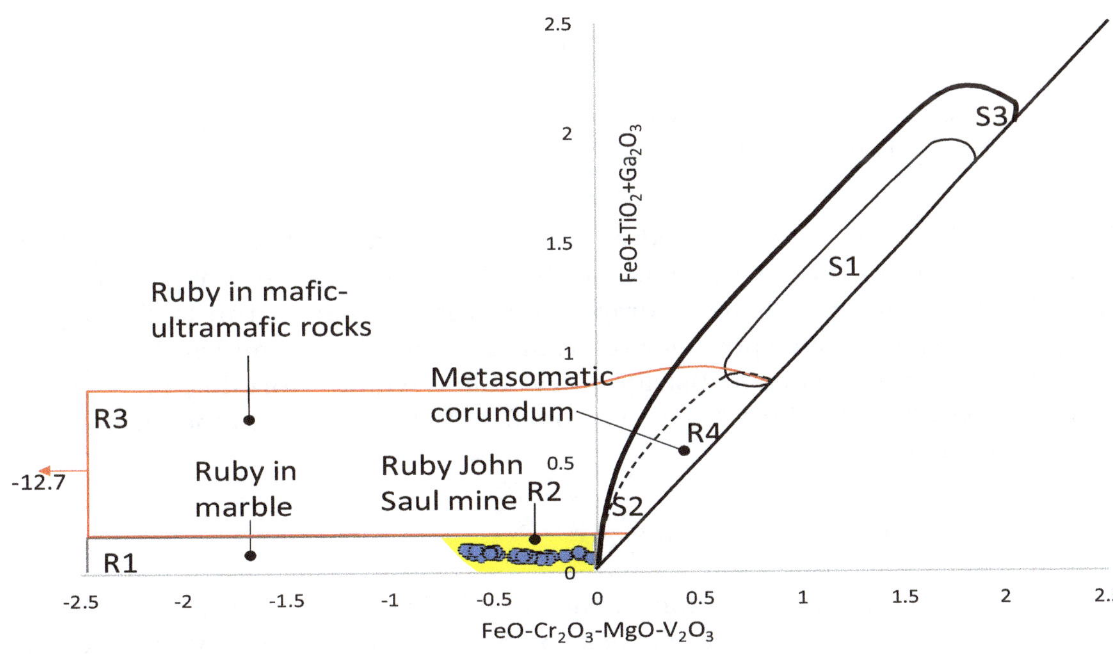

Figure 8. A FeO–Cr_2O_3–MgO–V_2O_3 vs. FeO–TiO_2–Ga_2O_3 diagram (in wt%) for rubies from the Snezhnoe deposit, modified after [34].

On the Cr/Ga vs. Fe/Ti discrimination diagram of [33], the Snezhnoe rubies clearly plot within the "metamorphic" field (Figure 9A). On the other hand, they are in the "transitional" and, partly, in the "magmatic" field on the Fe (μg/g) vs. Fe/Ti plot of [32,33] (Figure 9B).

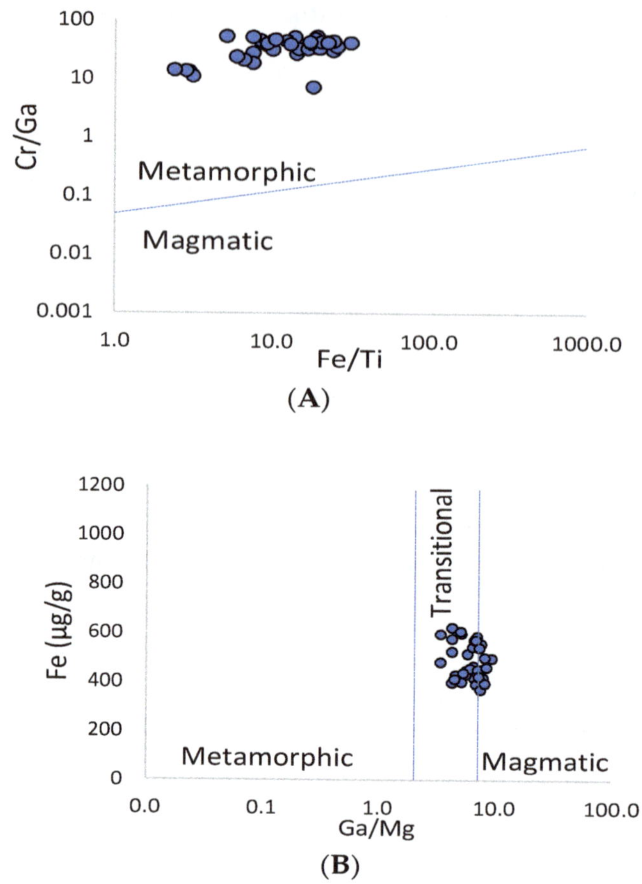

Figure 9. Cr/Ga vs. Fe/Ti plot of Snezhnoe ruby modified after [33] (**A**); Fe(µg/g) vs. Fe/Ti plot of Snezhnoe ruby modified after [32,33] (**B**).

The $\delta^{18}O$ isotopic composition is used as a proxy in corundum provenance determination distinguishing corundum of different genetic origins [35]. Two ruby grains previously analyzed using LA–ICP–MS for their trace element composition were chosen for rim-to-rim oxygen isotope analysis using SIMS. In total, eight spots were measured within these samples. The $\delta^{18}O$ rim-to-rim analyses show mostly a homogeneous distribution within the grains forming a narrow range from +15.1‰ to +15.3‰ with the standard deviation from 0.06‰ to 0.12‰ and the mean value of about +15.25‰ ± 0.07‰ (n = 8).

4.3. Petrochemical Features of the Ruby-Bearing Rocks

The whole-rock chemical composition of the ruby-bearing rocks varies considerably for all petrogenic elements (Table S12). These bulk-rock chemical variations correlate with changes in the mineral associations along strike.

Contents of the main petrogenic elements within the ruby-bearing rocks varied from 35 wt% to 98 wt% for Al_2O_3 (the most enriched in the Pamirs), up to 12 wt% for the sum of alkalis (Na_2O + K_2O) and from 3 wt% to 10 wt% for CaO. In the crustal normalized spectra, the ruby-bearing rocks are mostly enriched in three+ and four+-charged, and, in less extend, one+, two+ and five+-charged lithophile ions (Figure 10). However, they are slightly depleted in Si and Mg. The ruby-bearing rocks are also characterized by the joint presence of oxides and sulfides of iron, along with traces of nickel up to 2 wt%. All minerals found at the deposit contain Cr_2O_3 and V_2O_3 from hundredths to tenths parts of a percent, rarely, up to several percent. Cobalt and nickel were found in many minerals to hundredth parts of a percent. Minor elements within the ruby-bearing rocks exceed their average enrichment in the Earth's crust [36] by the orders of magnitude as follows: Ce (248), La (8.8), Ga (3.1), Sr (1.9), S (1.7), Cl (1.3) and Zn (1.3). The fluorine content is up to 4.6 wt% and this has played a significant role in increasing the mineral-forming activity of fluids.

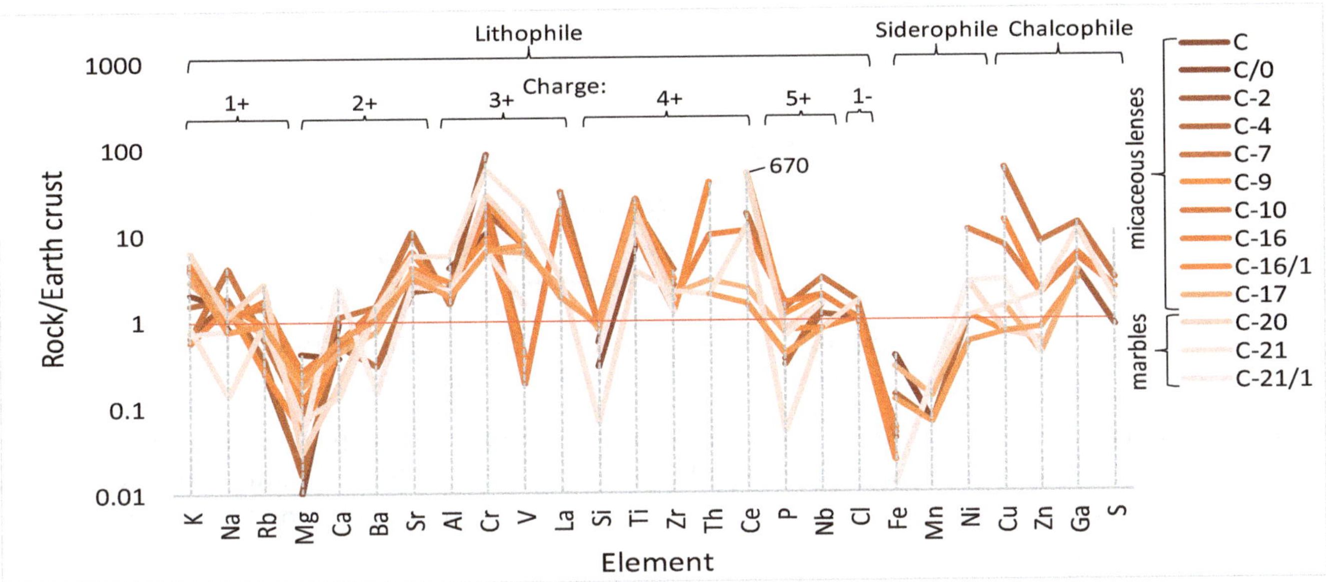

Figure 10. Crustal-normalized concentrations of lithophile (arranged by cation/anion charge), siderophile and chalcophile elements in the Snezhnoe deposit. The cerium ratio at 670 was cutoff. The values for the Earth's crust are from [36].

4.4. In Situ U–Pb LA–ICP–MS Rutile Dating, Zr-in-rutile Temperature and Raman Mapping of Rutile Inclusions

Rutile grains from the Sneznoe deposit varied from 200 × 200 μm to 600 × 600 μm in size with most showing a homogeneous composition. However, inclusions of apatite, Cr-bearing muscovite, margarite, zircon and calcite were found within rutile grains during SEM analyses. The following trace elements were detected in measured rutile grains (Table S13): V (1600–5000 μg/g), Cr (1360–5400 μg/g), Zr (1570–2650 μg/g), Nb (3000–6170 μg/g), Sn (170–240 μg/g), Sb (bdl–2.1), Hf (53–85 μg/g), Ta (210–400 μg/g), W (14–73 μg/g), Pb (bdl–68 μg/g), Th (bdl–52 μg/g) and U (14–46 μg/g). Molybdenum was measured but found to be below the detection limit (bdl varied from 0.3 to 0.9 μg/g).

The in situ LA–ICP–MS U–Pb analyses of the rutile grains (21 spots) were plotted on Concordia diagram in Figure 11. All the rutiles analyzed yielded discordant ages with lower intercepts on the Concordia at about 12 ± 1.5 Ma (2σ); MSWD (mean squared weighted deviation) is 9.7. The upper intercepts show an unreliable age at around 4.8–4.9 Ga likely due to contamination with common Pb; the point with $^{207}Pb/^{208}Pb$ ratio is about 0.8 on the Tera–Wasserburg diagram (Figure 11, [37]). However, the ages obtained may be applied with caution due to the re-opening of the U–Pb system in rutile.

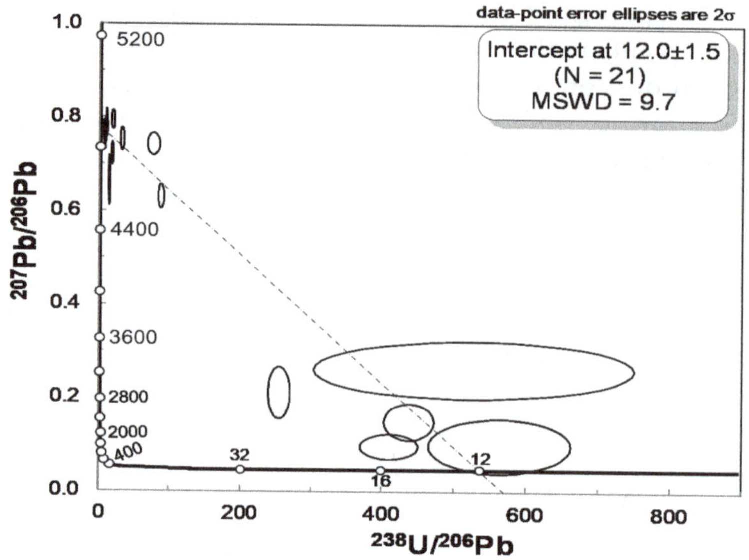

Figure 11. Tera–Wasserburg Concordia diagram based on the U–Pb LA–ICP–MS data for rutiles associated with ruby. The lower intercept age is shown in the legend.

Rutile grains were frequently found intergrown with zircon at the Snezhnoe deposit, and therefore could be used to calculate the apparent Zr-in-rutile temperature [38]. The calculations were performed using the following formula: $T\ (^{\circ}C) = \frac{4470 \pm 120}{(7.36 \pm 0.10) - \log\left(Zr, \frac{\mu g}{g}\right)} - 273$ [39]. The obtained temperatures were in the range from 830 ± 60 °C (Table S13).

Laser-induced luminescence measurements were applied for two rutile inclusions trapped within ruby grains; the same sample containing other rutile inclusions was used for U–Pb rutile dating. The rutile grains varied from about 10 × 20 μm to 10 × 70 μm in size. The Raman shift of Cr^{3+} R-lines in ruby photoluminescence spectra around rutile inclusions provided their minimum entrapment pressure at about 1–1.2 kbar (Figure 12) according to the calibration of [28].

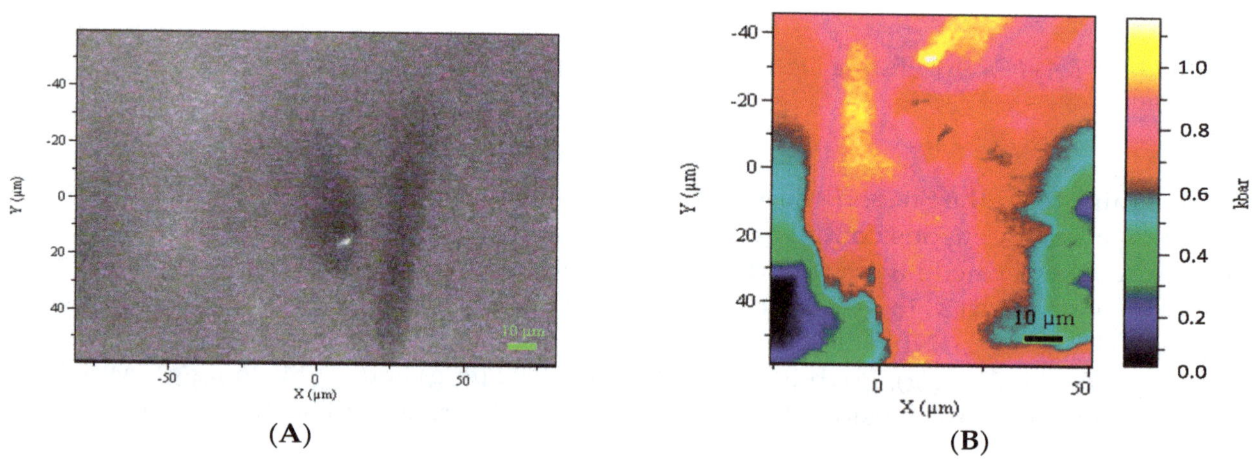

Figure 12. Photomicrograph of measured rutile inclusions within the ruby host (**A**); Raman map showing the shift of R-lines in laser-induced photoluminescence spectra expressed in pressure (in kbar) around rutile inclusions within the ruby host (**B**).

4.5. Rb–Sr and Sm–Nd Isotope Measurements

The ruby-bearing schists from Snezhnoe were studied by TIMS in order to obtain their Rb–Sr and Sm–Nd isotope systematics. For the analyses, single minerals fractions of plagioclase and phlogopite were extracted from the rock matrix. The plagioclase and phlogopite contained no visual secondary alteration, however, some of the phlogopite could be possibly partially replaced by microscopic chlorite group minerals (most likely clinochlore). The results for the ruby-bearing schists within marble host rocks (single mineral fraction of mica and feldspar along with bulk rock analyzes) provided one errorchron (the large scatter in Rb–Sr signatures defining significant ±errors and MSWD) with an age at 23.0 ± 1.6 Ma (MSWD = 37; Figure 13) and an initial $^{87}Sr/^{86}Sr$ ratio of 0.70791. The high MSWD value is likely linked to re-opening of the Rb–Sr system in phlogopite. Thus, 3 phlogopite samples provided one errorchron with a younger age of 12 ± 27 Ma (MSWD = 14) and an initial $^{87}Sr/^{86}Sr$ ratio of 0.7085.

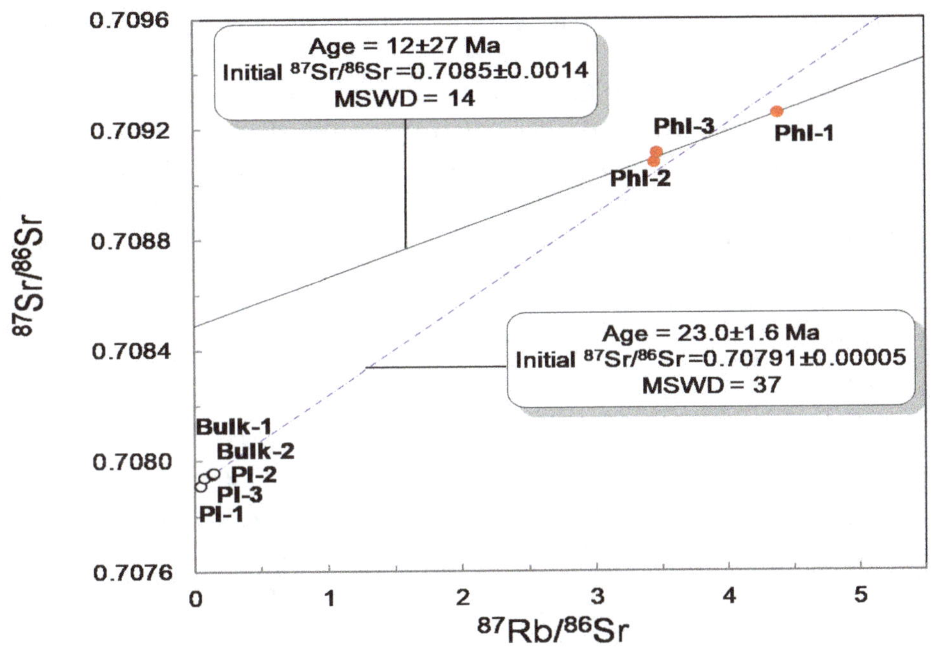

Figure 13. Two Rb–Sr errorchrons showing the ages of ruby-bearing rock at Snezhnoe (blue dotted line), with a likely alteration process affecting the Rb–Sr system in phlogopite (black solid line).

The epsilon notation value $\varepsilon Nd_{(20Ma)}$ for the bulk ruby-bearing rock was about −9.6 (Table S14).

5. Discussion

5.1. Hypotheses of Origin of the Snezhnoe Ruby-Bearing Rocks

Different hypotheses were proposed regarding the possible source for ruby-bearing rocks including metasomatic, hydrothermal or sedimentary-metamorphic scenarios. However, most of them remain controversial.

In the first scenario, Dufour et al. [9] proposed the formation of the ruby-bearing rocks through the metasomatic re-working of marbles intercalated with sandy-clay rocks, with the metasomatism induced likely by highly alkaline fluids. The desilication of sandy-clay rocks by fluids was followed by rapid enrichment in Al following the formation of biotite group minerals (according to our data, phlogopite) and muscovite. The Al_2O_3 excess resulted in the crystallization of corundum. This crystallization occurred during the final stage of regional metamorphism under the epidote–amphiboliteand greenschist facies with T = 600–650 °C, P = 4–6 kbar and X_{CO2} (mole fraction of CO_2) = 0.2–0.4 (Figure 14). The timing of metamorphism, and, hence, ruby formation, corresponded to the Miocene (N1) during which the last recorded regional metamorphic process in the area occurred. Kievlenko [40] also referred to the ruby-bearing rocks in the Snezhnoe deposit to being of metasomatic origin and formed as endo-skarns due to the desilication of Al–Si rocks by hydrothermal solutions. The crystallization temperature of rubies was estimated at 450–500 °C by measuring the homogenization temperature of fluid inclusions inside the rubies.

According to the second scenario [41], corundum mineralization formed by transport of Si, Al, Ti, Na, Cl and B in aqueous solutions migrated along fractures and layers in the marbles. Calcium and CO_2 were removed by the fluid-marble interaction. Terekhov et al. [42] shared a similar opinion on ruby formation. These authors pointed out that the source of aluminum, required for ruby formation, may be linked to the endogenous fluids formed in an alkaline magmatic chamber of unknown location.

In the third scenario [43,44], ruby and associated minerals within the marble host rocks were formed during metamorphic reworking of primary sedimentary rocks without significant involvement of any external material. This hypothesis was further developed by Litvinenko [45,46], Barnov [10] and Nasreddinov [12]. According to this scenario, ruby was formed during iso-chemical metamorphism of Proterozoic bauxite-like sediments hosted by limestone during amphibolite facies metamorphism. This scenario was further developed by Garnier et. al. [2], who pointed-out the significant role of evaporates hosted by marbles in the formation of ruby mineralization in several deposits located along the Alpine–Himalayan Belt. The source of elements required for their crystallization may be linked to the clay minerals. These clay minerals, hosted by the limestones, underwent the amphibolite facies metamorphism at T from 610 to 790 °C and P ~6 kbar. Ruby mineralization occurred during the retrograde stage at T = 620–670 °C and P = 2.6–3.3 kbar.

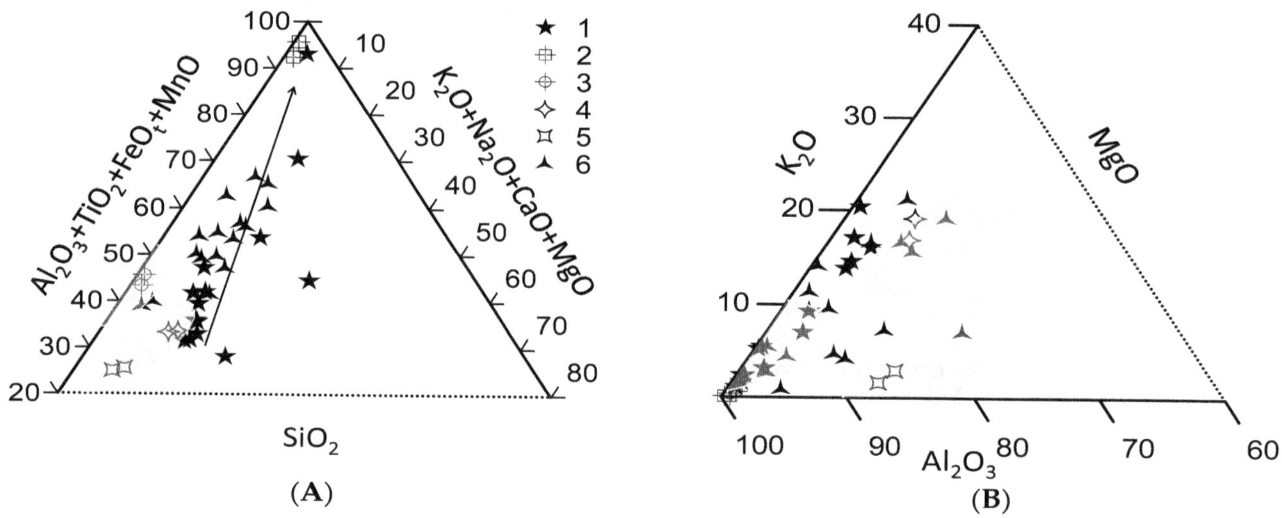

Figure 14. The Al_2O_3 + TiO_2 + FeOt + MnO—K_2O + Na_2O + CaO + MgO—SiO_2 (**A**) and K_2O—MgO—Al_2O_3 (**B**) ternary diagrams with the data of Snezhnoe ruby-bearing rocks from this study–1; Precambrian pelitic sediments: bauxite ores–2, kaolinitic clays–3 and green field, illitic clays–4 and orange field, montmorillonite clays–5 and blue field (numbers 2–5 are the mean values of sediments after [47], the pelitic rock fields are after [48]); Al-enriched metamorphic rocks of Muzkol series–6 are after [49]. The rocks under numbers 2 and 3 overlap in the panel B.

5.2. Origin of the Snezhnoe Ruby-Bearing Rocks

The following geological characteristics indicate the sedimentary origin of the protolith considering the interlayered occurrence of carbonate and ruby-bearing rocks tracing along the strike on several kilometers. The petrochemical data reveal very high Al_2O_3 content ranging from 35 wt% to 98 wt% coupled with high alkali concentrations (about 12 wt%) and a high proportion of CaO varying from 3 wt% to 10 wt%. The admixture of lithophile (Cr, V, Ti, U, etc.) and siderophile (Sn, Ni, etc.) elements coupled with Ga contents to 180 μg/g and REE elements up to several thousand μg/g common for sedimentary rocks (Table S8). On the Al_2O_3 + TiO_2 + FeOt + MnO—K_2O + Na_2O + CaO + MgO—SiO_2 ternary plot, the Shenzhoe ruby-bearing rocks overlap the fields of Precambrian illitic clays and bauxite ores with a distinct trend from the pelites toward the Al-enriched sediments (Figure 14). However, they are falling to illitic and kaolinitic clay fields in close proximity to lateritic bauxite values on K_2O—MgO—Al_2O_3 diagram (Figure 14).

The ruby geochemistry indicates a metamorphic origin with trace element ratios of Ga/Mg < 8.2, Fe/Mg < 51.2, Cr/Ga > 6.9 and Fe/Ti < 31.6 [32,33]. On the FeO–Cr_2O_3–MgO–V_2O_3 vs. FeO + TiO_2 + Ga_2O_3 plot of [34], Snezhnoe rubies overlap the area of "John Saul ruby mine" (Kenia) in the "ruby in marble" field in close proximity to the "metasomatic" corundum field. Meanwhile, the John Saul mine rubies can clearly be separated in terms of their mineral inclusions (exsolved rutile needles, etc.), geochemistry (Ga values varies from ~160 up to ~480 μg/g [5]) and formation time (533 ± 11 Ma by LA–ICP–MS U–Pb rutile dating [31]) from those of the Snezhnoe deposit. On the Cr/Ga vs. Fe/Ti discriminant diagram of [33], the Snezhnoe rubies plot in the "metamorphic" field (Figure 9a). However, they are in the "transitional" and partly "magmatic" fields on the Fe(μg/g) vs. Fe/Ti plot of [32,33] (Figure 9b), which is unlikely due to the geological data and mineral assemblage of the ruby-bearing rocks. The ruby–oxygen isotope composition varied in the narrow range of +15.25% ± 0.07‰. In a closed system at a temperature of around 800 °C the resulting corundum oxygen isotopic composition is controlled mainly by the protolith $\delta^{18}O$ values, however, it can also be buffered by the marble host rocks [50]. The obtained oxygen isotopic composition at ~+15.3‰ is slightly lower than that found in metamorphic marble-hosted corundums (from +16.3‰ to +23.0‰; [4]). However, this is within the range detected for pelitic and bauxite-like sediments with $\delta^{18}O$ values varying from about +15‰ to +20‰ [4,50]. Therefore, the obtained results confirm the third hypothesis indicating that the ruby-bearing rocks were formed from a sedimentary Al-enriched protolith re-worked during iso-chemical metamorphism.

According to the results of the Zr-in-rutile thermometer, the apparent crystallization temperatures were in the range of 830±60 °C. However, the application of this thermometer for rutiles crystallizing in the quartz-free environment with an unconstrained silicon activity (a_{SiO2}) may lead to significant temperature overestimation (up to 70 °C at 750 °C [51]). Therefore, the apparent Zr-in-rutile temperature should be recalculated to lower temperature of about 760 ±30 °C. This temperature estimation agrees with the range from 700 to 750 °C obtained in earlier studies [45] (Figure 15). Furthermore, this is close to the upper boundary of 600–650 °C detected previously by thermodynamic modeling of mineral equilibria in the system CaO–SiO2–Al2O3–Na2O–K2O [9], but well above the micro-thermometric estimates of 450–500 °C using fluid inclusions [40] (Figure 15). The Raman shift in the Cr^{3+} R-lines in ruby photoluminescence spectra targeted to areas around rutile inclusions resulted in estimates for the minimum pressure for rutile entrapment at about 1–1.2 kbar. Previous studies have suggested 4–6 kbar [9] and 8–9 kbar [45] required for the formation of ruby-bearing rocks (Figure 15). The lower pressures recorded by rutile entrapment in ruby could be caused by: (1) analysis of already decompressed areas due to the development of microscopic fractures in the ruby crystal or (2) measurement of the secondary rutile in such fractures formed by a replacement of the titanite inclusions originally trapped within the ruby host.

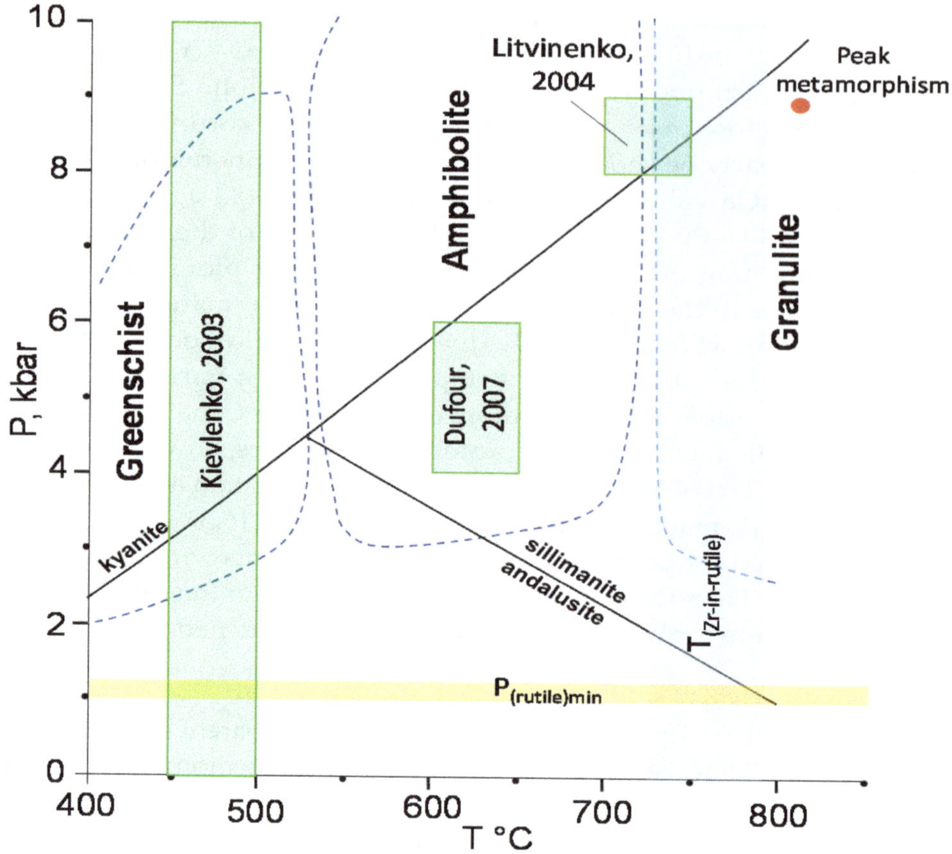

Figure 15. Temperature vs. pressure binary plot showing formation conditions of Snezhnoe ruby deposit recorded in rutiles: Zr-in-rutile thermometer $T_{(Zr\text{-}in\text{-}rutile)}$ (blue vertical field) and Raman mapping of rutile inclusions $P_{(rutile)min}$ (horizontal yellow line). The existing literature data by Kievlenko [40], Dufour [9] and Litvinenko [44,45] are shown as green fields. The peak metamorphism conditions are shown as red point after [18,19]. The stability fields of aluminosilicates (divided by black lines) are after [52]. Schematic fields of Greenschist, Amphibolite and Granulite metamorphic facies are shown for clarity.

The Rb–Sr isotopic measurements showed one errorchron with an age of 23 ± 1.6 Ma. However, these data should be used with caution due to the upper values in samples of phlogopite with different degrees of alteration to clinochlore. The phlogopite errorchron (Figure 13) as in the case for the lower intercept on the U–Pb Concordia (U–Pb rutile analyses in Figure 11) showed younger ages at about ca. 12 Ma falling well within the range of 7–17 Ma obtained previously during K/Ar muscovite dating [41]. This phenomenon suggests the re-opening of the Rb–Sr and K/Ar systems in mica group minerals along with U–Pb systematics in rutile linked likely to later hydrothermal process occurring at the deposit. This hydrothermal process led to the formation of several centimeter-thick veins filled with scapolite and mica group minerals clearly cross-cutting the ruby-bearing rocks. Meanwhile, these obtained Rb–Sr and U–Pb rutile ages are close to the Dufour et al. [9] estimations, while nowhere near the Budanova [19] observations suggesting a Precambrian age. These Rb–Sr ages are likely linked to the cooling stage and the period of relaxation after peak metamorphism during the Alpine–Himalayan Orogeny. However, they also fall in the time range for the other marble-hosted corundum deposits within this major tectonic structure [2,53] (Figure 16). The modeled age of the protolith T(DM)—model age—was estimated at about 1.3 Ga, later than the timing of the first metamorphic cycle in the area of the Snezhnoe deposit (1.9–1.6 Ga). However, this still confirms the hypothesis of a Proterozoic protolith for the ruby-bearing rocks.

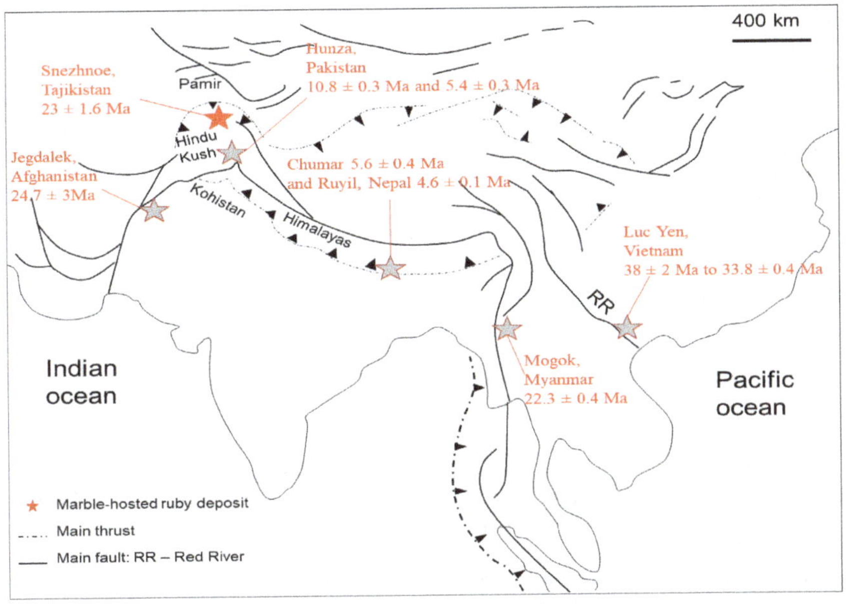

Figure 16. Tectonic map of Central and South East Asia with location of marble-hosted gem corundum deposits shown as gray stars (modified after [54]). Red star indicates the Snezhnoe ruby deposit with Rb–Sr age obtained in this study. The ages of ruby deposits are from [2,53].

6. Conclusions

1. The regional position of the Snezhnoe deposit is localized by the Muzkol–Rangkul anticlinorium within the Cimmerian zone of the Central Pamir. This tectonic structure comprises Precambrian metamorphic and igneous rocks that have undergone multi-stage metamorphism (1) under high-temperature amphibolite facies conditions (Precambrian) and, then, (2) from the melting zone in the central part of the anticlinorium through epidote–amphibolite to greenschist facies in the periphery (Mesozoic–Cenozoic). Ruby and associated minerals from the Snezhnoe deposit most likely formed at temperatures of 760 ± 30 °C. The minimum pressure for rutile entrapment was estimated at ~1–1.2 kbar. The Rb–Sr errorchron age (phlogopite—plagioclase—bulk rock) yielded 23 ± 1.6 Ma. However, the Rb–Sr systematics of the phlogopite was likely affected by later hydrothermal processes at Snezhnoe. Therefore, the Rb–Sr errorchron age for the ruby-bearing rocks should be used with caution.

2. The ratios of chromophore trace elements within Snezhnoe ruby are Ga/Mg < 8.2, Fe/Mg < 51.2, Cr/Ga > 6.9 and Fe/Ti < 31.6, all within the range for metamorphic corundum. On the $FeO–Cr_2O_3–MgO–V_2O_3$ vs. $FeO + TiO_2 + Ga_2O_3$ plot, the Snezhnoe rubies overlap those of the "John Saul ruby mine" in the "ruby in marble" field. On the Cr/Ga vs. Fe/Ti discriminant diagram, the Snezhnoe rubies plot within the "metamorphic" field. The ruby–oxygen isotope composition is consistently around +15.25% ± 0.07‰ well within the field for metamorphic and sedimentary rocks. The initial $^{87}Sr/^{86}Sr$ ratio is about 0.70791 and εNd is about -9.6 for bulk corundum-bearing rocks, confirming the geological data on the crustal origin of the ruby-bearing rocks at Snezhnoe. The possible protolith for these rocks was Proterozoic pelitic or bauxite-like sediments intercalated with carbonates.

3. The chemical composition of the ruby-bearing rocks varies considerably in terms of petrogenetic elements and is linked to the varying mineral associations along strike. Four mineral associations are distinguished: (1) ruby + scapolite (marialite) + muscovite + margarite + phlogopite; (2) ruby + plagioclase (albite–andesine) + muscovite + margarite; (3) ruby + muscovite + phlogopite + margarite; (4) ruby + calcite. The fourth mineral association is the most economically valuable. Meanwhile, the association of plagioclase + scapolite + phlogopite + muscovite is characteristic of the ruby-free zones. Mica group minerals were found in most ruby associations. Their specific features are traces of Cr and V linked to their green color being the main prospecting indicator for the ruby mineralization.

Supplementary Materials
Table S1: Description of the ruby-bearing samples used in this study; Table S2: Chemistry of muscovite (no. 1–4) chromium-bearing muscovite (no. 5–20) and fuchsite (no. 21–23); Table S3: Chemistry of margarite; Table S4: Chemistry of phlogopite; Table S5: Chemistry of plagioclase group minerals; Table S6: Chemistry of scapolite group minerals; Table S7: Chemistry of titanite; Table S8: Chemistry of rutile; Table S9: Chemistry of tourmaline group minerals; Table S10: Chemistry of corundum using EMPA (in wt.%); Table S11: Chemistry of corundum obtained by LA–ICP–MS (in µg/g); Table S12: Chemistry of ruby-bearing samples (wt.%); Table S13: Geochemistry of rutile grains used for in situ U–Pb LA–ICP–MS geochronology; Table S14: Rb–Sr and Sm–Nd isotope measurements of phlogopite, plagioclase and ruby-bearing rock from the Snezhnoe deposit.

Author Contributions: Conceptualization, A.K.L. and E.S.S.; methodology, T.H., E.S.S., Y.A.K. and T.L.; formal analysis, A.K.L., E.S.S., T.H., A.V.S., T.L. and T.V.R.; investigation, A.K.L. and E.S.S.; writing—original draft preparation, A.K.L. and E.S.S.; writing—review and editing, E.S.S., T.H., R.E.B., Y.A.K., A.V.S. and W.H.; supervision, R.E.B. and W.H.; funding acquisition, E.S.S. All authors have read and agreed to the published version of the manuscript.

Acknowledgments: We are grateful to our colleagues from MGRI (Mikhail Yu. Gurvitch), GEOKHI RAS (Lia N. Kogarko, Natalia N. Kononkova, Tatiana G. Kuzmina, Victor A. Turkov), Stefan Buhre (JGU Mainz), colleagues from Heidelberg University (Axel Schmitt, Mario Trieloff and Sonja Storm), Jasper Berndt (Westfälische Wilhelms-Universität Münster), Delia Rösel (TU Bergakademie Freiberg and University of Gothenburg), Natalia P. Gorbunova (Institute of Geology, Academy of Science of Tajikistan) for their help in sample preparations, assistance in analytical measurements and discussions on the topic of research. The authors thank the Editor of Minerals and Guest Editors Frederick Lin Sutherland and Khin Zaw for invitation to submit the manuscript for consideration and possible contribution in a special issue on "Mineralogy and geochemistry of ruby". We are also thankful to Ian Graham (University of New South Wales, Australia) and two anonymous reviewers for improving the text of an earlier version of this article.

References

1. Graham, I.; Sutherland, F.L.; Zaw, K.; Nechaev, V.; Khanchuk, A. Advances in our understanding of the gem corundum deposits of the West Pacific continental margins intraplate basaltic fields. *Ore Geol. Rev.* **2008**, *34*, 200–215. [CrossRef]
2. Garnier, V.; Giuliani, G.; Ohnenstetter, D.; Fallick, A.E.; Dubessy, J.; Banks, D.; Hoàng Quang, V.; Lhomme, T.; Maluski, H.; Pêcher, A.; et al. Marble-hosted ruby deposits from Central and Southeast Asia: Towards a new genetic model. *Ore Geol. Rev.* **2008**, *34*, 169–191. [CrossRef]
3. Stern, R.G.; Tsujimori, T.; Harlow, G.; Groat, L.A. Plate tectonic gemstones. *Geology* **2013**, *41*, 723–726. [CrossRef]
4. Giuliani, G.; Ohnenstetter, D.; Fallick, A.E.; Groat, L.; Fagan, A.G. The geology and genesis of gem corundum deposits. In *Geology of Gem Deposits*, 2nd ed.; Groat, L.A., Ed.; Mineralogical Association of Canada: Tucson, AZ, USA, 2014; pp. 29–112.
5. Sorokina, E.S.; Rösel, D.; Häger, T.; Mertz-Kraus, R.; Saul, J.M. LA-ICP-MS U–Pb dating of rutile inclusions within corundum (ruby and sapphire): New constraints on the formation of corundum deposits along the Mozambique belt. *Miner. Depos.* **2017**, *52*, 641–649. [CrossRef]
6. Saul, J.M. Transparent gemstones and the most recent supercontinent cycle. *Int. Geol. Rev.* **2017**, *60*, 889–910. [CrossRef]
7. Meng, F.; Shmelev, V.R.; Kulikova, K.V.; Ren, Y. A red-corundum-bearing vein in the Rai-Iz ultramafic rocks, Polar Urals, Russia: The product of fluid activity in a subduction zone. *Lithos* **2018**, *320*, 302–314. [CrossRef]
8. Dmitriev, E.A. Types of corundum mineralization in the Precambrian marbles of the Muzkol-Rangkulsky anticlinorium. In *Proceedings on Geology and Prospecting of the Gemstone Deposits of Tajikistan*; Klimkin, A.V., Ed.; Donish: Dushanbe, Tajikistan, 1987; pp. 34–36. (In Russian)
9. Dufour, M.S.; Kol'tsov, A.B.; Zolotarev, A.A.; Kuznetsov, A.B. Corundum-bearing metasomatic rocks in the Central Pamirs. *Petrology* **2007**, *15*, 151–167. [CrossRef]
10. Barnov, N.G. Geological Conditions of Localization and Prerequisites for Industrial Mineralization of Ruby in Marbles Using the Example of the Snezhnoye Deposit (Central Pamir). Unpublished Ph.D. Thesis, MGRI, Moscow, Russia, 2010. (In Russian).
11. Sorokina, E.S. Ontogeny and Quality of Gem Ruby from the Deposits of Central and South-East Asia. Unpublished Ph.D. Thesis, Fedorovsky All-Russian Research Institute of Mineral Resources, Moscow, Russia, 2011. (In Russian).

12. Nasriddinov, Z.Z. Geological and Mineralogical Features, Determining Quality of Rough from Snezhnoe Deposit. Unpublished Ph.D. Thesis, Russian State Geological Prospecting University, Moscow, Russia, 2013. (In Russian).
13. Barkhatov, B.P. *Tectonics of the Pamirs*; Leningrad University: Leningrad, USSR, 1963; 240p. (In Russian)
14. Litvinenko, A.K.; Nasreddinov, Z.Z.; Sorokina, E.S. Margarite—Rock-forming mineral from Snezhnoe ruby deposit, Central Pamir. *Proc. High. Educ. Establ. Geol. Explor.* **2015**, *4*, 80–82. (In Russian)
15. Sorokina, E.S.; Litvinenko, A.K.; Hofmeister, W.; Häger, T.; Jacob, D.E.; Nasriddinov, Z.Z. Rubies and Sapphires from Snezhnoe, Tajikistan. *Gems Gemmol.* **2015**, *51*, 160–175. [CrossRef]
16. Baratov, R.B. *Dismemberment of Stratified and Intrusive Rocks of Tajikistan*; Donish: Dushanbe, Tajikistan, 1976; 268p. (In Russian)
17. Budanov, V.I.; Budanova, K.T. Geological and petrological characteristics of the exposed crystalline basement. In *The Earth's Crust and the Upper Mantle of Tajikistan*; Donish: Dushanbe, Tajikistan, 1993; 275p. (In Russian)
18. Dufour, M.S.; Popova, V.A.; Krivets, T.N. *Alpine Metamorphic Complex of the Eastern Part of the Central Pamirs*; Leningrad University: Saint Petersburg, Russia, 1970; 126p.
19. Budanova, K.T. *Metamorphic Formations of Tajikistan*; Donish: Dushanbe, Tajikistan, 1991; 336p. (In Russian)
20. Litvinenko, A.K. *Geology of Snezhnoe Ruby Deposit, Central Pamirs*; Gornaya kniga: Moscow, Russia, 2020; 103p. (In Russian)
21. Jochum, K.P.; Nohl, U.; Herwig, K.; Lammel, E.; Stoll, B.; Hofmann, A.W. GeoReM: A new geochemical database for reference materials and isotopic standards. *Geostand. Geoanal. Res.* **2005**, *29*, 333–338. [CrossRef]
22. Jochum, K.P.; Weis, U.; Stoll, B.; Kuzmin, D.; Yang, Q.; Raczek, I.; Jacob, D.E.; Stracke, A.; Birbaum, K.; Frick, D.A.; et al. Determination of reference values for NIST SRM 610-617 glasses following ISO Guidelines. *Geostand. Geoanal. Res.* **2011**, *35*, 397–429. [CrossRef]
23. Luvizotto, G.; Zack, T.; Meyer, H.P.; Ludwig, T.; Triebold, S.; Kronz, A.; Münker, C.; Stockli, D.; Prowatke, S.; Klemme, S.; et al. Rutile crystals as potential trace element and isotope mineral standards for microanalysis. *Chem. Geol.* **2009**, *261*, 346–369. [CrossRef]
24. Axelsson, E.; Pape, J.; Berndt, J.; Corfu, F.; Mezger, K.; Raith, M.M. Rutile R632—A new natural reference material for U-Pb and Zr determination. *Geostand. Geoanal. Res.* **2018**, *42*, 319–338. [CrossRef]
25. Kooijman, E.; Berndt, J.; Mezger, K. U-Pb dating of zircon by laser ablation ICP-MS: Recent improvements and new insights. *Eur. J. Miner.* **2012**, *24*, 5–21. [CrossRef]
26. Mezger, K.; Hanson, G.N.; Bohlen, S.R. High-precision U–Pb ages of metamorphic rutile: Application to the cooling history of high-grade terranes. *Earth Planet. Sci. Lett.* **1989**, *96*, 106–118. [CrossRef]
27. Ludwig, K.R. *A User's Manual*; Berkeley Geochronology Center: Berkeley, CA, USA, 2009; p. 100.
28. Wanthanachaisaeng, B.; Häger, T.; Hofmeister, W.; Nasdala, L. Raman-und fluoreszenz-spektroskopische Eigenschaften von Zirkon-Einschlüssen in chrom-haltigen Korunden aus Ilakaka und deren Veränderung durch Hitzebehandlung. Z Dt. *Gemmol. Ges.* **2006**, *55*, 119–131. (In German)
29. Mariga, J.; Ripley, E.M.; Li, C.; McKeegan, K.D.; Schmidt, A.; Groove, M. Oxygen isotopic disequilibrium in plagioclase-corundum-hercynite xenoliths from the Voisey's Bay Intrusion, Labrador, Canada. *Earth Planet. Sci. Lett.* **2006**, *248*, 263–275. [CrossRef]
30. Sorokina, E.S.; Ozhogina, E.G.; Jacob, D.E.; Hofmeister, W. Some features of corundum ontogeny and the quality of ruby from Snezhnoe deposit, Tajikistan (the Eastern Pamirs). *Zapiski RMO (Proc. Russ. Mineral. Soc.)* **2012**, *141*, 100–108. (In Russian)
31. Sorokina, E.S.; Hofmeister, W.; Häger, T.; Mertz-Kraus, R.; Buhre, S.; Saul, J.M. Morphological and chemical evolution of corundum (ruby and sapphire): Crystal ontogeny reconstructed by EMPA, LA-ICP-MS, and Cr3+ Raman mapping. *Am. Mineral.* **2016**, *101*, 2716–2722. [CrossRef]
32. Peucat, J.J.; Ruffault, P.; Fritch, E.; Bouhnik-Le-Coz, M.; Simonet, C.; Lasnier, B. Ga/Mg ratio as a new geochemical tool to differentiate magmatic from metamorphic blue sapphires. *Lithos* **2007**, *98*, 261–274. [CrossRef]
33. Sutherland, F.L.; Abduriyim, A. Geographic typing of gem corundum: A test case from Australia. *J. Gemmol.* **2009**, *31*, 203–210. [CrossRef]
34. Giuliani, G.; Caumon, G.; Rakotosamizanany, S.; Ohnenstetter, D.; Rakototondrazafy, M. Classification chimique des corindons par analyse factorielle discriminante: Application à la typologie des gisements de rubis et saphirs. *Rev. Gemmol.* **2014**, *188*, 14–22.

35. Giuliani, G.; Fallick, A.E.; Garnier, V.; France-Lanord, C.; Ohnenstetter, D.; Schwarz, D. Oxygen isotope composition as a tracer for the origins of rubies and sapphires. *Geology* **2005**, *33*, 249–252. [CrossRef]
36. Yaroshevsky, A.A. Abundances of chemical elements in the Earth's crust. *Geochem. Int.* **2006**, *44*, 48–55. [CrossRef]
37. Kostitsyn, Y.A.; Belousova, E.A.; Silant'ev, S.A.; Bortnikov, N.S.; Anosova, M.O. Modern problems of geochemical and U-Pb geochronological studies of zircon in oceanic rocks. *Geochem. Int.* **2005**, *53*, 759–785. [CrossRef]
38. Zack, T.; Moraes, R.; Kronz, A. Temperature dependence of Zr in rutile: Empirical calibration of a rutile thermometer. *Contrib. Mineral. Petrol.* **2004**, *148*, 471–488. [CrossRef]
39. Watson, E.B.; Wark, D.A.; Thomas, J.B. Crystallization thermometers for zircon and rutile. *Contrib. Mineral. Petrol.* **2006**, *151*, 413–433. [CrossRef]
40. Kievlenko, E.Y. *Geology of Gems*; Soregaroli, A., Ed.; Ocean Pictures Ltd.: Littleton, CO, USA, 2003; 432p.
41. Dmitriev, E.A.; Ishan-Sho, G.A. Chromium-bearing muscovites in metamorphic and hydrothermal formations of the Pamirs. *Zapiski VMO (Proc. All-Union Mineral. Soc.)* **1987**, *116*, 690–697. (In Russian)
42. Terekhov, E.N.; Akimov, A.P. Tectonic setting and genesis of the gem corundum deposits of High Azia. *Lithosphere* **2013**, *5*, 122–140. (In Russian)
43. Rossovskiy, L.N.; Konovalenko, S.I.; Ananjev, S.A. Conditions of ruby formation in marbles. *Geol. Rudn. Mestorozhd.* **1982**, *24*, 57–66. (In Russian)
44. Rossovsky, L.N. Ruby and sapphire deposits of the Alpine-Himalayan fold belt and their formation conditions. In *Proceedings on Geology and Prospecting of Gemstone Deposits in Tajikistan*; Klimkin, A.V., Ed.; Donish: Dushanbe, Tajikistan, 1987; pp. 36–38. (In Russian)
45. Litvinenko, A.K. Nuristan-South Pamir province of Precambrian gems. *Geol. Ore Deposit.* **2004**, *46*, 263–268.
46. Litvinenko, A.K. The reconstruction of bauxite-like sediments in the Early Proterozoic metamorphites of the Central Pamir. In Proceedings of the Materials of the 5th All-Russian Lithologic Conference "the Types of Sedimentogenesis and Lithogenesis and Their Evolution in the History of the Earth", Ekaterinburg, Russia, 14–16 October 2008; pp. 428–430. (In Russian).
47. Predovsky, A.A. *Reconstruction of Conditions of Sedimentogenesis and Volcanism of Early Precambrian*; Nauka: Leningrad, USSR, 1980; 152p. (In Russian)
48. Golovenok, V.K. *High-Alumina Precambrian Formations*; Nedra: Leningrad, USSR, 1977; 267p. (In Russian)
49. Budanova, K.T. Mineralogical and petrochemical characteristics of high-alumina rocks in the eastern part of Central Pamir. *Proc. AS Repub. Tajikistan (Zap. AN Resp. Tadjikistan).* **1993**, *1*, 26–33. (In Russian)
50. Giuliani, G.; Fallick, A.E.; Rakotondrazafy, A.F.M.; Ohnenstetter, D.; Andriamamonjy, A.; Rakotosamizanany, S.; Ralantoarison, T.; Razanatseheno, M.M.; Dunaigre, C.; Schwarz, D. Oxygen isotope systematics of gem corundum deposits in Madagascar: Relevance for their geological origin. *Miner. Deposita* **2007**, *42*, 251. [CrossRef]
51. Ferry, J.M.; Watson, E.B. New thermodynamic models and revised calibrations for the Ti-in-zircon and Zr-in-rutile thermometers. *Contrib. Mineral. Petrol.* **2007**, *154*, 429–437. [CrossRef]
52. Bohlen, S.; Montana, A.; Kerrick, D.M. Precise determination of the equilibria kyanite=sillimanite and kyanite=andalusite and a revised triple point for Al_2SiO_5 polymorphs. *Am. Mineral.* **1991**, *76*, 677–680.
53. Phyo, M.M.; Wang, H.A.; Guillong, M.; Berger, A.; Franz, L.; Balmer, W.A.; Krzemnicki, M.S. U–Pb Dating of Zircon and Zirconolite Inclusions in Marble-Hosted Gem-Quality Ruby and Spinel from Mogok, Myanmar. *Minerals* **2020**, *10*, 195. [CrossRef]
54. Mattauer, M.; Matte, P.; Jolivet, J.L. A 3D model of the India-Asia collision at late scale. *Comptes Rendus* **1999**, *328*, 499–508.

5

The First Find of Cr_2O_3 Eskolaite Associated with Marble-Hosted Ruby in the Southern Urals and the Problem of Al and Cr Sources

Aleksander Kissin [1,2,*], Irina Gottman [1,2], Sergei Sustavov [2], Valery Murzin [1] and Daria Kiseleva [1]

1 A.N. Zavaritsky Institute of Geology and Geochemistry, Ural Branch of Russian Academy of Sciences, Vonsovskogo str., 15, 620016 Ekaterinburg, Russia; gottman@igg.uran.ru (I.G.); murzin@igg.uran.ru (V.M.); Kiseleva@igg.uran.ru (D.K.)

2 Department of Mineralogy, Petrology and Geochemistry, Ural State Mining University, Kuybysheva str.,30, 620144 Ekaterinburg, Russia; sustavov.s@ursmu.ru

* Correspondence: kissin@igg.uran.ru

Abstract: The results of the study of eskolaite associated with marble-hosted ruby found for the first time in the Kuchinskoe occurrence (Southern Urals) are presented. Here, eskolaite was located on the surface and near-surface regions of ruby crystals. Eskolaite diagnostics was confirmed by powder X-ray diffraction (URS-55). The morphology and chemical composition of eskolaite and associated ruby was studied using a JSM-6390LV scanning electron microscope and a Cameca SX 100 electron probe microanalyzer. The eskolaite crystals were hexagonal and tabular, up to 0.2 mm in size. Ruby mineralization was formed during prograde and retrograde dynamothermal metamorphism. The eskolaite associated with the prograde stage ruby contained Al_2O_3 (9.1–23.62 wt %), TiO_2 (0.52–9.66 wt %), V_2O_3 (0.53–1.54 wt %), FeO (0.03–0.1 wt %), MgO (0.05–0.24 wt %), and SiO_2 (0.1–0.21 wt %). The eskolaite associated with the retrograde stage ruby was distinguished by a sharp depletion in Ti and contained Al_2O_3 (12.25–21.2 wt %), TiO_2 (0.01–0.07 wt %), V_2O_3 (0.32–1.62 wt %), FeO (0.01–0.08 wt %), MgO (0.0–0.48 wt %), and SiO_2 (0.01–0.1 wt %). The associated rubies contained almost equal amounts of Cr_2O_3 (2.36–2.69 wt %) and were almost free from admixtures. The identification of the eskolaite associated with the marble-hosted rubies from the Kuchinskoe occurrence is a new argument in favor of introduction of Al and Cr into the mineral formation zone. The mineralization was localized in the metamorphic frame of the granite gneiss domes and was formed synchronously with them.

Keywords: gems; ruby; marble; eskolaite; Southern Urals

1. Introduction

The issue of Al and Cr sources in ruby (a corundum containing a chromium isomorphic impurity) deposits in calcite (dolomite) marbles is actively discussed in the scientific literature. Various ideas have been studied on this issue: (1) the lateritic weathering of an impure limestone [1–4]; (2) the lenses of bauxite-like sediments in limestones [5]; (3) the desilication of primary sedimentary rocks during regional alkali metasomatism [6]; (4) the introduction of Al and Cr by metamorphogenic fluids during rock granitization and dynamothermal metamorphism [7,8]; (5) the introduction of Al using the gas phase of deep fluids during alkaline magmatism [9]; (6) the introduction of Al as a result of decompression during deep tectonic erosion [10]; and (7) the redistribution of Al and Cr during the metamorphism of sedimentary limestones with evaporitic lenses [11,12]. A review of the views on this issue was given by Giluani et al. [13]. All researchers noted the presence of Cr-containing minerals in ruby-bearing marble, such as muscovite (fuchsite), phlogopite, margarite, pargasite, diopside, tourmaline, diaspore, rutile, sphene, and red spinel. In the Middle Ural Mountains (Urals, Russia), uvarovite with a grossular admixture (Cr_2O_3 up to 18.60 wt %) is found in ruby-bearing marbles [14]. The article reports on the discovery of chromium oxide in the intergrowth with ruby, which can be used in discussions about the sources of Al and Cr in ruby-bearing marbles.

Natural chromium oxide (Cr_2O_3) of supposedly hydrothermal-metasomatic genesis was discovered in the middle of the last century in the Outokumpu Cu-Co-Zn skarn deposit in Finland and was called eskolaite [15]. Hydrothermal-metasomatic eskolaite was also described in the chromitites of the Mariinsky and Bazhenovsky ophiolite massifs in the Middle Urals. In the Mariinsky emerald deposit, eskolaite formed inclusions in mariinskite (a chromium analogue of chrysoberyl) associated with fluorophlogopite, Cr-containing muscovite, and tourmaline [16]. In the Bazhenovsky massif, eskolaite was found in alumina chromitites [17] as well as in the form of inclusions in mariinskite [18].

The eskolaite of magmatic origin was found in the chromite ores of the Rai-Iz and Voikar–Syninsky hyperbasite massifs in the Polar Urals [19]. It formed as a result of fractionation of sulfide and chromospinelide ores from a silicate melt at temperatures 900–1100 °C The magmatic eskolaite was also described in magnetite-ilmenite ores of the Kusa gabbro intrusion (Southern Urals) [20].

Eskolaite was found in kimberlites [21–23]. The eskolaite intergrown with diamond and with picrochromite inclusions was reported for the Udachnaya kimberlite pipe (Yakutia) [24].

In the Cis-Baikal region, the eskolaite and karelianite (V_2O_3) of metamorphic genesis were revealed in the thin layers of graphite-enriched sillimanite-cordierite quartzite schists of the Olkhon series [25]. The grain size was 5–20 μm in intergrowth with rutile, $(Cr,V)_2Ti_3O_9$ olkhonskite, and $V_2Ti_3O_9$ schreyerite. In the metacarbonate rocks of the Slyudyanka series (Southern Cis-Baikal), the eskolaite was described in calcite-quartz-diopside rocks [26]. The eskolaite was accompanied by magnesiochromite and $(Cr,V)_2O_3$ karelianite-eskolaite. The authors of [26] believe that Cr and V were introduced into paleo-sediments by chemical deposition. Then, during prograde granulite facies metamorphism, Cr and V participated in the formation of magnesiochromite, eskolaite, and karelianite. The formation of eskolaite in metamorphic rocks is possible even at low chromium concentrations in the substrate, but under the conditions of isochemical metamorphism, Cr inertness, and the absence of its isomorphic incorporation into co-crystallizing minerals.

This article presents the results of the studies of eskolaite from the Kuchinskoe marble-hosted ruby occurrence in the Southern Urals, a mineral phase that has not been previously described in association with marble-hosted ruby.

2. Geological Setting

The Kuchinskoe marble-hosted ruby occurrence is located in the Kochkar anticlinorium (Southern Urals) (Figure 1). The anticlinorium runs in the submeridional direction for 140 km, with a width up to 28 km [27]. The tectonic boundaries are represented by thrusts dipping under the adjacent synclinorium structures. The anticlinorium is characterized by the presence of granite gneiss domes, which form the centres of dynamothermal metamorphism [28–30]. Crystalline schists, amphibolites, and marbles intruded by the dikes of granites and pegmatites occur in the metamorphic frame of the domes [31]. The metamorphism is syntectonic: prograde in the Carboniferous period and retrograde in the Permian-Triassic period. The level of metamorphism in the dome structures reached the conditions of the amphibolite facies and in the inter-dome structures reached the epidote-amphibolite facies (according to mineral paragenesis) [28,29].

The geology of the Kuchinskoe marble-hosted ruby occurrence was studied by one of the authors (A.K.) from 1979 to 1988 as a field geologist performing geological explorations. The ruby-bearing marbles of the Kuchinskoe occurrence are localized in the inter-dome structure. The marbles are banded, cleaved, and repeatedly recrystallized [32]. At the deposit: (1) coarse-grained calcite marble is the most prevalent, while (2) Mg-calcite marble (which can be schisted) is found locally, among which the bodies of (3) dolomite-calcite/calcite-dolomite marble are found. The marbles are split by dikes of granite and pegmatite. Type 1 and 2 marbles existed at the time of the dike introduction. The dike was not tectonically disturbed in the time intervals of its contact with these marbles. Consequently, the cleaving and schisting of these marbles took place before the dike introduction. There are no quenching zones in the dike and no marble recrystallization at the contact. This is explained by the similar temperatures of the embedded granite solution-melt and the host rocks. At the contacts with the massive two-carbonate marble, dikes are disturbed by low-amplitude shears and a zone of forsterite skarn 1.5–3.0 cm wide is formed along the marble. Consequently, two-carbonate marble was formed after the introduction of the dike. The dikes are genetically related to anatectic granites in the apical part of the granite gneiss domes. The time of their massive introduction records the time of stress relief and the drop in all-round pressure. Thus, 1 and 2 type marbles were formed during prograde metamorphism, while type 3 marble was formed during retrograde metamorphism [7,8,33,34].

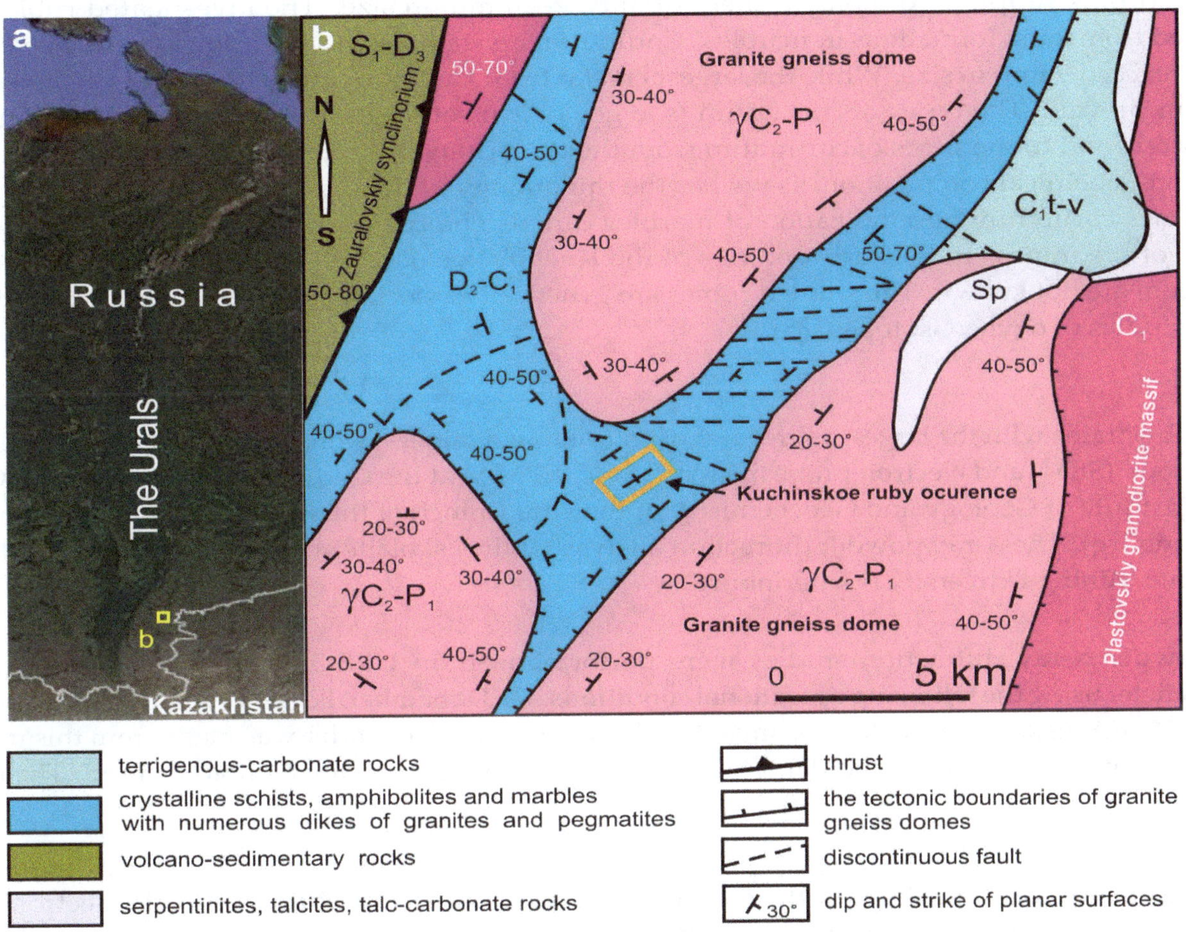

Figure 1. (**a**) The localization of the Kuchinskoe marble-hosted ruby occurrence and (**b**) a schematic geological map of the site.

Type 1 rubies are dark red crystals of tabular shape with the highest Cr_2O_3 content (1.22–2.91 wt %). They are found in the form of disseminated impregnation in Mg-calcite marble and are replaced by spinel in dolomite marble. They are accompanied by pyrite, anhydrite, and apatite. Type 2 rubies are pink, short-columnar, rounded crystals with the lowest Cr_2O_3 content (0.11–0.51 wt %). They are found in the form of disseminated impregnation in dolomite-calcite/calcite-dolomite marble and accompanied by rutile and pyrite, often replaced with pink spinel associated with graphite, forsterite, and norbergite. Type 3 rubies (more correctly called corundum) are represented by long-prismatic crystals and irregular-shaped grains of red, blue, violet, and white color, and are often polychromatic; the red corundum is close to type 1 rubies in terms of Cr_2O_3 contents (0.03–2.97 wt %). These corundums are confined to cleavage cracks in marbles and are accompanied by colorless phlogopite, pyrite, pyrrhotite, sphalerite, and other minerals. They are replaced by spinel associated with fluorite [7,8].

Karst formations filled with Paleogene–Neogene kaolin-montmorillonite clays mixed with quartz sand are widely developed in the area of marble distribution in the Kochkar anticlinorium. All formations were studied by drilling and tested during ruby prospecting. In karst sediments, ruby and related minerals (fuchsite, phlogopite, Cr-pargasite, red spinel, and others) are found only if ruby-bearing marbles are present under the karst. There are no signs of abrasion on the surfaces of ruby crystals and associated minerals, which indicates the absence of their transportation by water flows.

3. Materials and Methods

The authors had at their disposal several kilograms of crystals and the grains of red corundum (ruby and pink sapphire) of non-jewellery quality, 2–20 mm in size. The investigated rubies were extracted from karst formations in marbles. Corundum crystals from the Kuchinskoe marble quarries did not exceed 9 mm in size. While the largest crystal from karst formations reaches 5 cm, crystals up to 4 mm prevail. The types of corundum (see geological setting) from karst are easily diagnosed and correspond to the types of corundum from the underlying marbles. Their colors, transparency, inclusions, chemical composition, as well as the mineralogy of the intergrowths were also identical. Sometimes corundum mineralization of marbles can also be traced in the karst clays lying on the surface of the mines. In the surroundings of the Kuchinskoe occurrence, other primary sources of corundum are not known. The studied rubies are genetically associated with marbles, as is evidenced by the studies of oxygen isotopes [35].

Ruby grains with the largest aggregates of an unknown mineral were studied by scanning electron microscopy (SEM) and electron probe microanalysis (EPMA) at the Geoanalitik Collective Use Center of the Institute of Geology and Geochemistry of the Ural Branch of the Russian Academy of Sciences (Ekaterinburg). The X-ray powder diffraction analysis of the eskolaite structure was performed at the Ural State Mining University (Ekaterinburg).

Powder X-ray diffraction studies were performed on an URS-55 X-Ray powder diffraction spectrometer using the following operational conditions: RKD chamber, $Fe_{K\alpha+\beta}$ radiation, voltage 30 kV, current 10 mA. Eskolaite powder was mixed with rubber glue. A cylinder was made from this mixture, which was fixed in the sample holder. The analysis was carried out using a photo method. The results are shown in Table 1.

The chemical composition of the minerals was studied on a CAMECA SX 100 electron probe microanalyzer (Cameca, Gennevilliers, France). The analyses were carried out with an accelerating voltage of 15 kV, a current of 10 nA for Al, 20 nA for Cr, and 150 nA for Fe, Ti, Mg, Si, and V, with an electron beam diameter of 1–3 μm on the sample. The following standard samples were used: Cr_2O_3, Al_2O_3, Fe_2O_3, MgO, TiO_2, V_2O_5, and SiO_2. To identify all the peaks, the most intense Kα lines were used. The superposition of the V Kα line on Ti Kβ was taken into account. Al, Mg, Ga were measured using TAP crystals, chromium–LPET, titanium and vanadium-PET, and iron-LLIF crystal. The duration of the intensity measurement at the peaks of the analytical lines and the background on each side of the peak were: for the major elements—Al and Cr—10 and 5 s, respectively; for Fe, Ti, Mg, Si, Al, Cr 30 and 15 s; and for V 60 and 30 s. Standard deviation of the elemental content (wt %) was less than 0.5 for Cr, less than 0.6 for Al, less than 0.01 for Fe, Ti, Mg, Si, and V, 0.02 for Ga, less than 0.06 for Al (in eskolaite), and 0.05 for Cr (in corundum). Three polished sections of ruby with eskolaite were studied (Table 2). Sample 1 was a type 1 ruby (analyses 1 and 2) with 3 inclusions (analyses 3–5). Sample 2 was a type 3 ruby (analyses 6–9) with 6 inclusions (analyses 10–15). Sample 3 was a type 3 ruby (analyses 16–18) with 5 inclusions (analyses 19–23).

Table 1. Eskolaite X-Ray diffraction data.

Eskolaite JCPDS			Eskolaite with Ruby	
d (Å)	I/I_1	*hkl*	*I*	$d_{\alpha/n}$
3.633	74	012	6	3.64
2.666	100	104	10	2.67
2.480	96	110	10	2.48
2.264	12	006	1	2.26
2.176	38	113	3	2.16
2.048	9	202	-	-
1.816	39	024	4	1.811
1.672	90	116	10	1.663 *
1.579	13	122	2	1.565
1.465	25	214	3	1.462
1.4314	60	300	7	1.426 *
1.2961	20	1.0.10	1	1.286
1.2398	17	220	1	1.233
1.2101	7	306	-	-
1.1731	14	128, 312	-	-
1.1488	10	0.210	3	1.144
1.1239	10	134	-	-
1.0874	17	226	10	1.083
1.0422	16	2.1.10	8	1.038
X-Ray diffraction pattern contains 9 lines not included in the Table				
$a_o = 4.954$ $c_o = 13.584$			$a_o = 4.94$ $c_o = 13.51$	

* Reflections used for unit cell parameters calculation.

Table 2. The chemical composition of eskolaite and associated ruby of the prograde and retrograde metamorphism according to X-ray microanalysis.

No	Object	Oxide (wt %)								Crystal Chemical Formula
		TiO_2	FeO	V_2O_3	MgO	Al_2O_3	SiO_2	Cr_2O_3	Total	
		Prograde Metamorphism								
1	Ruby 1	bdl	bdl	bdl	bdl	95.95	0.07	2.91	98.93	$Cr_{0.04}Al_{1.96}O_3$
2	Ruby 1	bdl	bdl	bdl	bdl	97.51	bdl	2.47	99.98	$Cr_{0.03}Al_{1.97}O_3$
3	Esk	9.66	bdl	1.54	0.24	9.1	0.1	78.21	98.9	$Cr_{1.52}Al_{0.26}V_{0.03}Ti_{0.18}Mg_{0.01}O_3$
4	Esk	0.52		0.53	0.15	23.62	0.14	73.3	98.37	$Cr_{1.34}Al_{0.64}V_{0.01}Mg_{0.01}O_3$
		Retrograde Metamorphism								
5	Ruby 3	bdl	bdl	0.04	bdl	97.12	0.04	2.16	99.36	$Cr_{0.03}Al_{1.97}O_3$
6	Ruby 3	bdl	bdl	0.11	bdl	97.39	0.07	1.78	99.35	$Cr_{0.02}Al_{1.98}O_3$
7	Ruby 3	bdl	bdl	bdl	bdl	96.2	0.03	2.97	99.20	$Cr_{0.04}Al_{1.96}O_3$
8	Ruby 3	bdl	bdl	bdl	bdl	95.79	0.03	2.67	98.49	$Cr_{0.04}Al_{1.96}O_3$
9	Esk	bdl	0.08	1.37	0.16	12.25	bdl	83.76	97.64	$Cr_{1.62}Al_{0.35}V_{0.03}O_3$
10	Esk	bdl	0.08	0.32	0.08	16.15	0.1	82.08	98.82	$Cr_{1.54}Al_{0.45}V_{0.01}O_3$
11	Esk	bdl	bdl	0.62	bdl	17.81	bdl	80.34	98.8	$Cr_{1.49}Al_{0.49}V_{0.02}O_3$
12	Esk	bdl	bdl	0.56	0.08	18.5	bdl	80.23	99.41	$Cr_{1.48}Al_{0.51}V_{0.01}O_3$
13	Esk	0.07	bdl	1.62	bdl	19.11	bdl	78.69	99.52	$Cr_{1.45}Al_{0.52}V_{0.03}O_3$
14	Esk	bdl	0.06	0.93	0.12	21.2	bdl	77.6	99.96	$Cr_{1.41}Al_{0.57}V_{0.02}O_3$
15	Ruby 3	bdl	bdl	bdl	bdl	97.13	0.03	2.64	99.80	$Cr_{0.04}Al_{1.96}O_3$
16	Ruby 3	bdl	bdl	bdl	bdl	97.48	0.07	2.40	99.95	$Cr_{0.03}Al_{1.97}O_3$
17	Ruby 3	bdl	bdl	bdl	bdl	96.97	0.03	2.40	99.40	$Cr_{0.03}Al_{1.97}O_3$
18	Esk	0.02	0.09	0.73	0.17	18.0	bdl	79.4	98.41	$Cr_{1.48}Al_{0.50}V_{0.01}Mg_{0.01}O_3$
19	Esk		0.02	0.39	0.12	26.02	bdl	73.74	100.3	$Cr_{1.30}Al_{0.69}V_{0.01}O_3$
20	Esk	bdl	0.02	1.5	0.14	24.59	bdl	72.95	99.2	$Cr_{1.31}Al_{0.66}V_{0.03}O_3$
21	Esk	bdl	0.02	1.94	0.26	13.39	bdl	82.08	97.69	$Cr_{1.57}Al_{0.38}V_{0.04}Mg_{0.01}O_3$
22	Esk		bdl	0.6	0.21	25.61	bdl	73.75	100.18	$Cr_{1.30}Al_{0.68}V_{0.01}Mg_{0.01}O_3$

bdl: the values below the detection limit.

The sections of ruby grains with eskolaite as well as ruby grain surface morphology were studied using scanning electron microscopy (SEM). Back scattered electron images were obtained by a JSM-6390LV electronic scanning microscope (JEOL, Akishima, Tokyo, Japan) with an INCA Energy 450 X-Max 80 energy dispersive spectrometer (Oxford Instruments, Abingdon, Great Britain). Operating conditions were an acceleration voltage of 20 kV and an exposure time of 5 ms per pixel.

4. Results

SEM indicated that the dark colored crusts on a ruby were represented by Cr_2O_3 eskolaite with Al, Fe, Ti, and V admixtures. The eskolaite was represented by an aggregate of small tabular crystals (Figures 2–4) up to 10 microns in size, rarely up to 0.2 mm. More often, crystals of diaspore, spinel, rutile, and green tourmaline were observed on the surface of ruby crystals. The growth of eskolaite crystals on corundum is chaotic and tends to develop along and within surface imperfections on the ruby crystal faces and on their near-surface part. X-ray diffraction studies confirmed the diagnosis of the mineral (Table 1). The results of the X-ray microanalysis of eskolaite and associated ruby are given in Table 2. The admixtures of Al, Ti, V, and sometimes Fe, widely varying in content, are common for eskolaite.

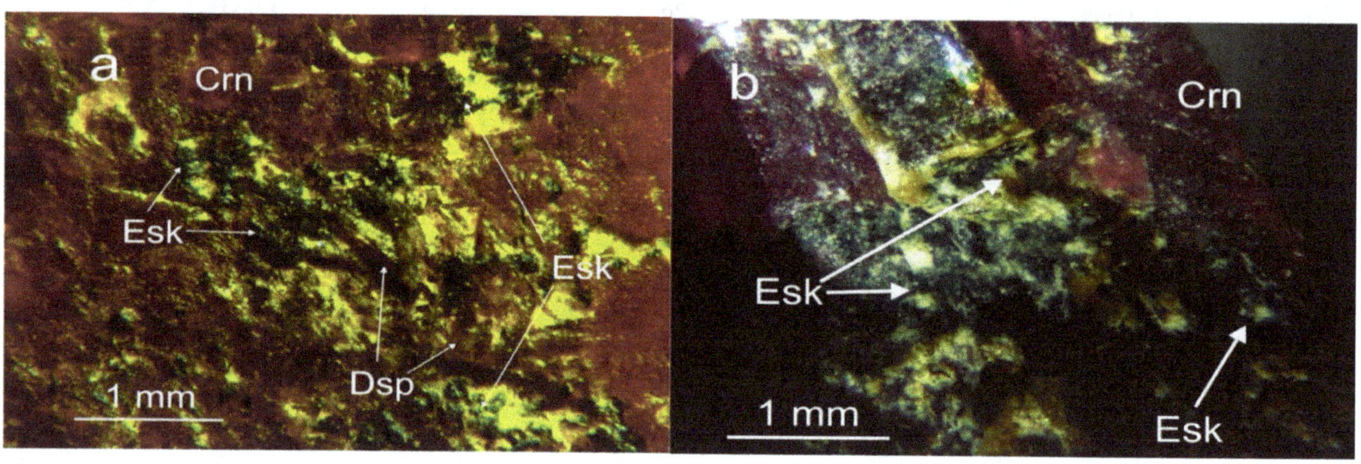

Figure 2. Eskolaite (Esk) and diaspore (Dsp) on the (**a**) pinacoidal and (**b**) rhombohedral crystal faces of the prograde stage ruby (Crn).

Figure 3. The aggregation of eskolaite crystals (Esk) on the surface of a ruby (Crn) crystal: (**a**) general view and (**b**) a close-up of the yellow fragment depicted in Figure 3a.

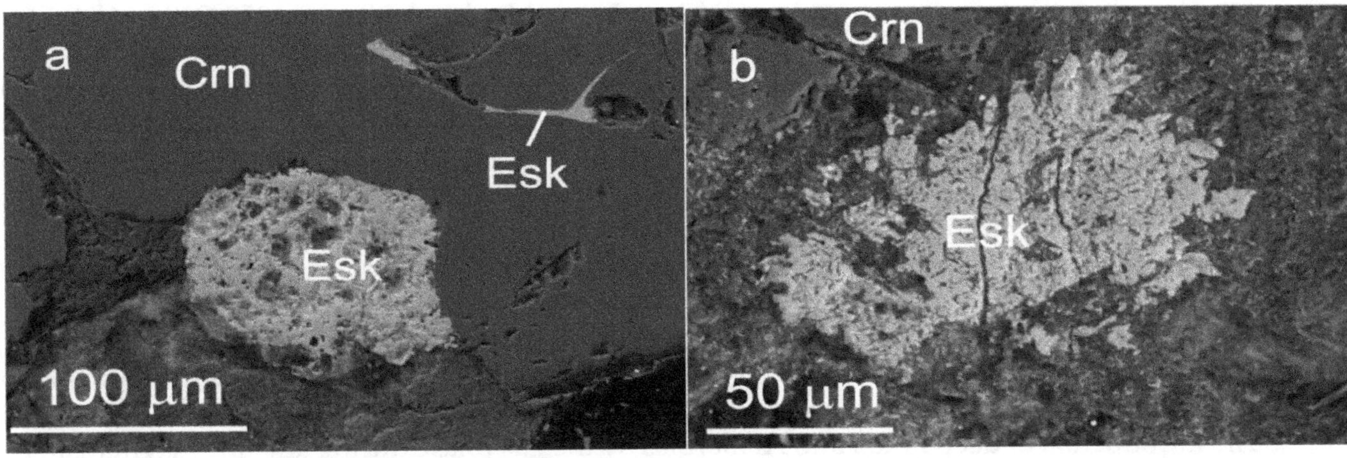

Figure 4. Back-scattered electron (BES) image of the eskolaite (Esk) (**a**) on type 1 ruby and (**b**) on type 3 ruby.

5. Discussion

5.1. PT-Conditions for the Formation of Ruby and Eskolaite Association in the Kuchinskoe Occurrence

At the peak of the regional metamorphism of carbonate rocks in the Kuchinskoe marble-hosted ruby occurrence, the temperature reached 660 °C (calcite-dolomite geothermobarometer [36]) at a pressure of 2.6–1.2 kbar [7]. The temperature of type 3 ruby formation is estimated as 600–550 °C, p = 2.2–1.9 kbar, and p_{CO_2} = 1.4–0.4 kbar. PT conditions for the formation of type 1 ruby have not yet been evaluated. However, the temperature was probably in the range of 550–600 °C and the pressure could reach 3–4 kbar. These PT conditions for the formation of ruby-bearing marbles of the Kuchinskoe occurrence are in good agreement with the results of the studies of regional metamorphism in the Kochkar anticlinorium using mineral paragenesis (Garnet + Biotite + Plagioclase + Quartz ± Staurolite ± Sillimanite ± Cordierite): 570–640 °C, 3–5 kbar [28]; 590–620 °C, 2.5–3.0 kbar [29]; and 500–620 °C, 3.0–4.0 kbar (in inter-dome structures) and 620–700 °C (in domes) [30]. The pressure during the retrograde metamorphism is estimated by the depth of formation of the Kochkar anticlinorium, which is 4.0–4.5 km (about 1.5 kbar) [28]. At the Kuchinskoe occurrence, the eskolaite is only associated with high-chromium rubies, which possibly indicates the genetic relationship between these minerals. To estimate the temperature of eskolaite formation, a geothermometer was proposed based on the experimental data on the stability of corundum-eskolaite solid solution at various PT parameters [37]. The molar fractions of Cr_2O_3 eskolaite crystals (X_{Cr2O3}) on the rubies of prograde and retrograde stages, according to their crystal-chemical formulas, are 0.65–0.8 (see Table 2). Such X_{Cr2O3} values for eskolaite, subject to their homogeneity, make it possible to estimate the formation temperature of this mineral by the corundum-eskolaite geothermometer as 700–850 °C Such a high temperature excludes the formation of eskolaite in karst conditions. At the same time, it is higher than the estimate of the temperatures of metamorphism in the Kochkar anticlinorium (500–670 °C). The mismatch in the temperature parameter of the ruby and eskolaite formation is still difficult to explain. This discrepancy may be due to the influence of oxygen fugacity, pH, or other factors.

5.2. Aluminum and Chromium Sources in the Formation of Ruby and Eskolaite in Marbles

1. The following geological data dispute the sedimentary origin of Al and Cr in ruby-containing marbles [1–5]:

(i) there is no evidence of sedimentary lamination in ruby-containing marbles;

(ii) in the Kochkar anticlinorium, the length of a strip of carbonate rocks is measured in the tens of kilometers, with a width of up to 2.7 km. The marbles of this strip contain an admixture of Al and Cr and are in the same zone of metamorphism, but ruby mineralization occurs very locally;

(iii) the length of the areas of ruby mineralization in marbles rarely exceeds 250–300 m along the flow cleavage and their width does not exceed 80 m. The vertical scale of mineralization has not been established, but is the greatest in small bodies. Thus, ruby mineralization forms flattened lenticular columnar bodies and this form of ruby-bearing marbles contradicts the concept of stratified sediments;

(iv) the formation of rubies in marble occurred both at the prograde and retrograde metamorphism, and was not one-act (see geological setting): 2–3 types of ruby can be seen in one marble sample, which requires the changes of conditions during their formation.

But we did not exclude the participation of Al and Cr of sedimentary origin in the formation of rubies, provided that they were dissolved by metamorphogenic fluids.

2. Desilication of sedimentary terrigenous-carbonate rocks during regional alkaline metasomatism [6] could take place. But there are no signs of marble desilication in the Kuchinskoe occurrence. In addition, the formation of rubies occurred during the prograde and retrograde metamorphism.

3. The introduction of deep fluids into Al marbles by the gas phase during alkaline magmatism [9] or decompression during deep tectonic erosion of the Earth's crust [9] cannot be applied for the Kuchinskoe marble-hosted ruby occurrence since there is no alkaline magmatism and deep tectonic erosion here.

4. The redistribution of Al and Cr during metamorphism of sedimentary limestones with evaporate lenses [10,11] does not contradict facts (i–iii), but does not explain fact (iv). Mg-metasomatism in the marbles of the Kochkar anticlinorium manifested twice: (1) through organogenic limestones at an early prograde metamorphism with the formation of fine-grained dolomites and (2) through Mg-calcite and calcite marbles at an early retrograde metamorphism, with the formation of medium-grained dolomite-calcite marbles. Mg metasomatism of the prograde metamorphism is probably caused by the destruction of evaporites where metasomatic dolomites of this stage are enriched in Al, Cr, V, Cu, Co, and REE. By contrast, two-carbonate marbles are depleted in these elements.

5. The introduction of Al and Cr by metamorphogenic fluids during rock granitization [7,8] does not contradict the facts (i–iv). By "granitization" we mean a high temperature metasomatic process that brings the initial rock closer to granite in chemical composition through the introduction of SiO_2, Na, K, H_2O, and the removal of Mg, Fe, Ca and other femic components. As a possible cause of granitization, we considered the deformation of the continental crust during Late Paleozoic collision [30,34,38], the times of tectonic deformations, dome formation, and metamorphism in the Kochkar anticlinorium coincide [28–30]. They culminated in the introduction of anatectic granites and pegmatites. The rocks of the dome structures are enriched with fluorophilic elements (Be, Li, Sn, Ta, Nb, and others) and the rocks of the inter-dome structures are rich in Fe, Mg, Ca, Ti, and others. The substrate was volcanic-sedimentary rocks, distributed around the Kochkar anticlinorium [27].

6. The solution for the problem of the Cr and Al sources for the formation of red corundum in the hyperbasite Rai-Iz massif (the Polar Urals) was recently considered [39]. The authors [39] believe that the vein was the product of the interaction between a subduction-zone-derived fluid and mantle wedge peridotite. The Cr content of the red corundum was potentially derived from accessory chromian spinel in the peridotite. Popov et al. [40] considered that the chromium-bearing spinels of the Bazhenovsky ophiolite complex are the main chromium sources for the coloring of emeralds and alexandrites of the Mariinsky deposit (Middle Urals), located 15 km west. We did not comment on these statements, since only ruby-containing marbles were considered in this article.

7. In Mg-calcite (ruby-bearing) marble, the content of Al and Cr is low at 0.11–0.13 wt % Al_2O_3 and 0.00096–0.0014 wt % Cr_2O_3 [8]. Marbleized bituminous organogenic limestones in the marginal parts of the anticlinorium (about 5 km from the Kuchinskoe occurrence) contain 3-times greater the amount of Al and Cr. The primary chemical composition of limestones changed significantly during the prograde and initial retrograde metamorphism [8,32,34]. Evaporitic lenses, if present in limestone (as suggested in [11,12]), could also undergo these processes.

We believe that ruby and eskolaite in the Kuchinskoe occurrence were formed by the hydrothermal-metasomatic process. Eskolaite was observed only on the surface of ruby crystals or in their near-surface region. This is probably due to the formation of eskolaite only at the final stage of the ruby crystal growth. Chromium zonal distribution in ruby crystals was not observed. Corundum enrichment (or depletion) in chromium in the contact zone with eskolaite was sometimes present, but is not a rule. The reason for this has not yet been established. Probably, the chromium content in corundum is limited by crystallization conditions rather than by chromium activity. This corresponds to the observed facts: the spinel and ruby have maximum Cr contents at the prograde stage, minimal Cr concentration at transitional stage, and again enriched in Cr at the retrograde stage. Sazonov [41] noted that Al^{3+} and Cr^{3+} have amphoteric properties and migrate under the similar conditions in hydrothermal metasomatic processes. In this case, we can assume that Al and Cr were introduced into the marble by high-temperature fluid. The metamorphism of sedimentary carbonate rocks in the Kochkar anticlinorium was not isochemical [7,8] and Cr was a component of many accessory minerals in marbles. Consequently, eskolaite could not form under the conditions of isochemical metamorphism. Due to the evolution of metamorphic conditions, the introduction of Al into the system ceased, but Cr still entered, forming crystalline crusts on the ruby.

6. Conclusions

The identification of eskolaite in association with rubies in the marbles of the Kuchinskoe occurrence is a new argument in favor of the introduction of Al and Cr into the mineral formation zone. Here, eskolaite was associated only with high-chromium rubies of the prograde and retrograde metamorphism, which confirms the migration of these elements in similar conditions. The formation of eskolaite occurs at the final stage of the ruby crystal growth and later. The eskolaite of prograde stage contains an admixture of titanium, which is explained by its low activity and inability to form its own minerals. The occurrence of rutile and sphene at the retrograde stage explains the low titanium content in the recent eskolaite. The parameters and morphology of the mineralized zones also indicate the introduction of Al and Cr by metamorphogenic fluids.

Author Contributions: Conceptualization, investigation and writing—original draft preparation, A.K.; writing-review and "Results" and "Discussion" section editing, supervision, V.M.; Powder X-Ray diffraction analysis and structure solution, S.S.; SEM and EMPA analyses, I.G.; writing—review, "Materials and Methods" section editing and translation, D.K. All authors have read and agreed to the published version of the manuscript.

Acknowledgments: The Editor of *Minerals* and Guest Editors are thanked for organizing the Special Issue on "Mineralogy and Geochemistry of Ruby and Allied Assemblages" and extending an invitation to submit a contribution for consideration. This particularly includes Frederick Lin Sutherland and Khin Zaw. We thank the Reviewers for the important comments and constructive suggestions, which helped us to improve the quality of the manuscript.

References

1. Bank, H.; Okrusch, M. Über Rubin—Vorkommen in Marmoren von Hunza (Pakistan). *Z. Dt. Gemmol. Ges.* **1976**, *25*, 67–85.
2. Harlow, G.E.; Bender, W. A study of ruby (corundum) compositions from the Mogok Belt, Myanmar: Searching for chemical fingerprints. *Am. Mineral.* **2013**, *98*, 1120–1132. [CrossRef]
3. Okrusch, M.; Bunch, T.E.; Bank, H. Paragenesis and petrogenesis of a corundum bearing marble at Hunza (Kashmir). *Miner. Depos.* **1976**, *11*, 278–297. [CrossRef]
4. Rossovskiy, L.N.; Konovalenko, S.I.; Ananjev, S.A. Conditions of ruby formation in marbles. *Geologiya Rudnyh Mestorozhdeniy* **1982**, *24*, 57–66. (In Russian)
5. Litvinenko, A.K. The reconstruction of bauxite-like sediments in the Early Proterozoic metamorphites of the Central Pamir. In Proceedings of the Materials of the 5th All-Russian Lithologic Conference "The Types of Sedimentogenesis and Lithogenesis and Their Evolution in the History of the Earth", Ekaterinburg, Russia, 14–16 October 2008; pp. 428–430. (In Russian).
6. Dufour, M.S.; Kol'tsov, A.B.; Zolotarev, A.A.; Kuznetsov, A.B. Corundum-containing metasomatites of the Central Pamir. *Petrologiya* **2007**, *15*, 160–177. (In Russian) [CrossRef]

7. Kissin, A.Y. *Marble-hosted Ruby Deposits (by the Example of Urals)*; UB RAS Publishing House: Sverdlovsk, Russia, 1991. (In Russian)
8. Kissin, A.Y.; Murzin, V.V.; Tomilina, A.A.; Pritchin, M.E. Ruby-sapphire-spinel mineralization in the marbles of the Middle and Southern Urals: Geology, mineralogy, genesis. *Geol. Ore Depos.* **2016**, *58*, 385–402. [CrossRef]
9. Terekhov, E.N.; Kruglov, V.A.; Levitsky, V.I. Rare earth elements in corundum bearing metasomatites and associated rocks of the Eastern Pamir. *Geochimiya* **1999**, *3*, 238–250. (In Russian)
10. Terekhov, E.N.; Akimov, A.P. Tectonic position and genesis of jewelry corundum deposits in High Asia. *Litosfera* **2013**, *5*, 122–140. (In Russian)
11. Spiridonov, E.M. Gemstone deposits of the former Soviet Union. *J. Gemmol.* **1998**, *26*, 111–124. [CrossRef]
12. Garnier, V.; Giuliani, G.; Ohnenstetter, D.; Fallick, A.E.; Dubessy, J.; Banks, D.; Vinh, H.Q.; Lhomme, T.; Maluski, H.; Pêcher, A.; et al. Marble-hosted ruby deposits from Central and Southeast Asia: Towards a new genetic model. *Ore Geol. Rev.* **2008**, *34*, 169–191. [CrossRef]
13. Giuliani, G.; Ohnenstetter, D.; Fallick, A.E.; Groat, L.; Fagan, J. The geology and genesis of gem corundum deposits. In *Geology of Gem Deposits*; (Short Course Series); Raeside, E.R., Ed.; Mineralogical Association of Canada: Québec, QC, Canada, 2014; pp. 29–112.
14. Kisin, A.Y.; Murzin, V.V.; Batalina, A.A.; Tomilina, A.V. Green garnet from the Alabashka and Lipovka ruby occurrences (Middle Urals). *Vestnik UrO RMO* **2014**, *11*, 52–59. (In Russian)
15. Kuova, O.; Vuorelainen, Y. Eskolaite, a new chromium mineral. *Am. Mineral.* **1958**, *43*, 1098–1105.
16. Pautov, L.A.; Popov, M.P.; Erokhin, Y.V.; Khiller, V.V.; Karpenko, V.Y. Mariinskit $BeCr_2O_4$—A new mineral, chrysoberyl analog. *Zap. RMO* **2012**, *6*, 43–62. (In Russian)
17. Erokhin, Y.V. Chromite mineralization of the Bazhenovsky ophiolite complex (Middle Urals). *Litosfera* **2006**, *3*, 160–165. (In Russian)
18. Erokhin, Y.V.; Khiller, V.V.; Zoloev, K.K.; Popov, M.P.; Grigoryev, V.V. Mariinskit from the Bazhenovsky ophiolite complex—The second find in the world. *Dokl. Earth Sci.* **2014**, *455*, 441–443. (In Russian)
19. Moloshag, V.P.; Alimov, V.Y.; Anikina, E.V.; Gulyaeva, T.Y.; Vakhrusheva, N.V.; Smirnov, S.V. Accessory mineralization of chromitites of alpinotype hyperbasites of the Urals. *Zap. VMO.* **1999**, *2*, 71–83. (In Russian)
20. Bocharnikova, T.D.; Voronina, L.K. The first find of eskolaite in the magnetite-ilmenite ores of the Kusa gabbro intrusion. *Ezhegodnik–2007 Proc. IGG UB RAS.* **2008**, *155*, 226–229. (In Russian)
21. Sobolev, N.V. *Deep Inclusions in Kimberlites and the Problem of the Upper Mantle Composition*; Nauka (Siberian Branch): Novosibirsk, Russia, 1974; 264p. (In Russian)
22. Sobolev, V.K.; Zherdev, P.Y.; Kolodko, A.A. Eskolaite—The first find in kimberlites. *Mineral. Zhurnal* **1993**, *15*, 83–85. (In Russian)
23. Schulze, D.J.; Flemming, R.L.; Shepherd, P.H.M.; Helmstaedt, H. Mantle-derived guyanaite in a Cr-omphacitite xenolith from Moses Rock diatreme, Utah. *Amer. Miner.* **2014**, *99*, 1277–1283. [CrossRef]
24. Logvinova, A.M.; Wirth, R.; Sobolev, N.V.; Seryotkin, Y.V.; Yefimova, E.S.; Floss, C.; Taylor, L.A. Eskolaite associated with diamond from the Udachnaya kimberlite pipe, Yakutia, Russia. *Amer. Miner.* **2008**, *93*, 685–690. [CrossRef]
25. Koneva, A.A.; Suvorova, L.F. Rare chromium and vanadium oxides in the metamorphic rocks of the Priolkhonye (Western Baikal region). *Zap. VMO.* **1995**, *124*, 52–61. (In Russian)
26. Reznitsky, L.Z.; Sklyarov, E.V.; Karmanov, N.S. Eskolaite in the metacarbonate rocks of the Slyudyanka series (Southern Baikal). *Dokl. Earth Sci.* **1998**, *362*, 657–661. (In Russian)
27. *Geology of the USSR. T. XII. Perm, Sverdlovsk, Chelyabinsk and Kurgan regions. Part I. Geological description. Book 2*; Nedra: Moscow, Russia, 1969; 304p. (In Russian)
28. Keilman, G.A. *Migmatite Complexes of Mobile Belts*; Nedra: Moscow, Russia, 1974; 200p. (In Russian)
29. Boltyrov, V.B.; Pystin, A.M.; Ogorodnikov, V.N. Regional metamorphism of rocks in the northern framing of the Sanarsky granitoid massif on the South Urals. In *Geologiya Metamorficheskikh Kompleksov Urala (Geology of Metamorphic Complexes of the Urals)*; SGI: Sverdlovsk, Russia, 1973; pp. 53–66. (In Russian)
30. Ogorodnikov, V.N.; Sazonov, V.N.; Polenov, Y.A. Kochkarsky ore district (Southern Urals). In *Minerageny of Suture Zones of the Urals*; UGGGA: Ekaterinburg, Russia, 2004; 216p. (In Russian)

31. Talantsev, A.S. *Pocket Pegmatites of the Urals*; Nauka: Moscow, Russia, 1988; 144p. (In Russian)
32. Kisin, A.Y. Deformation macrostructures in carbonate rocks of granite-gneiss complexes of Urals. *Lithosfera* **2007**, *1*, 90–108. (In Russian)
33. Kissin, A.J. Ruby and Sapphire from the Southern Ural Mountains, Russia. *Gems Gemol.* **1994**, *30*, 243–252. [CrossRef]
34. Kisin, A.Y.; Koroteev, V.A. *Block Folding and Ore Genesis*; IGG UrO RAN: Ekaterinburug, Russia, 2017; 349p.
35. Vysotskiy, S.V.; Nechaev, V.P.; Kissin, A.Y.; Yakovenko, V.V.; Ignat'ev, A.V.; Velivetskaya, T.A.; Sutherland, F.L.; Agoshkov, A.I. Oxygen isotopic composition as an indicator of ruby and sapphire origin: A review of Russian occurrences. *Ore Geol. Rev.* **2015**, *68*, 164–170. [CrossRef]
36. Talantsev, A.S. *Geothermobarometry according to Dolomite-Calcite Paragenesis*; Science: Moscow, Russia, 1981; 136p. (In Russian)
37. Chatterjee, N.D.; Leistner, H.; Terhart, L.; Abraham, K.; Klaska, R. Thermodynamic mixing properties of corundum-eskolaite, α-$(Al,Cr^{3+})_2O_3$, crystalline solid solutions at high temperatures and pressures. *Amer. Miner.* **1982**, *67*, 725–735.
38. Kissin, A.Y.; Koroteev, V.A. Stress Gradients as the Driving Force of Mass Movement in General Crustal Folding. *Dokl. Earth Sci.* **2009**, *424*, 7–10. [CrossRef]
39. Meng, F.; Shmelev, V.R.; Kulikova, K.V.; Ren, Y. A red-corundum-bearing vein in the Rai–Iz ultramafic rocks, Polar Urals, Russia: The product of fluid activity in a subduction zone. *Lithos* **2018**, *320–321*, 302–314. [CrossRef]
40. Popov, M.P.; Sorokina, E.S.; Kononkova, N.N.; Nikolaev, A.G.; Karampelas, S. New Data on the Genetic Linkage of the Beryl and Chrysoberyl Chromophores of the Ural's Emerald Mines with Chromium-Bearing Spinels of the Bazhenov Ophiolite Complex. *Dokl. Earth Sci.* **2019**, *486*, 630–633. [CrossRef]
41. Sazonov, V.N. *Chromium in the Hydrothermal Process (on the Example of the Urals)*; Nauka: Moscow, Russia, 1978; 287p. (In Russian)

6

Identification of Opaque Sulfide Inclusions in Rubies from Mogok, Myanmar and Montepuez, Mozambique

Wim Vertriest * and Aaron Palke

Gemological Institute of America, 5355 Armada Drive, Carlsbad, CA 92008, USA; apalke@gia.edu
* Correspondence: wvertrie@gia.edu

Abstract: The red variety of corundum owes its color and strong fluorescence to the presence of Cr, as well as traces of Fe. The latter can reduce the fluorescence and thus impact the appearance of the final gem. Gem quality rubies are rarely available for scientific study and even less common in their rough form. Opaque inclusions in rubies are often removed during faceting and remain unidentified. This study aims to identify opaque inclusions in rubies from the two most common origins seen in the high end market today: Mogok, Myanmar and Montepuez, Mozambique. Using electron probe microanalaysis (EPMA) the inclusions were identified as sphalerite and pyrrhotite in Mogok rubies. The paragenesis of Myanmar, marble-related rubies is fairly well understood and no Fe-rich minerals apart from sulfides have been identified. Opaque inclusions in Mozambican rubies are a complex mix of Fe-Cu-Ni sulfides with exsolution textures. These inclusions are interpreted to be small amounts of sulfide melt trapped during corundum formation. The different sulfide phases crystallized from this entrapped melt and some phases experienced later exsolution during cooling. The formation of amphibole-related, Mozambican rubies is not well understood, but it is obvious that very different processes are at work compared to the marble-related Myanmar ruby deposits.

Keywords: ruby; sulfides; Mogok; Montepuez; gemology; corundum

---

1. Introduction

Ruby has always been one of the most desired gemstones. Historically, the finest rubies were found in deposits throughout Central and South-East Asia [1]. In the last decades, many ruby deposits have been found in East Africa, introducing a slightly different type of ruby to the market that originated from geologically different deposits [2–14].

Marble-related rubies, such as the Myanmar stones, are famous for being extremely fluorescent and having very low Fe concentrations [15]. Amphibole-related rubies, like the Mozambican stones, show more variable fluorescence, ranging from almost none to intensities that nearly match marble-related stones [7]. The red gem variety of corundum owes its color and fluorescence to trace amounts of Cr. Some rubies from Mogok have a considerable V content that can modify the color to be more purple [16]. The presence of Fe lowers the fluorescence of the ruby.

Inclusion studies are a critical part of gemological research, because included crystals are easy to observe and describe in transparent minerals. The only required equipment to see them is a gemological microscope and adequate lighting tools. Gemologists rely heavily on inclusions to determine treatments and geographic origin of gem quality corundum [17–19]. However, in most cases their studies are limited to visual observation. Since most gemstones have a high value, destructive analysis is rarely an option. This means that the chemical analysis of inclusions is only possible when they reach the polished surface of the gem. Spectroscopic techniques such as FTIR and Raman spectroscopy offer a solution, with Raman spectroscopy being the most ideal candidate to identify crystal inclusions [20]. For most common inclusions in finished gemstones, this is sufficient. In many cases, these crystals are transparent/translucent carbonate, phosphate, oxide or silicate minerals [1,21].

In the case of rough rubies, an additional type of inclusion can be frequently seen. These are dark, opaque, inclusions with a metallic luster. These dark colored inclusions are rarely retained in the facetted gems, because their high contrast with the transparent red bodycolor creates eye visible distractions that lower the value significantly.

On top of that, they react poorly to heat treatment [2,17,18]. Heating rubies is a standard practice in the gem trade to improve the color or clarity of the gemstone. These metallic inclusions often expand during this process. Expansion fractures commonly surround the crystal, often filled with black opaque residue, making the inclusion even more visible and distracting. Thus, facetters aim to remove these inclusions before polishing the gem.

The goals of this study are to identify these opaque inclusions and assess how their presence may explain the differences in trace element chromophores and visual appearance between the Myanmar and Mozambique rubies. Given the affinity for S to bind with Fe and create sulfides, the presence of S during corundum formation may potentially be an important factor for the final fluorescence and appearance of the gem. The formation of amphibole related rubies is not fully understood. An analysis of the opaque sulfide inclusion could provide additional insight in the environment that allowed the formation of amphibole-related rubies.

2. Materials and Methods

2.1. Materials

For this study, rubies from Mogok, Myanmar and Montepuez, Mozambique were used (Table 1). These sources were chosen because they represent the majority of high-end rubies seen in the gem trade nowadays. Mogok, Myanmar is the most coveted ruby source in the world, due to its long history and the fabled gemstones that were sourced from its mines [1]. Montepuez Mozambique on the other hand is a very new source where rubies were only discovered in 2009, but its large volumes and high quality quickly made it the dominant supplier for the global ruby trade [2].

All samples are part of GIA's (Gemological Institute of America) Colored Stone Reference Collection (GIA CSRC) which consists of samples collected during GIA Field expeditions to gemstone mining locations around the world [22]. The selected samples were prepared and documented at GIA's laboratory in Bangkok, Thailand.

Every ruby selected contained opaque, mineral inclusions with metallic luster. In one case, the opaque mineral is not present as an inclusion. This sample (A6), from the Baw Lon Gyi mining area in Mogok, Myanmar, has an opaque mineral in the matrix attached to the ruby, rather than as an inclusion.

Table 1. Samples used during this study.

GIA CSRC Sample Number	Number	Country of Origin	Area	Mining Area	
100321364740	A1	Myanmar	Mogok	Baw Lon Gyi	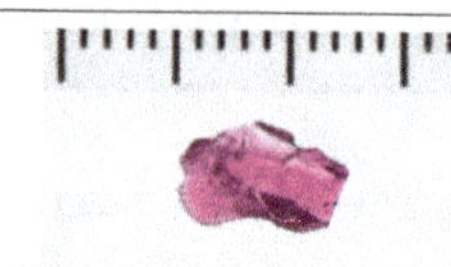
100315602328	A2	Myanmar	Mogok	Baw Lon Gyi	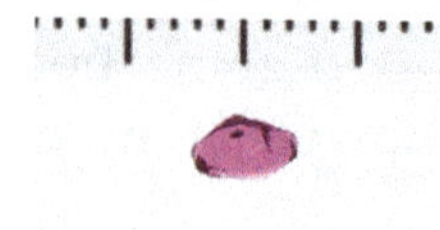
100322689933	A3	Myanmar	Mogok	Baw Lon Gyi	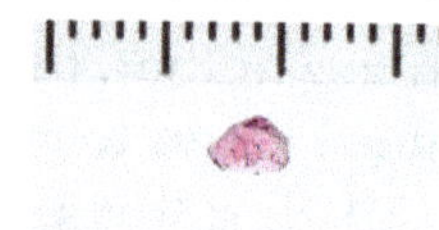
100318470512	A4	Myanmar	Mogok	Sin Kwa	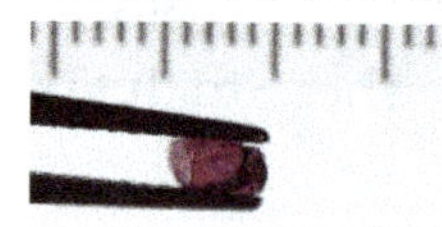
100309935280	A5	Myanmar	Mogok	Sin Kwa	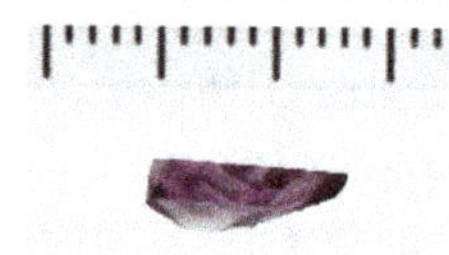
100327146806	A6	Myanmar	Mogok	Baw Lon Gyi	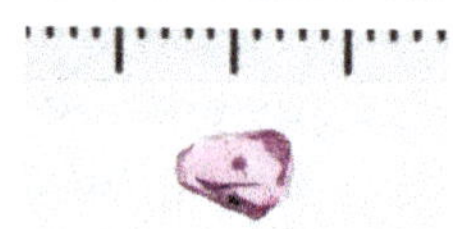
100319160887	B1	Mozambique	Montepuez	Mugloto	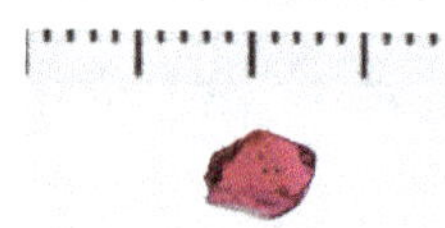
100327146596	B2	Mozambique	Montepuez	Mugloto	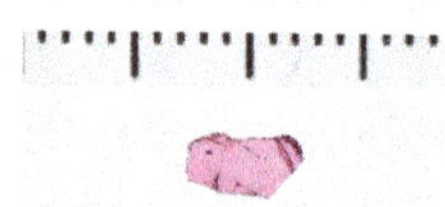
100319160890	B3	Mozambique	Montepuez	Mugloto	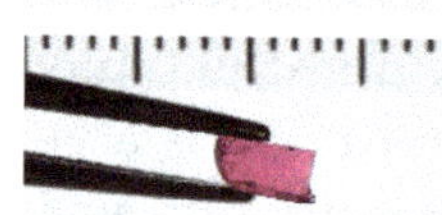
100319160927	B4	Mozambique	Montepuez	Mugloto	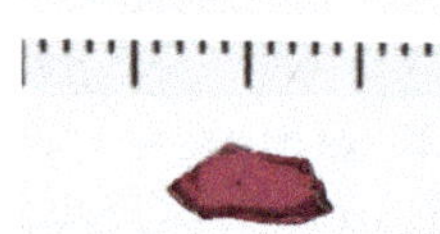

2.2. *Methods*

2.2.1. Photomicrography

Photomicrographs of internal features were captured at different magnifications with Nikon SMZ 18 and Nikon SMZ 1500 systems (Nikon, Tokyo, Japan), using darkfield, brightfield, diffused, and oblique illumination, together with a fiber-optic light source when necessary. It should be noted that the microscope magnification power was taken into consideration when calculating the field of view (FOV) information in the captions.

2.2.2. Inclusion Analysis

Backscattered electron (BSE) images were obtained using a ZEISS 1550VP field emission SEM (ZEISS, Oberkochen, Germany) and a JEOL 8200 electron microprobe (JEOL, Tokyo, Japan) with AsB (angle selective backscatter) and solid-state BSE detectors, respectively. Quantitative elemental microanalyses were carried out using the JEOL 8200 electron microprobe, operated at 10 kV (for smaller interaction volume) and 5 nA in focused beam mode. Analyses were processed with the CITZAF correction procedure [23] using the Probe for electron probe microanalaysis (EPMA) software from Probe Software, Inc. (Eugene, OR, USA).

The EPMA analysis was set up to measure 8 selected elements: S, As, Cu, Fe, Ni, Zn, Pb and Co. Analysis was calibrated using the following standards (Table 2).

Table 2. Details of the standards used during electron probe microanalaysis (EPMA) analysis.

Elements	Standard Number	Mineral	Mineral Code
Cu	St 129	Copper	P-1012
S	St 1830	MnS syn	P-698
Ni, S	St 1831	NiS syn	P-699
Zn, S	St 1832	ZnS syn	P-700
Pb, S	St 1833	PbS syn	P-701
As, S	St 1835	As_2S_3 syn	P-703
Co, S	St 1838	Co_9S_8 syn	P-706
Fe, S	St 1853	FeS_2 mac	P-1004

EPMA and BSE analysis were performed at the Geological and Planetary Science Division Analytical Facility at Caltech in Pasadena, California.

3. Geological Setting

Corundum only forms in natural environments that are depleted in Si. At high concentrations of silicon, Al will preferentially be incorporated in other minerals such as micas or feldspars. Consequently, corundum is often associated with Si-free carbonates or silicate minerals, with a relatively high Al/Si ratio such as plagioclase, amphibole or pyroxene [24]. For the formation of the ruby, it is necessary to incorporate trace amounts of Cr to act as a chromophore, as well as the right thermobarometric conditions to allow the growth of large crystals.

3.1. Geologogical Setting and Formation of Mogok Rubies

The Mogok Stone Tract, the source of many fine gems, is located in the northern part of the Mogok Metamorphic Belt. This metamorphic belt crosses Myanmar from north to south and forms the western margin of the Shan-Thai block [25–27]. The Mogok area is dominated by amphibolite to granulite facies marbles and paragneisses, as well as a variety of intrusive rocks and derived orthogneisses [28,29]. Recent data show a spatial relation between the gem corundum formation and intrusions, although the exact timing and the impact of the intrusive rocks on gem formation are not yet fully understood. It is suggested that Jurassic intrusions and associated alterations of the host rock were affected by regional granulite facies metamorphism related to the Himalayan Orogeny, lasting from 68–21 Ma. During this metamorphism, rubies and sapphires were formed in the marbles and metaskarns [30].

The rubies formed in metamorphosed carbonate platforms that contained some contamination of clays, organic material and evaporites (e.g., anhydrite and salts). The clays provided the necessary elements to build ruby (Al and Cr), while the molten salts allow these usually immobile elements to migrate through the marble. The molten salts and fluids associated with ruby formation are derived from the protolith and are also enriched in carbonate, sulfur and Al [31,32].

This results in layers of fairly pure marble with large, well-crystallized calcite interlayered, with silicate layers often containing amphiboles and micas [33]. Ruby crystals occur disseminated in the marble and are associated with accessory phlogopite, muscovite, scapolite, spinel, titanite, pyrite and graphite [34].

Throughout Central and Southeast Asia, several marble related deposits are found that are related to the Cenozoic Indo-Asian collision and share a similar formation history [35]. This includes Luc Yen, Vietnam; Murghab, Tajikistan; Hunza, Pakistan; Mong Hsu, Myanmar; Jegdalek, Afghanistan; etc. [32].

The Myanmar rubies from the Mogok valley are mined from hard rock deposits in the marbles, or from secondary mines inside karst features or alluvial deposits.

3.2. Geological Setting and Formation of Mozambican Rubies

Mozambican rubies are found in the Southern part of the Neoproterozoic Mozambique belt, which is closely linked with the Pan African orogeny [6–8,36]. This suture zone represents the collision between east and West Gondwana [37,38].

The geology of Mozambican rubies is not well understood. The deposit was only discovered in 2009 and most of the rubies come from secondary deposits. The primary deposits in the area are so weathered that they are barely recognizable.

The ruby-bearing rocks are amphibolites hosted in the Montepuez nappe of the Cabo Delgado complex [39]. The Montepuez complex is made up of orthogneiss (granitic to amphibolitic composition) and paragneiss (quartzite, meta-arkose, biotite gneiss and amphibole gneiss,) [40].

There is some evidence of intrusions seen in the Montepuez area in the form of pegmatites and granitic intrusions [7,40].

The rubies are closely related to the bodies with an amphibolite-rich composition, although they are extremely weathered [8]. To date, there is only one study published that describes the petrography of Mozambican ruby host rocks. They contain significant amounts of plagioclase and calcic amphibole next to spinel and corundum, although the silicates are often weathered to clays and micas [41]. Whole-rock geochemical analysis shows that the rocks have a composition similar to picrobasalt in the TAS diagram [41,42]. While this classification is traditionally used for magmatic rocks, it is not yet proven that the protolith of the Mozambican ruby host rock is of magmatic origin. However, the geochemical analysis is at least suggestive that it is derived from a basaltic magma [41].

At the moment, the genesis of Mozambican ruby and amphibole-related ruby is not fully understood. The current model states that amphibole and corundum formed during peak metamorphism, which reacted to plagioclase and spinel during retrograde metamorphism. According to P-T calculation based on a combination of mineral isopleths, pseudosections and geothermometers, the rubies formed at temperatures around 550–600 °C and pressures around 10.5–11 kbar [41]. These conditions place ruby formation in the amphibolite facies at high pressure and lower temperature. Terrain observations and geological surveys have shown the presence of migmatites and other features that partial melting took place in the amphibolites and gneisses [40,41].

4. Inclusion Analysis

4.1. Literature Studies

The most commonly used reference works for gemological inclusion studies describe nearly all opaque inclusions in marble-related ruby as pyrrhotite and pyrite [21,43]. The first main study on Mozambican rubies identifies them as chalcopyrite [7]. Nowadays, they are often described based on appearance as pyrite or chalcopyrite, for Myanmar and Mozambique respectively. Inclusions in gemstones are usually well described in the gemological literature. Characteristics such as color, transparency and shape are easy to observe and often allow for preliminary identification. Raman spectroscopy is often used to further identify the crystal inclusions, but this is not always conclusive. More advanced techniques are rarely employed by gemologists. Most of the gemstone inclusion knowledge stems from the three volume Photo-Atlas series, which documents and describes the internal world of gemstones [21]. The following is a list of crystal inclusions documented in rubies from Mogok, Myanmar.

1. Zircon, spinel, rutile prisms, mica, garnet, corundum [44]
2. Sphene (titanite), calcite [45]
3. Apatite, olivine, pargasite, sphalerite [46]
4. Dolomite, pyrrhotite, pyrite [21]
5. Sulfur [47]
6. Scapolite, graphite [48]
7. Vesuvianite [49]

The deposits around Montepuez, Mozambique were only recently discovered and there is not as much information is available [6,7]. In literature, the following inclusions have been described.

8. Amphibole and muscovite [7,39,50]
9. Chalcopyrite [2,7]
10. Orange, irregular masses identified as chromite [7]

However, based on the authors' experience, several additional, less common inclusions like tourmaline, feldspar, zicorn and anhydrite also occur in rubies from this deposit.

This study focusses on the black opaque inclusions that are found in many rough rubies, which are typically described as pyrrhotite/pyrite and chalcopyrite, for Myanmar and Mozambique respectively.

4.2. *Analytical Results*

4.2.1. Imaging and Description

All studied inclusions are opaque and have a metallic luster. The apparent color of the inclusions varies from black to red, which are colors not typically associated with most common crystallized metallic minerals. This is because they are found as inclusions within a bright red ruby matrix, where the color of the host mineral obscures the color of the actual inclusion.

Most of the crystals in the Myanmar rubies have sharp angles and well-formed surfaces (A1, A2, A4, A5), while sample A2 shows a very clear step pattern on its surface (Figure 1). Sample A4 has a clear (pseudo?-) hexagonal outline. None of these samples are associated with fractures or show any evidence that they could be secondary.

The inclusions in Mozambican rubies are very similar in color and luster to the Myanmar ones (Figure 2). Their outlines are more spherical but still exhibit flat crystal faces, indicating that all of these crystals are primary inclusions. Sample B1 is associated with a fracture, but this does not extend to the exterior of the stone.

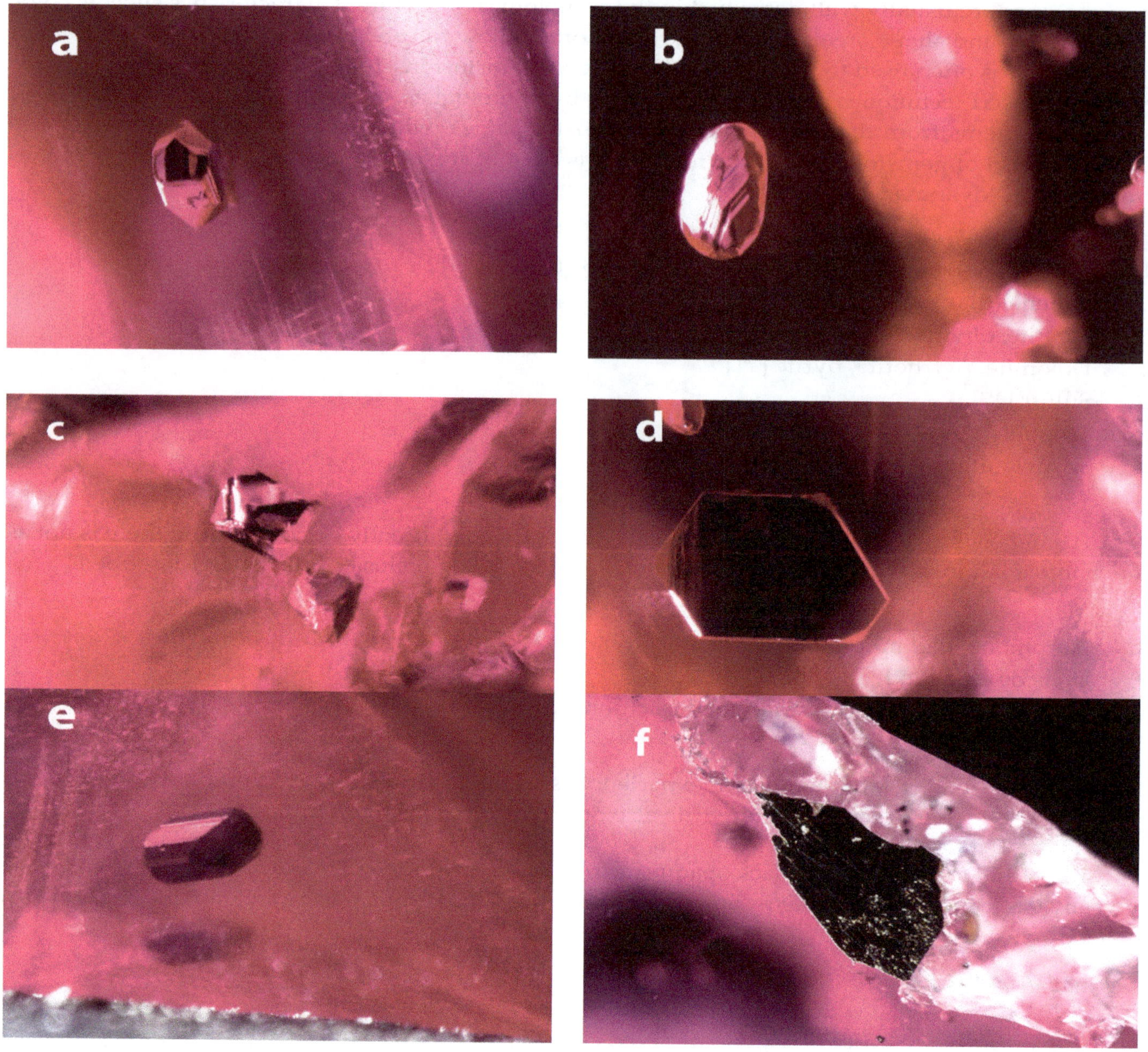

Figure 1. Images of the inclusions in the Mogok rubies. (**a**) Opaque inclusion in A1, showing well-formed crystal faces. FOV: 1.2 mm; (**b**) Opaque inclusion in A2 has more rounded appearance due to the terraced angular features. FOV: 1.01 mm; (**c**) Opaque inclusions in A3 exhibit almost a bipyramidal shape reminiscent of a cubic crystal structure. FOV: 1.2 mm; (**d**) Opaque inclusion in A3 has a very flat, (pseudo?) hexagonal outline. FOV: 0.85 mm; (**e**) Opaque inclusion in A5 shows well developed crystal faces and a slightly elongated blocky crystal shape. FOV: 0.85 mm; (**f**) The opaque minerals associated with sample A6 are an irregular mass, attached to the ruby and surrounded by carbonates. FOV: 1.44 mm.

We also checked a number of samples from GIA's CSRC, to see how abundant those inclusions are. In 54 rough samples from Mogok, we found 46 samples that contained at least one metallic opaque inclusion. In Mozambican rubies, metallic inclusions are much less common, with only 37 out of 109 Mozambican rubies showing this type of inclusion.

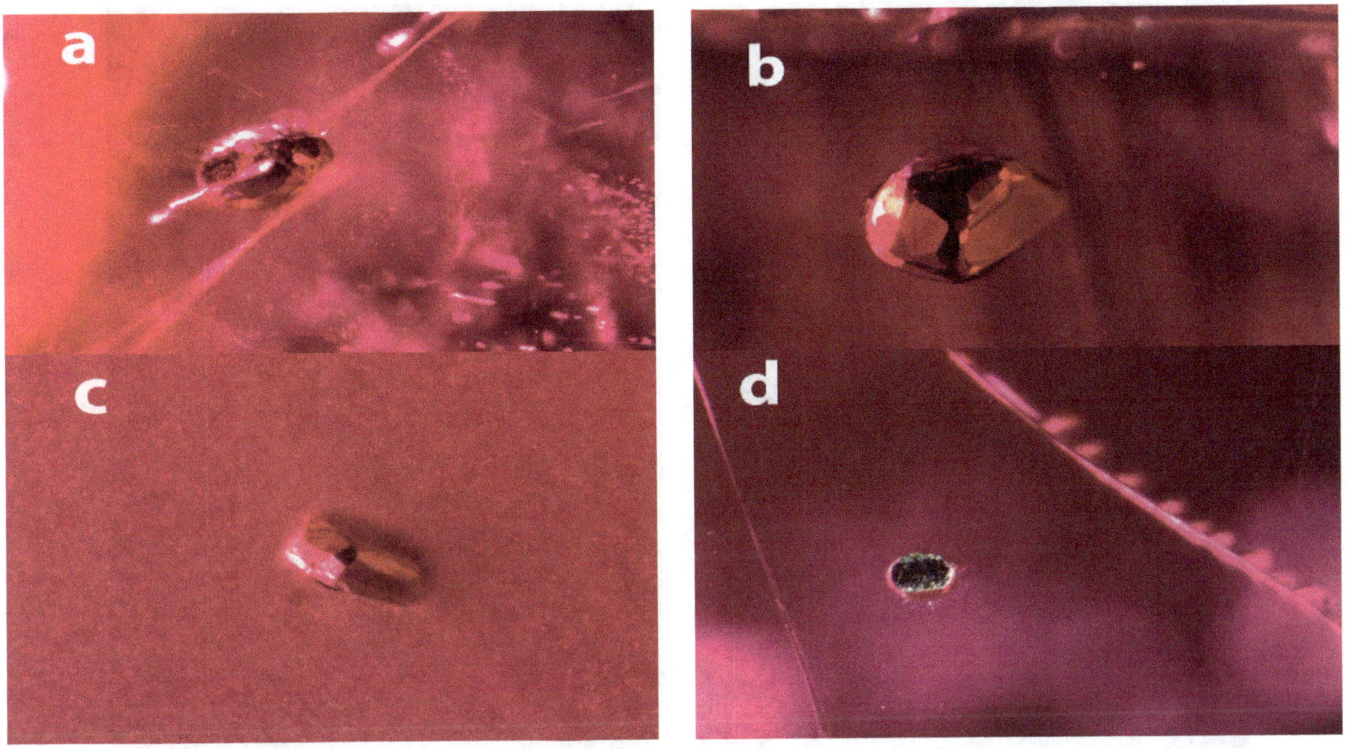

Figure 2. Inclusions in Mozambican rubies. (**a**) The inclusion in sample B1 is more egg shaped, but shows many small flat crystal faces. It is surrounded by a fracture that is reminiscent of an expansion fracture, which might be an indication of a temperature increase during its geological history after the corundum formed. FOV: 1.50 mm;(**b**) The opaque inclusion in sample B2 show many larger crystal faces that do not seem regularly distributed. The overall appearance is rather spherical. FOV: 0.78 mm; (**c**) The inclusions in sample B3 are similar to B2 but appears flatter. FOV: 1.2 mm; (**d**) The inclusion in sample B4 was cut through during sample fabrication and shows the brassy color around the edges. FOV: 0.72 mm.

4.2.2. BSE Imaging

Back-scattered electron imaging shows that the inclusions in Mogok rubies are mainly single phases (Figure 3). Polish lines and pits were caused by the sample preparation process. Some samples show fractures that were already present, but might have been enlarged during polishing. In some cases, the inclusion is not entirely preserved and some parts were removed during the fabrication process. One sample (A4) shows some obvious zoning with a darker area, showing a herringbone texture. The sample with the sulfide located in the attached matrix (Figure 3f) shows a subtle transition between phases.

The Mozambican ruby inclusions show much more complex features (Figure 4). All samples have clearly separated phases. Sample B1 shows three clearly separated phases based on the intensity in BSE images. They are sometimes separated by fractures at the phase contact. In samples B1 and B2, the darkest phase also shows clear exsolution textures. The exsolution pattern in sample B2 is more expressed than it is in sample B1. The homogenous phases without exsolution patterns are also smaller and confined to the fringes of the inclusion. Samples B3 and B4 lack any obvious exsolution patterns, but still show very clear phase separations.

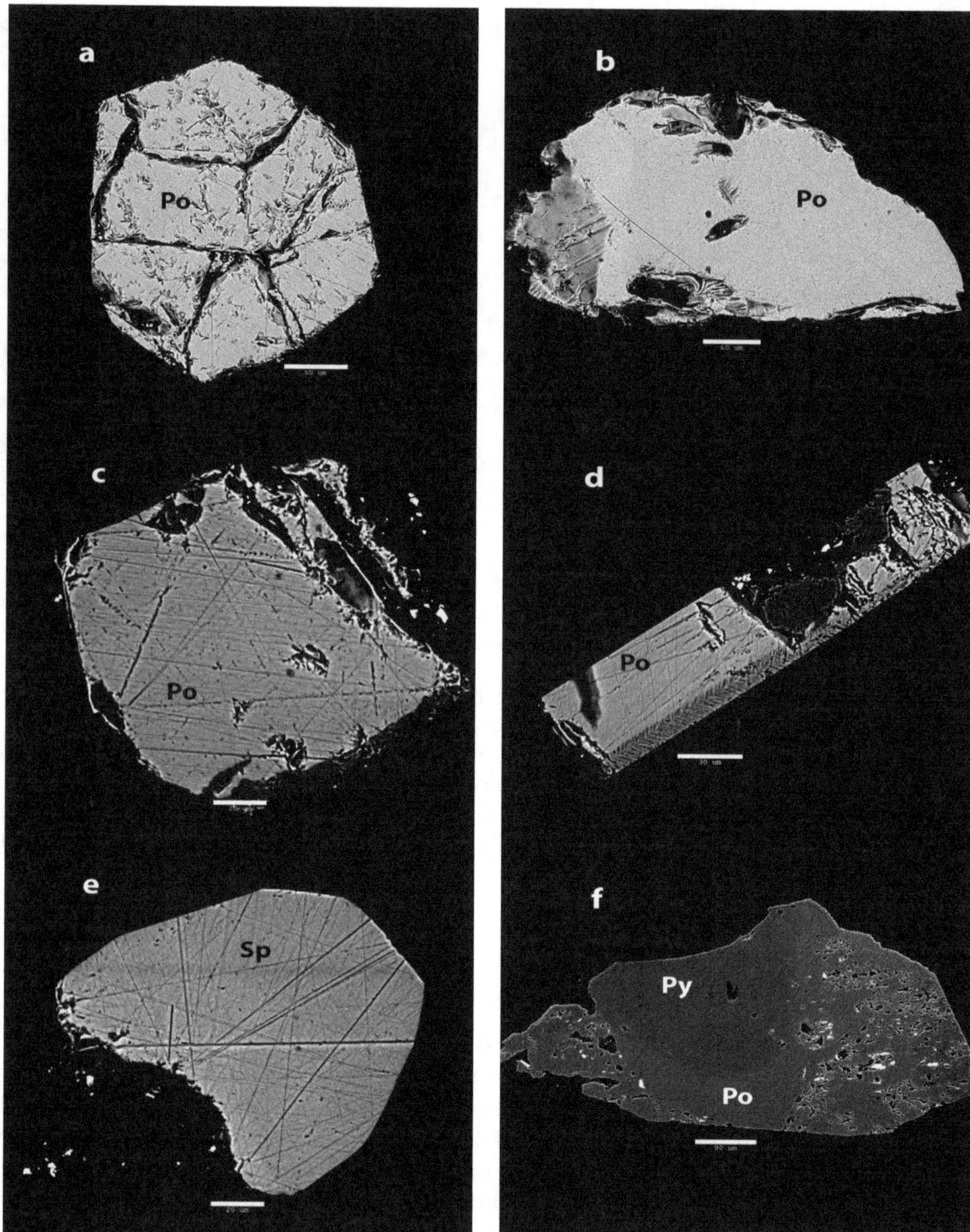

Figure 3. BSE images of Mogok rubies. (**a**) Inside sample A1, the inclusion shows hexagonal outline and large internal cracks. Scale bar: 50 μm; (**b**) The inclusion in sample A2 shows an irregular outline with a homogenous interior. Scale bar: 60 μm; (**c**) The inclusion in sample A3 shows straight outline with numerous polish lines. Scale bar: 20 μm; (**d**) Sample A4 also shows a lot of polish lines, but most notable is the herringbone texture on the side. Scale bar: 30 μm; (**e**) The inclusion in sample A5 has a well-defined outline on the top right. The bottom left has been removed during polishing. Apart from polishing lines, this inclusion appears very homogenous. Scale bar: 20 μm; (**f**) The opaque minerals attached to sample A6 show many pits but most important is the subtle zoning in the central part. The ruby is attached to the lower part. Scale bar: 90 μm.

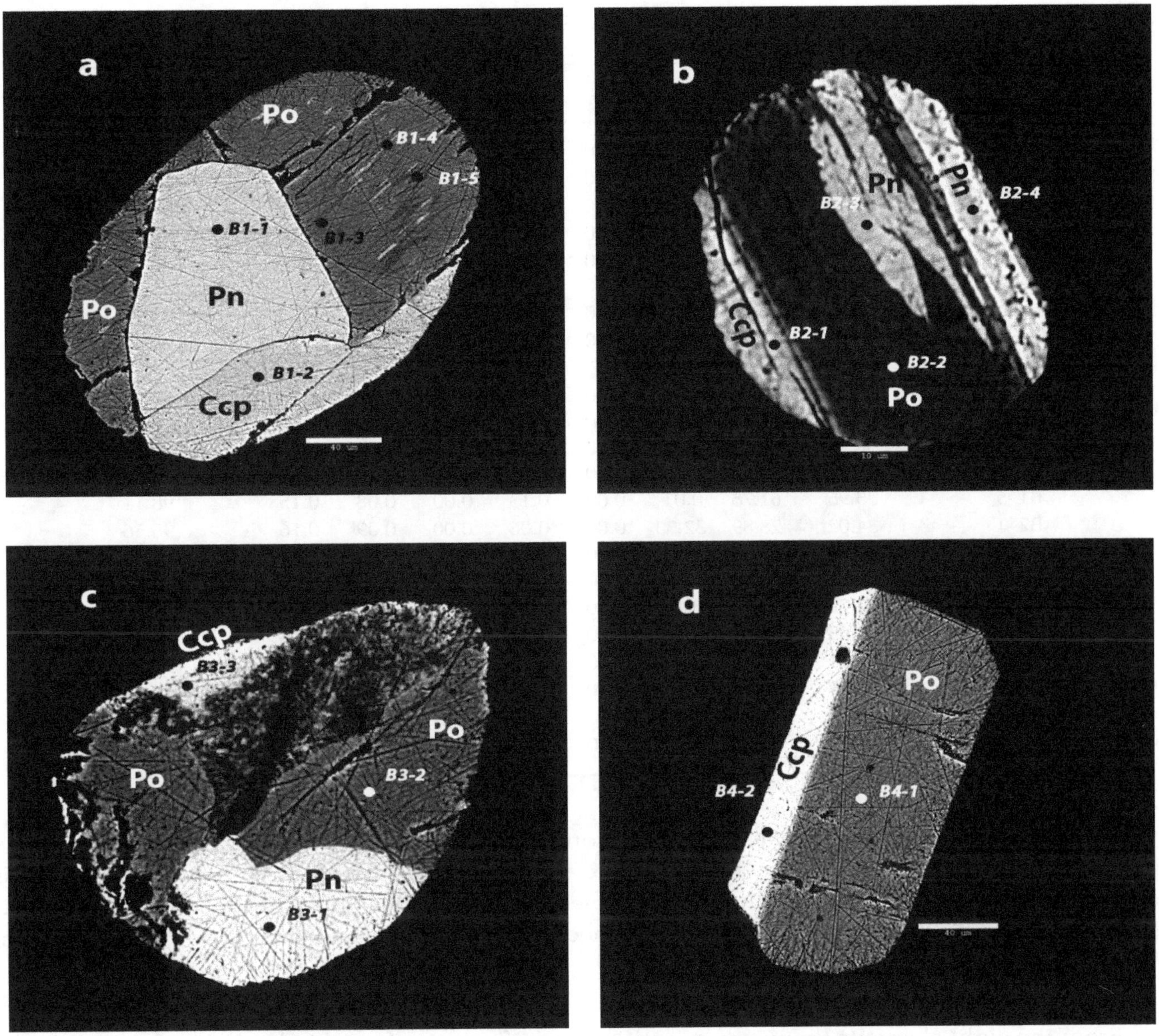

Figure 4. Back scattered electron images of the sulfide inclusions in Mozambican ruby. (**a**) The inclusion in sample B1 shows three main phases, of which one has various exsolutions. Scale bar: 40 μm; (**b**) In sample B2, there are three phases with two phases that show exsolution patterns. Scale bar: 10 μm; (**c**) In sample B3, there are three distinct phases. The darker phase shows deeper polish lines and large pits, which might indicate a more unstable mineral. Scale bar: 20 μm; (**d**) B4 shows two clearly separated phases without any exsolution textures. Scale bar: 40 μm.

4.2.3. EPMA Analysis

During the EPMA analysis, eight selected elements were quantified (Tables 3 and 4). All of the inclusions turned out to be sulfides, which is consistent with the previous work [7,21]. Most of the analyses resulted in an elemental total of 97–98% for the inclusions in Mogok rubies. The missing percentages are most likely explained by surface oxidation of the samples. However, it cannot be excluded that an element was not measured during our analysis, which was limited to eight elements. Notably, in sample A4, the elemental total measured in the section with the herringbone pattern was significantly lower.

Table 3. EPMA results: Elemental percentages and the elemental totals.

Mogok	S	Fe	Cu	As	Co	Ni	Zn	Pb	Elemental Totals
A1	38.66	59.14	0.00	0.00	0.13	0.03	0.01	0.20	98.19
A2	38.58	59.28	0.00	0.00	0.14	0.06	0.10	0.16	98.31
A3	38.68	59.51	0.01	0.00	0.12	0.15	0.10	0.18	98.75
A4	38.72	58.91	0.00	0.00	0.11	0.15	0.06	0.20	98.14
A4—herringbone	41.96	43.18	0.00	0.00	0.09	0.58	0.03	0.18	86.01
A5	33.21	10.70	0.00	0.00	0.02	0.00	53.95	0.06	97.94
A6—dark phase	53.34	46.76	0.00	0.00	0.07	0.00	0.03	0.24	100.45
A6—light phase	38.75	59.49	0.00	0.00	0.12	0.22	0.04	0.18	98.80
Mozambique									
B1–1	32.82	28.74	0.05	0.00	1.22	34.43	0.05	0.16	97.46
B1–2	34.77	29.27	32.55	0.00	0.05	0.00	0.02	0.17	96.85
B1–3	36.22	60.97	0.01	0.00	0.14	0.02	0.08	0.19	97.63
B1–4	34.22	40.43	0.04	0.00	0.17	22.52	0.09	0.15	97.62
B1–5	35.95	61.78	0.01	0.00	0.13	0.00	0.08	0.15	98.11
B2–1	34.02	28.89	32.33	0.00	0.06	0.00	0.09	0.16	95.56
B2–2	39.14	58.43	0.13	0.00	0.12	0.16	0.14	0.16	98.27
B2–3	33.19	30.28	0.07	0.00	1.22	32.51	0.21	0.14	97.61
B2–4	32.72	29.16	0.11	0.00	1.27	34.54	0.00	0.16	97.96
B3–1	32.65	29.02	0.00	0.00	1.14	34.75	0.02	0.18	97.75
B3–2	38.11	58.95	0.07	0.00	0.12	0.40	0.18	0.15	97.99
B3–3	33.47	29.49	32.07	0.00	0.07	0.00	0.10	0.13	95.31
B4–1	38.63	58.50	0.00	0.00	0.13	0.93	0.11	0.19	98.49
B4–2	34.02	29.84	32.18	0.00	0.07	0.00	0.03	0.18	96.31

In the case of the Mozambican ruby inclusions, there was a larger spread in elemental totals. The minerals that are dominated by Fe-Ni-S have elemental totals between 97% and 98%, which is comparable to the numbers seen for the Mogok rubies. Samples with significant Cu content have lower elemental totals (95–97%), which could be explained by a more intense alteration or a minor element that is missing.

From the eight elements that were quantified, only five were found to be major elements in the minerals. In every sample, the Co and Pb content is low enough to consider them trace elements. Arsenic was not detected in any of the samples.

The analysis shows the presence of pyrrhotite $Fe_{(0.88)}S$ and sphalerite $(Zn_{0.8}, Fe_{0.18})S$ in the Mogok rubies (Table 4). Additionally, in the matrix sample, pyrrhotite $Fe_{(0.88)}S$ and pyrite FeS_2 are found together.

The inclusions in Mozambican samples are more complex. They show three main mineral components: chalcopyrite, pentlandite and pyrrhotite. These are well separated and do not show any mixing. The darker pyrrhotite phase often shows different flame-like exsolution patterns. The exsolved minerals consist of Fe-Ni sulfides and pyrrhotite, with slightly elevated Fe contents.

Table 4. Molar percentages normalized to sulfur.

Mogok	S	Fe	Cu	Co	Ni	Zn	Mineral Name	Mineral Formula
A1	1.00	0.88	0.00	0.00	0.00	0.00	pyrrhotite	$Fe_{0.88}S$
A2	1.00	0.88	0.00	0.00	0.00	0.00	pyrrhotite	$Fe_{0.88}S$
A3	1.00	0.88	0.00	0.00	0.00	0.00	pyrrhotite	$Fe_{0.88}S$
A4	1.00	0.87	0.00	0.00	0.00	0.00	pyrrhotite	$Fe_{0.87}S$
A5	1.00	0.18	0.00	0.00	0.00	0.80	sphalerite	$(Fe_{0.18}, Zn_{0.80})S$
A6—dark phase	1.00	0.50	0.00	0.00	0.00	0.00	pyrite	FeS_2
A6—light phase	1.00	0.88	0.00	0.00	0.00	0.00	pyrrhotite	$Fe_{0.88}S$
Mozambique								
B1–1	1.00	0.50	0.00	0.02	0.57	0.00	pentlandite	$(Fe_4, Ni_{4.56} Co_{0.16})S_8$
B1–2	1.00	0.48	0.47	0.00	0.00	0.00	chalcopyrite	$Cu_{0.47}Fe_{0.48}S$
B1–3	1.00	0.97	0.00	0.00	0.00	0.00	pyrrhotite	$Fe_{0.97}S$
B1–4	1.00	0.68	0.00	0.00	0.36	0.00	?Fe-rich pentlandite?	
B1–5	1.00	0.99	0.00	0.00	0.00	0.00	pyrrhotite	$Fe_{0.99}S$
B2–1	1.00	0.49	0.48	0.00	0.00	0.00	chalcopyrite	$Cu_{0.48}Fe_{0.49}S$
B2–2	1.00	0.86	0.00	0.00	0.00	0.00	pyrrhotite	$Fe_{0.86}S$
B2–3	1.00	0.52	0.00	0.02	0.53	0.00	pentlandite	$(Fe_{4.16}, Ni_{4.24} Co_{0.16})S_8$
B2–4	1.00	0.51	0.00	0.02	0.58	0.00	pentlandite	$(Fe_{4.04}, Ni_{4.64}, Co_{0.16})S_8$
B3–1	1.00	0.51	0.00	0.02	0.58	0.00	pentlandite	$(Fe_{4.04}, Ni_{4.64}, Co_{0.16})S_8$
B3–2	1.00	0.89	0.00	0.00	0.01	0.00	pyrrhotite	$Fe_{0.89}S$
B3–3	1.00	0.51	0.48	0.00	0.00	0.00	chalcopyrite	$Cu_{0.48}Fe_{0.51}S$
B4–1	1.00	0.87	0.00	0.00	0.01	0.00	pyrrhotite	$Fe_{0.87}S$
B4–2	1.00	0.50	0.48	0.00	0.00	0.00	chalcopyrite	$Cu_{0.48}Fe_{0.50}S$

5. Discussion

Based on the EPMA analyses, it is clear that the inclusions in Mozambican rubies have not been correctly identified by previous methods. Originally described as pyrite and chalcopyrite in Mogok and Mozambican rubies respectively, they were now identified as pyrrhotite and Fe-Ni-Cu-sulfides.

Pyrrhotite has a very narrow stability field at surface conditions, but is the most stable iron-sulfide at higher temperatures (>743 °C). Fluid inclusion studies of marble-related ruby deposits in Central Asia and northern Vietnam provide formation temperatures between 620 and 670 °C, with pressures between 2.6 and 3.3 kbar [32].

Various studies have been done to determine P-T conditions of gneisses in the Mogok Metamorphic Belt, which stretches more than 1000km and has shown a range of temperatures from 625 to 950 °C [51,52]. For the Mogok deposit, there are no temperature estimates available for the marbles. However, various studies have determined the formation temperatures of the gneisses in the Mogok stone tract and these yielded significantly higher temperatures. Formation temperatures of quartz-dominated gneisses are reported to be 750–790°C [53]. However, it is important to note that the prevalent genetic model for marble-hosted rubies places the ruby formation in the retrograde path [32]. These temperatures are significantly higher than the formation conditions of other marble related deposits and most likely represent peak conditions of the host rock, which was reached before ruby formation. Under the higher peak conditions, pyrrhotite is stable, making it the most likely Fe-S mineral that would later be incorporated as a protogenetic inclusion in the ruby [32].

The presence of sphalerite has been previously documented, but it is described as a more transparent phase [46]. It is a common mineral and frequently encountered during low grade metamorphism of sedimentary rocks. The observed sphalerite inclusion most likely formed during metamorphism of the platform carbonates and their associated clay sediments, although it cannot be excluded that it formed from external hydrothermal fluids.

The EPMA analysis of the herringbone structure in sample A4 only totals 86.01%. The most likely explanation is that this patterned zone consists of a different mineral or an intergrowth of different minerals. Our analysis was set up to quantify eight elements that are commonly found in sulfides, but it is possible that this zone incorporates elements that we did not analyze. Although a different mineral is the most likely explanation, alteration due to the fabrication process cannot be excluded until this inclusion is further analyzed.

The paragenesis of Mogok rubies is pretty well understood and it is interesting to note that the only mineral that incorporates Fe, are these sulfides. Thus, the presence of sulfur might be an important factor for the formation of low-Fe, and thus, more fluorescent rubies. Marble-related ruby forming systems have a significant amount of evaporites, including sulfates. During increasing metamorphism, these sulfates will reduce, due to reactions with organic matter [54]. This reaction provides an ample supply of reduced, reactive sulfur to react with all Fe and form Fe-sulfides before corundum forms. The presence of S-rich fluid inclusions and native sulfur found in marble related ruby inclusions supports this model [31,55,56]. The same model could also explain the extreme fluorescence witnessed in red/pink spinel from marble related deposits. Gem spinel shares a very similar formation model with ruby in marbles, apart from the fact that it has a significant Mg component and most likely formed at slightly different P-T conditions [34,57,58]. A similar mechanism where sulfur is responsible for Fe removal during formation of high quality gemstones has been proposed for the Colombian emeralds [59]. It seems that, in both situations, the reaction of sulfates with organic matter provided sufficient reduced sulfur to trap much of the Fe present in the system in sulfide phases [59]. This results in Fe-poor, Cr-rich emeralds and rubies, creating a strong color and fluorescence. While this mechanism is similar between these deposits, the rest of the geological conditions are very different.

Although the inclusions in Mozambican amphibole-related rubies are visually very similar to the ones from Mogok's marble-related rubies, they are chemically very different. This was expected, given that their geological history and host rocks are very different.

The composition of the inclusions in Mozambican rubies strongly resembles a sulfide melt trapped during formation. The texture of these inclusions is suggestive of phases crystallizing from a melt individually, rather than exsolving from one originally homogeneous inclusion [60]. For instance, certain crystallized phases in these inclusions do show very obvious exsolution textures (Figure 4a,b), which are distinctly different from the overall textural relationships observed between the main sulfide phases. Such sulfide melts can form at higher temperatures, precipitating multiple sulfide phases upon cooling [61,62]. The overall consistency of these inclusions between the different samples indicates that they formed as a separate and homogeneous phase, possibly a partial sulfide melt, and were later incorporated in the ruby. The rounded shape of the inclusions is also more suggestive of a molten phase. Conversely, in Mogok rubies, the more regular, almost euhedral, outlines suggest that these inclusions were trapped as solid crystals.

During cooling, the individual components in such a sulfide melt would have separated and crystallized into the different phases (chalcopyrite, pentlandite, pyrrhotite). The creation of sulfide melts through partial melting has been documented at temperatures as low as 300 °C, but this applies only to a limited amount of sulfide minerals, often with considerable amounts of As [62]. The most common Fe-Ni-Cu sulfide system in nature consists of chalcopyrite and monosulfide solid-solution (FeS-NiS). Early experimental work suggested that significant quantities of melt are produced only at around 850 °C [63]. However, some experiments have shown that first melting may initiate at about 760 °C in this system [64,65]. The mineralogy of sulfide crystallization in these references is consistent with that seen in the Mozambican ruby inclusions with chalcopyrite and Fe-Ni sulfides, that have completely separated from what could have been a single homogeneous partial melt at high temperature. The flame-like patterns observed also suggests exsolution of the Fe-Ni sulfide phase during cooling, that formed the patterns seen now. The main potential concern with this hypothesis is that the melting temperatures of this system are significantly higher than the 550–650 °C reported for the ruby host rock in previous studies [41,66]. However, these pressure-temperature calculations should be approached with some caution. These PT estimates are based on models that assume equilibrium between various phases found associated with corundum in the Mozambique deposit. However, the Mozambican ruby host rock is extensively weathered and in most cases completely reduced to clay and sedimentary detritus. It is often hard to say if the minerals that are seen in thin sections, and used in geothermobarometry, are actually related to the ruby formation. For instance, it is possible (if not likely) that the kyanite and Fe-carbonate used to determine a temperature of 650 °C and pressure of 9.5 kbar, are alteration products formed during retrograde metamorphism and/or interaction with fluids rich in Si and carbonates [66]. The same applies to the stability field used to estimate a temperature estimate of 550–600 °C, based on a plagioclase-garnet-diopside-pargasite-clinochlore-quartz-corundum assemblage [41], especially as clinochlore is a common alteration product when amphibole weathers. It would be highly unlikely that these minerals all formed together, especially quartz and corundum, which cannot coexist under ordinary geological conditions. The presence of an inclusion derived from partial sulfide melt rich in Ni-Cu-S, as we have proposed here, would suggest that reported formation temperatures of around 550–600 °C are underestimated.

Given that the data on Mozambican ruby is limited, the data on other amphibole related deposits in East-Africa can provide limited insight. The ruby deposit in Winza, Tanzania has reported P-T conditoons of 750–850 °C and 9.5–12 kbar [67]. Similar conditions are reported for Vohibory, Madagascar, although the temperature range is narrower at 750–800 °C, with pressure of 9–11.5kbar [68]. While these conditions are not directly related to the Mozambican ruby deposit, they can provide an indication for the formation of amphibole-related rubies in East Africa. These temperatures are more in line with the likely temperatures for the formation of a partial sulfide melt. The effect of various parameters, such as water content, oxygen fugacity, sulfur fugacity, the involvement of other minerals and the potential for lowering the initial melting temperature of a free sulfide melt by addition of other sulfide components, are not well understood and could potentially have a significant impact on formation conditions of these potential sulfide melt inclusions [62]. It has to be noted that these conditions are unknown in most ruby forming systems.

In Mogok, the presence of abundant sulfur allows for the formation of low-Fe ruby, because much of the Fe is locked in pyrrhotite. The sulfide inclusions in Mozambican rubies are more complex, as is their associated mineral assemblage. Although it is not well studied, many of the observed minerals contain a significant Fe component, which indicates that the Fe-balance in the amphibole-related ruby forming system has more controlling factors than those in marble-related systems. In the Mozambican rubies, with Fe-Cu-Ni sulfides, it is clear that simply an availability of abundant sulfur is not the controlling factor for the Fe concentration in rubies.

This is reflected in the wider variety of iron contents seen in rubies from Montepuez, which ranges from 414 to 4000 ppmw [2,6]. This variability in Fe concentrations obviously results in much more variable fluorescence in Mozambican rubies compared to Mogok rubies.

6. Conclusions

Opaque inclusions are common in many rubies forming in different geological environments. The identification of these crystals is usually based on their visual characteristics, which are sometimes misleading. Chemical analysis has shown that the inclusions in Mogok rubies are dominantly homogenous pyrrhotite and with one inclusion identified as sphalerite. Pyrite is found in the marble that hosts these rubies, but was not found as an inclusion. The presence of these sulfides and the presence of abundant sulfur in the marble could be an important geological indicator for strong fluorescent rubies. Fluorescence is deemed an important quality factor in ruby valuation and is quenched by the presence of Fe. In marble hosted ruby deposits, there is abundant sulfur present to trap all iron in sulfide minerals, allowing the formation of low-Fe rubies.

The opaque inclusions in Mozambican rubies from Montepuez look very similar to the ones found in Mogok in the microscope, but they are very different compositionally and mineralogically. Analysis of these stones shows a complex mix of Fe-Cu-Ni-sulfides with clearly separated phases, as well as exsolution patterns. Their compositions and textures are reminiscent of melt inclusions forming from a sulfide partial melt. To form such inclusions, the temperatures should be higher than the reported P-T conditions, meaning that the temperatures reported in literature might be too low. More studies are required to better understand the genesis of these inclusions and its ruby host.

Author Contributions: Conceptualization and methodology, W.V. and A.P.; writing—original draft preparation, W.V.; writing—review and editing, W.V. and A.P. All authors have read and agreed to the published version of the manuscript.

Acknowledgments: The authors would like to thank Charuwan Khowpong, Sasithorn Engniwat and Suwasan Wongchacree for sample analysis and inclusion photography. We thank Mike Breeding (GIA) for helpful comments on the manuscript. We also thank two anonymous peer reviewers for their help in improving this manuscript. We would also like to thank Lin Sutherland and Khin Zaw for extending the invitation to submit a manuscript to this Special Issue on ruby geology and geochemistry, as well as their editorial input. Chi Ma, director of the GPS analytical facility at Caltech, Pasadena, CA was of great assistance during the setting up of the EPMA analysis.

References

1. Hughes, R.W.; Manorotkul, W.; Hughes, E.B. *Ruby and Sapphire: A Gemologist's Guide*; Lotus Publishing: Chichester, UK; London, UK, 2017.
2. Vertriest, W.; Saeseaw, S. A decade of Ruby from Mozambique: A review. *Gems Gemol.* **2019**, *55*. [CrossRef]
3. Pardieu, V. Update on Mozambique ruby mining and trading. *Gems Gemol.* **2017**, *53*, 3.
4. Vertriest, W.; Pardieu, V. Update on gemstone mining in Northern Mozambique. *Gems Gemol.* **2016**, *52*, 4. [CrossRef]
5. Hughes, R.W. Red Rain: Mozambique Ruby Pours into the Market. Available online: http://www.lotusgemology.com/index.php/library/articles/316-red-rain-mozambique-ruby-pours-into-the-market (accessed on 10 February 2020).
6. Hsu, T.; Lucas, A.; Pardieu, V. *Mozambique: A Ruby Discovery for the 21st Century*; GIA: Carlsbad, CA, USA, 2014.
7. Pardieu, V.; Sangsawong, S.; Muyal, J.; Chauviré, B.; Massi, L.; Sturman, N. *Rubies from the Montepuez Area (Mozambique)*; GIA: Carlsbad, CA, USA, 2013.
8. Pardieu, V.; Jacquat, S.; Bryl, L.P.; Senoble, J.B. Rubies from northern Mozambique. *InColor* **2009**, *12*, 32–36.
9. Vincent, P.; Rakotosona, N. Ruby and Sapphire Rush Near Didy, Madagascar. Available online: http://www.giathai.net/pdf/Didy_Madagascar_US.pdf (accessed on 22 February 2020).
10. Rakontondrazafy, A.F.M.; Giuliani, G.; Ohnenstetter, D.; Fallick, A.E.; Rakotosamizanay, S.; Andriamamonjy, A.; Ralantoarison, T.; Razanatseheno, M.; Offant, Y.; Garnier, V.; et al. Gem corundum deposits of Madagascar: A review. *Ore Geol. Rev.* **2008**, *34*, 135–154.
11. Pardieu, V.; Wise, R.W. Ruby boom town. *Colored Stone* **2006**, *19*, 30–33.
12. Hughes, R.W.; Pardieu, V.; Schorr, D. Sorcerers & Sapphires: A Vsit to Madagascar. Available online: https://lotusgemology.com/index.php/library/articles/161-sorcerers-sapphires-a-visit-to-madagascar (accessed on 22 February 2020).
13. Schwarz, D.; Schmetzer, K. Rubies from the Vatomandry area, eastern Madagascar. *J. Gemmol.* **2001**, *27*, 409–416. [CrossRef]
14. Abduriyim, A.; Kitawaki, H. New geological origin: Ruby from Winza of Tanzania. *Gemmology* **2008**, *8*, 4–7.
15. Keller, P.C. The rubies of Burma: A review of the Mogok Stone Tract. *Gems Gemol.* **1983**, *19*, 209–219. [CrossRef]
16. Zaw, K.; Sutherland, L.; Yui, T.-F.; Meffre, S.; Thu, K. Vanadium-rich ruby and sapphire within Mogok Gemfield, Myanmar: Implications for gem color and genesis. *Miner. Depos.* **2015**, *50*, 25–39. [CrossRef]
17. Saeseaw, S.; Kongsomart, B.; Atikarnsakul, U.; Khowpong, C.; Vertriest, W.; Soonthorntantikul, W. *Update on "Low-Temperature" Heat Treatment of Mozambican Ruby: A Focus on Inclusions and FTIR Spectroscopy*; GIA: Carlsbad, CA, USA, 2018.
18. Pardieu, V.; Saeseaw, S.; Detroyat, S.; Raynaud, V.; Sangsawong, S.; Bhurisom, T.; Engniwat, S.; Muyal, J. *"Low Temperature" Heat Treatment of Mozambique Ruby*; GIA: Carlsbad, CA, USA, 2015.
19. Palke, A.C.; Saeseaw, S.; Renfro, N.D.; Sun, Z.; McClure, S.F. Geographic origin determination of ruby. *Gems Gemol.* **2019**, *55*, 580–613. [CrossRef]

20. Kiefert, L.; Karampelas, S. Use of the Raman spectrometer in gemmological laboratories: Review. *Spectrochim. Acta A* **2011**, *80*, 119–124. [CrossRef] [PubMed]
21. Gübelin, E.J.; Koivula, J.I. *Photoatlas of Inclusions in Gemstones*; ABC Edition: Zürich, Switzerland, 1986; p. 532.
22. Vertriest, W.; Palke, A.C.; Renfro, N.D. Field gemology: Building a research collection and understanding the development of gem deposits. *Gems Gemol.* **2019**, *55*, 580. [CrossRef]
23. Armstrong, J.T. CITZAF-A package of correction programs for the quantitative electron microbeam X-ray-analysis of thick polished materials, thin-films and particles. *Microbeam Anal.* **1995**, *4*, 177–200.
24. Garnier, V.; Giuliani, G.; Ohnenstetter, D.; Schwarz, D. Les gisements de corindon: Classification et genèse. *Règne Miner.* **2004**, *55*, 7–47.
25. Iyer, L.A.N. The geology and gem-stones of the Mogok stone tract, Burma. *Geol. Surv. India-Mem.* **1953**, *82*, 8–100.
26. Themelis, T. *Gems and Mines of Mogok*; Self-published: Bangkok, Thailand, 2008.
27. Chhibber, H.L. *The Geology of Burma*; MacMillan and Co.: London, UK, 1934; p. 538.
28. Thu, Y.K.; Win, M.M.; Enami, M.; Tsuboi, M. Ti–rich biotite in spinel and quartz–bearing paragneiss and related rocks from the Mogok metamorphic belt, central Myanmar. *J. Miner. Petrol. Sci.* **2016**, *111*, 270–282.
29. Thu, K. *The Igneous Rocks of the Mogok Stone Tract: Their Distributions, Petrography, Petrochemistry, Sequence, Geochronology and Economic Geology*; University of Yangon: Yangon, Myanmar, 2007.
30. Searle, M.; Garber, J.; Hacker, B.; Htun, K.; Gardiner, N.; Waters, D.; Robb, L. Timing of syenite-charnockite magmatism and ruby and sapphire metamorphism in the mogok valley region, Myanmar. *Tectonics* **2020**, *39*, 39. [CrossRef]
31. Giuliani, G.; Dubessy, J.; Ohnenstetter, D.; Banks, D.; Branquet, Y.; Feneyrol, J.; Fallick, A.E.; Martelat, J.-E. The role of evaporites in the formation of gems during metamorphism of carbonate platforms: A review. *Miner. Depos.* **2017**, *53*, 1–20. [CrossRef]
32. Garnier, V.; Giuliani, G.; Ohnenstetter, D.; Fallick, A.E.; Dubessy, J.; Banks, D.; Hoang, V.Q.; Lhomme, T.; Maluski, H.; Pecher, A.; et al. Marble-hosted ruby deposits from Central and Southeast Asia: Towards a new genetic model. *Ore Geol. Rev.* **2008**, *34*, 169–191. [CrossRef]
33. Kievlenko, E.Y. *Geology of Gems*; Ocean Pictures Ltd.: Littleton, CO, USA, 2003.
34. Giuliani, G.; Ohnenstetter, D.; Fallick, A.; Groat, L.; Fagan, A.J. Chapter 2: The geology and genesis of gem corundum deposits. In *Geology of Gem Deposits*, 2nd ed.; Groat, L., Ed.; Mineralogical Association of Canada: Tucson, AZ, USA, 2014.
35. Garnier, V.; Maluski, H.; Giuliani, G.; Ohnenstetter, D.; Schwarz, D. Ar-Ar and U-Pb ages of marble-hosted ruby deposits from central and southeast Asia. *Can. J. Earth Sci.* **2006**, *43*, 509–532. [CrossRef]
36. Bingen, B.; Jacobs, J.; Viola, G.; Henderson, I.H.C.; Skar, O.; Boyd, R.; Thomas, R.J.; Solli, A.; Key, R.M.; Daudi, E.X.F. Geochronology of the Precambrian crust in the Mozambique Belt in NE Mozambique, and implications for Gondwana assembly. *Precambrian Res.* **2009**, *170*, 231–255. [CrossRef]
37. Meert, J.G. A synopsis of events related to the assembly of eastern Gondwana. *Tectonophysics* **2003**, *362*, 1–40. [CrossRef]
38. Muhongo, S. Anatomy of the Mozambique Belt of eastern and southern Africa: Evidence from Tanzania. *Gondwana Res.* **1999**, *2*, 369–375. [CrossRef]
39. Pardieu, V.; Thanachakaphad, J.; Jacquat, S.; Senoble, J.B.; Bryl, L.P. *Rubies from the Niassa and Cabo Delgado Regions of Northern Mozambique*; GIA Laboratory: Carlsbad, CA, USA, 2009; pp. 5–16.
40. Boyd, R.; Nordgulen, O.; Thomas, B.; Bingen, B.; Bjerkgård, T.; Grenne, T.; Henderson, I.; Melezhik, V.; Sandstad, J.; Solli, A.; et al. The geology and geochemistry of the East African orogen in northeastern Mozambique. *S. Afr. J. Geol.* **2010**, *113*, 87–129. [CrossRef]
41. Fanka, A.; Sutthirat, C. Petrochemistry, mineral chemistry, and pressure–temperature model of corundum-bearing amphibolite from Montepuez, Mozambique. *Arab. J. Sci. Eng.* **2018**, *43*, 3751–3767. [CrossRef]
42. Lebas, M.J.; Lemaitre, R.W.; Streckeisen, A.; Zanettin, B. A chemical classification of volcanic-rocks based on the total alkali silica diagram. *J. Petrol.* **1986**, *27*, 745–750.
43. Gübelin, E.J. On the nature of mineral inclusions in gemstones—Parts 1 and 2. *Gems Gemol.* **1969**, *13*, 42–56.
44. Gübelin, E.J. *Inclusions as a Means of Gemstone Identification*; GIA: Los Angeles, CA, USA, 1953; p. 220.
45. Gübelin, E.J. On the nature of mineral inclusions in gemstones. *J. Gemmol.* **1969**, *11*, 149–192. [CrossRef]
46. Gübelin, E.J. *Internal World of Gemstones*; reprinted 1983 ed.; ABC Verlag: Zürich, Switzerland, 1973; p. 234.
47. Fritsch, E.; Rossman, G.R. New technologies of the 1980s: Their impact on gemology. *Gems Gemol.* **1990**, *26*, 64–75. [CrossRef]

48. Kammerling, R.C.; Scarratt, K.; Bosshart, G.; Jobbins, E.A.; Kane, R.E.; Gubelin, E.J.; Levinson, A.A. Myanmar and its gems—An update. *J. Gemmol.* **1994**, *24*, 3–40. [CrossRef]
49. Renfro, N.; Koivula, J. Vesuvianite in burmese ruby. *Gems Gemol.* **2017**, *51*, 469.
50. Leelawathanasuk, T.; Pisutha-Arnond, V.; Atichat, W.; Sutthirat, C.; Wathanakul, P.; Sriprasert, B. Some characteristics of "Mozambique ruby". In *31st International Gemmological Congress—Abstracts*; IGC: Arusha, Tanzania, 2009; pp. 33–34.
51. Searle, M.P.; Noble, S.R.; Cottle, J.M.; Waters, D.J.; Mitchell, A.H.G.; Hlaing, T.; Horstwood, M.S.A. Tectonic evolution of the Mogok metamorphic belt, Burma (Myanmar) constrained by U-Th-Pb dating of metamorphic and magmatic rocks. *Tectonics* **2007**, *26*, 26. [CrossRef]
52. Yonemura, K.; Osanai, Y.; Nakano, N.; Adachi, T.; Charusiri, P.; Zaw, T. EPMA U-Th-Pb monazite dating of metamorphic rocks from the Mogok Metamorphic Belt, central Myanmar. *J. Mineral. Petrol. Sci.* **2013**, *108*, 184–188. [CrossRef]
53. Myat Phyo, M.; Franz, L.; Capitani, C.; Balmer, W.; Krzemnicki, M.; Christian, C. Petrology and PT-conditions of quartz- and nepheline-bearing gneisses from Mogok Stone Tract, Myanmar. In Proceedings of the 15th Swiss Geoscience Meeting, Davos, Switzerland, 17–18 November 2017. [CrossRef]
54. Goldstein, T.; Aizenshtat, Z. Thermochemical sulfate reduction a review. *J. Therm. Anal. Calorim.* **1994**, *42*, 241–290. [CrossRef]
55. Giuliani, G.; Dubessy, J.; Banks, D.; Hoang, V.Q.; Lhomme, T.; Pironon, J.; Garnier, V.; Phan, T.T.; Pham, L.V.; Ohnenstetter, D.; et al. CO_2-H_2S-S_8-AlO(OH)-bearing fluid inclusions in ruby from marble-hosted deposits in Luc Yen area, North Vietnam. *Chem. Geol.* **2003**, *194*, 167–185. [CrossRef]
56. Hoang, V.Q.; Giuliani, G.; Phan, T.T.; Pham, L.V. Fluid inclusion on ruby from the Yen Bai Province. In *Geo- and Material-Science on Gem Minerals of Vietnam*; GIA: Carlsbad, CA, USA, 2003; pp. 136–144.
57. Phyo, M.M.; Bieler, E.; Franz, L.; Balmer, W.; Krzemnicki, M.S. Spinel from mogok, myanmar—A detailed inclusion study by raman microspectroscopyand scanning electron microscopy. *J. Gemmol.* **2019**, *36*, 418–435. [CrossRef]
58. Malsy, A.; Klemm, L. Distinction of gem spinels from the Himalayan mountain belt. *Chimia* **2010**, *64*, 741–746. [CrossRef]
59. Ottaway, T.; Wicks, F.; Bryndzia, L.; Spooner, E. Formation of the Muzo hydrothermal emerald deposit in Colombia. *Nature* **1994**, *369*, 552–554. [CrossRef]
60. Bard, J.P. *Microtextures of Igneous and Metamorphic Rocks*; Springer Netherlands: Dordrecht, The Netherlands, 1986.
61. Tomkins, A.G.; Pattison, D.R.M.; Frost, B.R. On the initiation of metamorphic sulfide anatexis. *J. Petrol.* **2006**, *48*, 511–535. [CrossRef]
62. Frost, B.; Mavrogenes, J.; Tomkins, A. Partial melting of sulfide ore deposits during medium- and high-grade metamorphism. *Can. Mineral.* **2002**, *40*, 1–18. [CrossRef]
63. Craig, J.R.; Kullerud, G. The Cu-Fe-Ni-S system. *Carnegie Inst. Wash. Yearb.* **1968**, *66*, 413–417.
64. Peregoedova, A.; Ohnenstetter, M. Collectors of Pt, Pd and Rh in a S-poor Fe-Ni-Cu sulfide system at 760 °C: Experimental data and application to ore deposits. *Can. Mineral.* **2002**, *40*, 527–561. [CrossRef]
65. Sugaki, A.; Kitakaze, A. High form of pentlandite and its thermal stability. *Am. Mineral.* **1998**, *83*, 133–140. [CrossRef]
66. Institute, S.S.G. *New Rubies from Montepuez, Mozambique*; GIA: Carlsbad, CA, USA, 2010; p. 17.
67. Schwarz, D.; Pardieu, V.; Saul, J.M.; Schmetzer, K.; Laurs, B.M.; Giuliani, G.; Klemm, L.; Malsy, A.K.; Erel, E.; Hauzenberger, C.; et al. Rubies and sapphires from Winza, central Tanzania. *Gems Gemol.* **2008**, *44*, 322–347. [CrossRef]
68. Nicollet, C. Saphirine et staurotide riche en magnésium et chrome dans les amphibolites et anorthosites à corindon du Vohibory Sud, Madagascar. *Bull. Mineral.* **1986**, *109*, 599–612. [CrossRef]

7

U–Pb Dating of Zircon and Zirconolite Inclusions in Marble-Hosted Gem-Quality Ruby and Spinel from Mogok, Myanmar

Myint Myat Phyo [1,2], Hao A.O. Wang [2,*], Marcel Guillong [3], Alfons Berger [4], Leander Franz [1,*], Walter A. Balmer [2] and Michael S. Krzemnicki [2]

1 Department of Environmental Sciences, Mineralogy and Petrology, University of Basel, Bernoullistrasse 30, 4056 Basel, Switzerland; myint.myatphyo@ssef.ch

2 Swiss Gemmological Institute SSEF, Aeschengraben 26, 4051 Basel, Switzerland; info@collectionbalmer.com (W.A.B.); director@ssef.ch (M.S.K.)

3 Institute of Geochemistry and Petrology, ETH Zurich, Clausiusstrasse 25, 8092 Zürich, Switzerland; marcel.guillong@erdw.ethz.ch

4 Institute of Geological Sciences, University of Bern, Baltzerstrasse 1+3, 3012 Bern, Switzerland; alfons.berger@geo.unibe.ch

* Correspondence: hao.wang@ssef.ch (H.A.O.W.); leander.franz@unibas.ch (L.F.)

Abstract: The Mogok area in Myanmar (Burma) is known since historic times as a source for some of the finest rubies and spinels in the world. In this study, we focus on in-situ U–Pb geochronological analyses of zircon and zirconolite, either present as inclusions in gem-quality ruby and spinel or as accessory minerals in ruby- and spinel-bearing marble and adjacent granulite facies gneisses. The age determination was carried out using both laser ablation inductively coupled plasma time-of-flight mass spectrometry (LA-ICP-TOF-MS) and sector-field mass spectrometry (LA-ICP-SF-MS). In addition, we present multi-element data (REE) of zircon and zirconolite collected with LA-ICP-TOF-MS to further characterize these inclusions. Most of the studied zircon grains display growth zoning (core/rim) regardless if as inclusion in gemstones, or as accessory mineral in host rock samples. U–Pb dating was conducted on both core and rim of zircon grains and revealed most ages ranging from ~200 Ma in the core to ~17 Ma in the rim. The youngest U–Pb ages determined from the rim of zircon inclusions in gem-quality ruby and spinel are 22.26 ± 0.36 Ma and 22.88 ± 0.72 Ma, respectively. This agreement in U–Pb ages is interpreted to indicate a simultaneous formation of ruby and spinel in the Mogok area. In ruby- and spinel-bearing marble from Bawlongyi, the youngest zircon age was determined as 17.11 ± 0.22 Ma. Furthermore, U–Pb age measured on the rim of zircon grains in a biotite-garnet gneiss reveals a Late Oligocene age (26.13 ± 1.24 Ma), however older ages up to Precambrian age were also recorded in the cores of zircon as accessory minerals from this gneiss. These old ages point to a detrital origin of the analysed zircon cores. Although non-matrix matched standard was applied, zirconolite U–Pb age results are narrower in distribution from ~35 Ma to ~17 Ma, falling within the range of zircon ages. Based on results which are well in accordance with previous geochronological data from the Mogok Metamorphic Belt (MMB), we deduce that gem-quality ruby and spinel from Mogok probably formed during a granulite-facies regional metamorphic event in Oligocene to Early Miocene, related to post collision tectonics of the Eurasian and Indian plates. Our data not only provide key information to understand the formation of gem-quality ruby and spinel in the so-called Mogok Stone Tract, but also provide assisting evidence when determining the country of origin of gemstones in gemmological laboratories.

Keywords: ruby; spinel; Mogok; geochronology; U–Pb dating; zircon; zirconolite

1. Introduction

Ruby, the chromium-bearing red variety of corundum (Al_2O_3) and chromium-bearing red spinel ($MgAl_2O_4$) are among the world's most valued gemstones. In the gem trade, the price of a ruby and spinel is not only defined by their size and quality, but also to a large part by their geographic (country of) origin, notably when they are from a historically important and famed gem deposit. Ruby and spinel of gem-quality from marble-hosted deposits are known from a number of localities, such as the Mogok area (Figure 1) and Mong Hsu in Myanmar, the Hunza Valley in Pakistan, Jegdalek in Afghanistan, Murgab in Tajikistan, Luc Yen in Vietnam and the Morogoro region in Tanzania. From these deposits, the Mogok area is considered by far the most important and historically reputed source for ruby and spinel in the trade. This explains, why in literature corundum (ruby and sapphire) and to a lesser extent spinel from Mogok have been extensively studied, either gemmologically and in general [1–4], or focusing specifically on trace elements [5,6], stable isotopes [7–10], solid inclusions [11] and fluid inclusions [12]. In geochronology, a number of studies were published about the Mogok area or further gem deposits worldwide, either investigating minerals related to gemstone formation or inclusions in rough and faceted gemstones [13–21].

Figure 1. Faceted gem-quality ruby (0.369 ct) and rough ruby crystal within marble from Mogok area, Myanmar. Photo: M.M.P.

In recent years, U–Pb dating of inclusions (e.g., zircon, titanite, etc.) in gemstones using laser ablation inductively coupled plasma mass spectrometry (LA-ICP-MS) has become more popular due to easier access to instrumentation, minimum sample preparation, smaller laser crater size and improved instrument sensitivity [22,23]. The U–Pb ages of inclusions found in a gemstone indicate the maximum possible formation age of that gemstone. As such, these radiometric ages may provide gemmological laboratories with additional evidence when carrying out country of origin determination as a commercial service to the gem trade. This is especially helpful when separating ruby, sapphire and spinel related to the Himalayan orogenesis (e.g., Mogok area in Myanmar) from those related to the much older Pan-African orogeny (e.g., Morogoro area in Tanzania) [17,19,24–26].

For the Mogok area, only a limited amount of geochronological data from inclusions in gemstones and from accessory minerals in host rocks have been published so far: zircon inclusion in ruby ~32 Ma [16], titanite in ruby ~32 Ma [14], zircon in sapphire ~27 Ma [24], zircon in granitic host rocks ~16.8 Ma [26]. More geochronological data from this gem-rich area is thus still essential in order to get a better understanding of its complex geology and to be able to compile a genetic model for the formation of gemstones in the Mogok area.

In this paper, we present new U–Pb ages of zircon and zirconolite, found as inclusions in gem-quality ruby and spinel or as accessory minerals in host rock samples from the Mogok area. In addition, we present rare earth element data (REE) of selected zircon and zirconolite grains for further characterization. Our data contributes to a better understanding of the complex geological processes resulting in the formation of gem-quality ruby and spinel in the studied area. Furthermore, our data may assist gemmological laboratories specifically in country of origin determination.

2. Regional Geological Setting

The Mogok Metamorphic Belt (MMB), further described in early studies [27,28], is a large sickle-shaped belt of regionally metamorphosed rocks stretching N-S from central to upper Myanmar as the southern continuation of the Himalayan syntaxis [29]. It forms the western margin of the Shan-Thai Block (Sibumasu Block) and is adjacent to the north-south trending right-lateral strike-slip Sagaing Fault, related to the youngest Cenozoic metamorphic events [30]. The Mogok-Mandalay-Mergui Belt extends to the Mergui area in the southern part of Myanmar [31,32]. In the so-called Mogok Stone Tract [27], located at the northeastern end of the MMB, there is an accumulation of world-renowned gem deposits, namely of rubies, sapphires and spinels [3,28]. The formation of those gemstones is related to late-stage Cenozoic metamorphic events within this geological unit [30,33,34]. The Mogok area mainly consists of upper amphibolite to granulite facies marbles, schists and gneisses [35]. Amongst these, the Mogok gneisses [27] in the southern part of the Mogok area are considered to be the oldest rock units in this area of Myanmar [36]. Gem-quality ruby and spinel are found in the Mogok area in both primary deposits in marbles and secondary deposits in alluvial and eluvial placers, as well as accumulated placers in karstic sinkholes and caverns [37].

The MMB has been extensively studied with a main focus on regional geology, geochemistry and geochronology [30,33,38–40]. In terms of geochronology, the ruby- and spinel-bearing marbles from the Mogok area are stratigraphically interpreted in previous studies [27,28] as being metamorphosed limestones of Precambrian age. Searle and Haq considered the Mogok metasediments to be originally Precambrian to Jurassic rocks of the Shan Plateau [41], whereas Clegg described them as mid-Cretaceous limestones and Jurassic sediments [42]. From the Precambrian to Cenozoic, the MMB experienced several magmatic and metamorphic events, which are commonly summarized as: Precambrian to Palaeozoic regional metamorphism and magmatism indicated by Ordovician age on zircon from foliated leucogranite [2] and Late Cambrian age on biotite gneiss [21] as well as Jurassic magmatism on biotite augen gneiss [40]. The latest, high-grade regional metamorphic event in the MMB occurred in Paleogene to Early Neogene due to the collision of the Sibumasu Block with the West Burma Block subsequent to the collision of the Indian and the Asian plates [43,44] and was followed by an intense episode of post-tectonic magmatism [2,26,45].

3. Materials and Methods

3.1. Zircon and Zirconolite

Prior to analyses, all gemstone samples (ruby and spinel) were polished to expose zircon and zirconolite inclusions. For host rocks, samples were first embedded in epoxy, before being polished.

The zircon and zirconolite inclusions were identified using a standard gemmological microscope (StereoZoom7, Cambridge Instruments, Cambridge, UK) and a polarization microscope (Leica DMLS, Wetzlar, Germany), two Raman micro-spectrometers (Renishaw inVia equipped with an argon-ion laser at 514.5 nm; Swiss Gemmological Institute SSEF, Basel, Switzerland; Bruker Senterra with a Nd-YAG laser at 532 nm; Department of Environmental Sciences, Mineralogy and Petrology, University of Basel, Basel, Switzerland), and a scanning electron microscope (FEI Nova Nano SEM 230, Nano Imaging Lab, University of Basel, Basel, Switzerland) equipped with an energy-dispersive spectrometer (EDS) and a back-scattered electron detector (BSE). In order to avoid any contamination, the samples were not coated. More details about the analytical conditions applied for Raman and SEM analyses are described in our previous publication on spinel inclusions from Mogok [11].

Zircon ($ZrSiO_4$) is found as inclusion in gem-quality ruby and spinel and as accessory mineral in three host rock samples (ruby- and spinel-bearing marble from Bawlongyi area, granulite-facies gneiss from Aunglan Taung area in central Mogok and granulite-facies gneiss from Kinn area in western Mogok, see Table 1).

Table 1. List of samples from Mogok investigated for this study. The scale bars represent 5 mm length for gemstones and 5 cm for host rocks.

Sample Name	Locality	Host Material	Analysed Grain	Weight (ct)	Sample Photo
SSEF96663_A	Mansin		4 Zircons and 5 Zirconolites	2.411	
SSEF96666_A	Kadotetat	Ruby	1 Zircon	2.431	
SSEF96666_B			1 Zircon	1.270	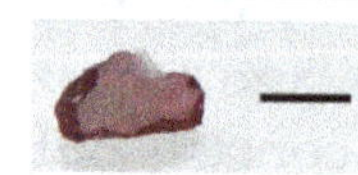
SSEF96667_A	Kyatpyin		1 Zircon	1.615	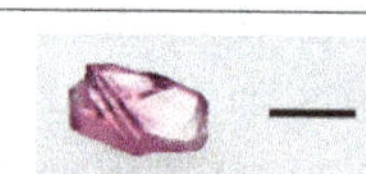
SSEF96668_A	Kyaukpoke		1 Zircon	0.268	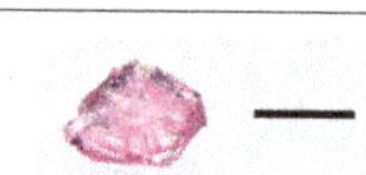
SSEF92720_D	Kyauksaung		2 Zircons	0.405	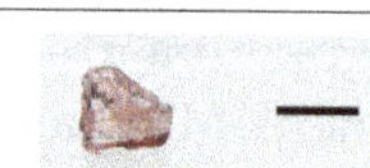
SSEF92725_D	Kyauksin	Spinel	2 Zircons	0.673	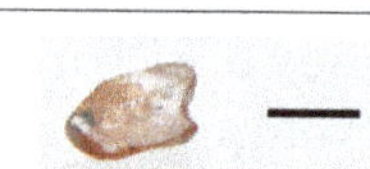
SSEF92725_E			1 Zircon	0.940	
BLG_12	Bawlongyi	Ruby- and Spinel-bearing Marble	32 Zircons and 9 Zirconolites		
ALT_03	Aunglan Taung	Gneiss	40 Zircons		
K_01	Kinn		24 Zircons		

Only 5 from a total of 85 ruby samples as well as 3 from 100 spinels contained zircon inclusions suitable for dating in terms of size and proximity to the sample's surface. Most of the observed zircon inclusions in the examined gemstone samples however were either too small (<10 µm) or located too deep inside the sample to be reached by laser ablation and therefore could not be considered for U–Pb dating in this study. In total, 8 zircon inclusions from ruby and 5 zircon inclusions from spinel were analysed for this study. The size of the analysed zircon grains in ruby range from 50 to 300 µm in diameter. In contrast to this, the zircon grains in spinel are smaller with a maximum diameter of 50 µm, with some inclusions being as small as 10 µm. In addition to the zircon inclusions from gemstones, we have studied zircon present as accessory mineral in host-rocks. In the ruby- and spinel-bearing marble from Bawlongyi area (Figure 2), we analysed a total of 32 zircons, occurring either as individual grains or intergrown with zirconolite. Furthermore, we analysed 40 zircons found as accessory mineral in a garnet-orthopyroxene gneiss from Aunglan Taung and 24 zircon grains in biotite-garnet gneiss from the Kinn area (Figure 2). The size of the accessory zircons in the Aunglan Taung gneiss is 35 µm to 50 µm, while those from Kinn gneiss are even smaller with a size of 10 µm to maximum 35 µm [46].

Zirconolite, ideally $CaZrTi_2O_7$, is known in several polytypoids, such as zirconolite-3*O* (orthorhombic), zirconolite-3*T* (trigonal), zirconolite-2*M* (monoclinic) and non-crystalline (metamict) or undetermined polytypoid [47]. This latter option explains why we were not able to identify zirconolite in our study using Raman spectroscopy. In fact, they all revealed featureless (amorphous) Raman spectra. However, by analysing the major and minor elemental composition using SEM equipped with an EDS detector, we were able to positively identify the zirconolite grains present in our samples. We observed two types of zirconolite: (1) individual inclusion of 40 µm to 100 µm size in ruby from Mansin, and (2) intergrown with zircon in Bawlongyi marble (see Table 1). Here, zirconolite forms either a 10–50 µm thick rim around zircon or occurs as small (maximum 5 µm) amoeboid inclusions within zircon.

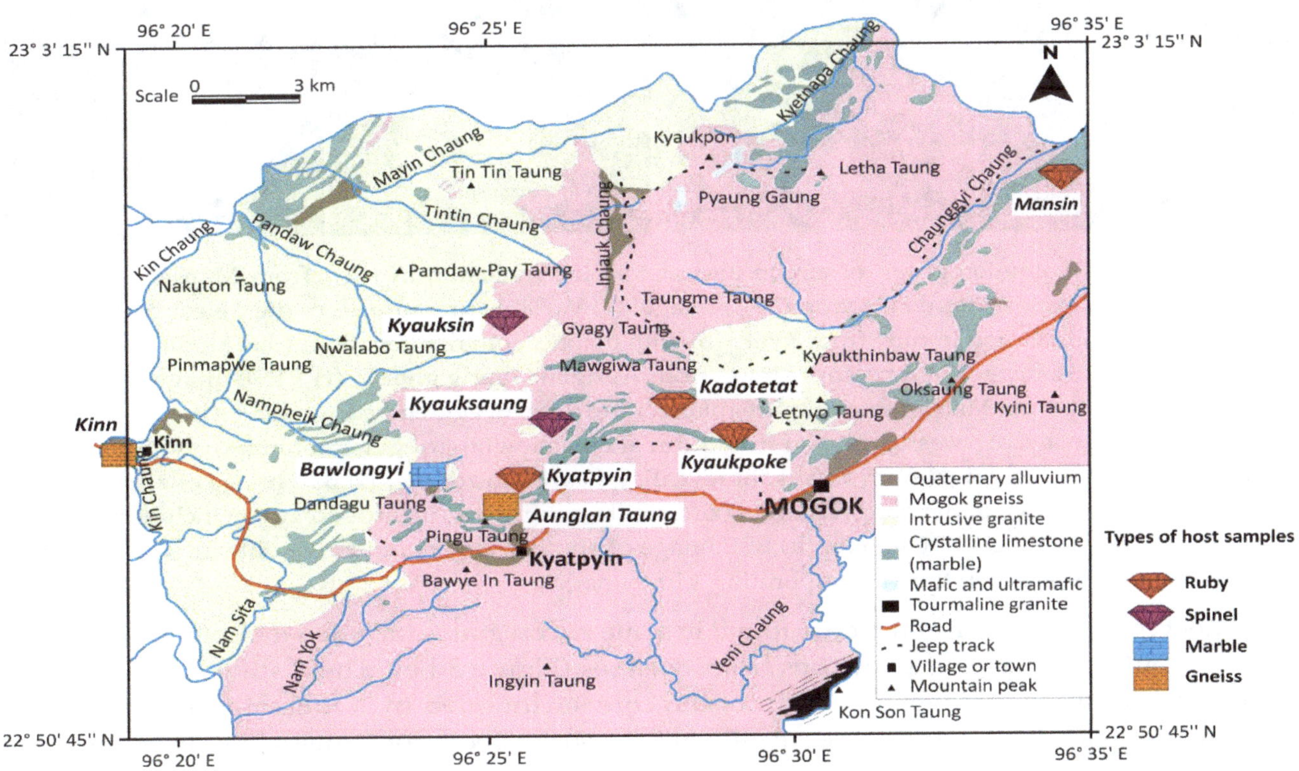

Figure 2. Sample locations plotted on the geological map of the Mogok area (modified after [28] and [48] with permission).

To our knowledge, this study provides the first detailed characterization of zirconolite from the Mogok area. A detailed list of the zircon and zirconolite analysed and their occurrences are given in Table 1 and Supplementary Data.

3.2. Growth Zonation of Zircon and Zirconolite

Zircon often shows complex growth zonation patterns. This may strongly influence radiometric dating if an analysis is conducted across growth zones of different ages [49]. In order to visualize such zoning in our zircon (and zirconolite) grains (Figure 3a,b) and to determine the appropriate position to carry out U–Pb dating, we additionally collected variable pressure secondary electron (VPSE) images using a ZEISS EVO 50 SEM (Institute of Geological Sciences, University of Bern, Bern, Switzerland). These images reveal three different types of patterns: (a) zircon with a characteristic core/rim growth zonation, (b) zircon showing an irregular zoning pattern and (c) zircon with no zonation detectable by VPSE imaging.

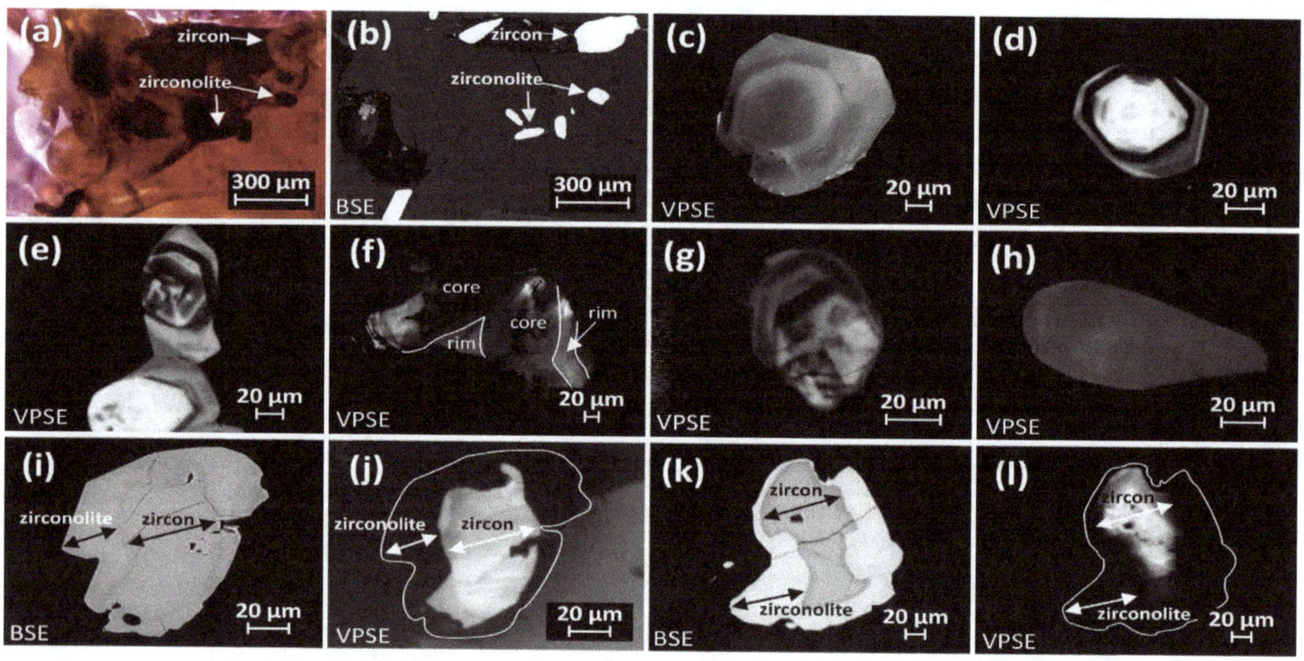

Figure 3. (**a**) Photomicrograph of zircon and zirconolite inclusion in ruby from Mansin, and (**b**) corresponding back-scattered electron (BSE) image. (**c**) Variable pressure secondary electron (VPSE) image of zircon inclusion in ruby from Kyatpyin with distinct core/rim zonation. (**d**,**e**) Similar core/rim zonation in VPSE images of accessory zircon from Aunglan Taung gneiss. (**f**) VPSE image of zircon inclusion in ruby from Mansin showing irregular zoning with an aggregated core and a homogeneous rim. (**g**) VPSE image of accessory zircon from Kinn gneiss showing irregular zoning. (**h**) VPSE image of zircon inclusion in spinel from Kyauksin revealing nearly no zoning. (**i**) BSE image of zircon and zirconolite intergrown in Bawlongyi marble (zirconolite occurs as a rim around zircon inclusion), and (**j**) their corresponding VPSE image. (**k**) BSE image of another zircon and zirconolite intergrowth in Bawlongyi marble and (**l**) their corresponding VPSE image.

Distinct core/rim growth zonation is found in some zircon grains included in rubies from Kyatpyin (Figure 3c), but also in accessory zircons from gneisses (Figure 3d,e). The distinct core/rim growth zonation may indicate a magmatic origin of the core surrounded by a metamorphic rim. Such a zonation pattern is often observed, specifically in accessory zircon from gneiss [50]. Irregular zoning is found in zircon inclusions in ruby from the Mansin area (Figure 3f) revealing aggregated textures in the core and more homogeneous rim sections, but also in zircons present as accessory minerals in the studied gneisses (Figure 3g). BSE and VSPE images showing no (significant) growth zoning patterns are observed for zircon inclusions in spinel (Figure 3h), but also for the accessory zircons in

Bawlongyi marble (Figure 3i–l). This is also true for all investigated zirconolite grains. They show a homogeneous VPSE reaction and no zoning features, regardless if present as individual inclusion in ruby from Mansin or intergrown with zircon in Bawlongyi marble (Figure 3j,l).

3.3. LA-ICP-MS Instrumentations

For in-situ radiometric U–Pb dating and the multi-element characterization (specifically REE) of the zircon and zirconolite grains, a combined approach of two different mass spectrometer setups was applied to benefit from the advantages of each of these systems: (1) LA-ICP time-of-flight MS (LA-ICP-TOF-MS, Swiss Gemmological Institute SSEF, Basel, Switzerland) and (2) sector-field MS (LA-ICP-SF-MS, Institute of Geochemistry and Petrology, ETH Zurich, Zurich, Switzerland). Single detector SF-MS has about 15 times higher sensitivity compared to TOF-MS [51], hence a distinctly smaller laser spot can be used for analysis, minimizing potential overlaps of age zones, especially when analysing zoned zircon. Since the SF-MS data acquisition is sequential, only a limited number of masses can be collected. In contrast to this, TOF-MS allows simultaneous acquisition of the full mass spectrum, hence it is ideal for constant monitoring of elemental variations (e.g., when passing from one mineral to another), especially when the laser ablates into small mineral intergrowth. The drawback of this second method is its relatively lower sensitivity, thus requiring a larger laser ablation spot (e.g., 35 μm). Obviously, this last condition (laser spot size) of LA-ICP-TOF-MS is challenging, especially when analysing small zircon and zirconolite with narrow and/or complex growth zoning. Therefore, we applied LA-ICP-SF-MS for small and complex zoned zircon and for zirconolite, using a small spot size of only 13 to 19 μm and integration times of 10 ms (^{202}Hg, ^{208}Pb, ^{232}Th, ^{235}U, ^{238}U), 20 ms (^{204}Pb), 75 ms (^{207}Pb) and 90 ms (^{206}Pb), resulting in a total sweep time of 0.243 s (including MS settling time) for these eight selected isotopes of interest. In doing so, we were able to determine the different ages of the core and the rim of zoned zircon separately. The LA-ICP-TOF-MS system was then used for a detailed elemental and isotopic characterization of selected zircon and zirconolite grains, allowing the simultaneous acquisition of the full mass spectrum ($^{7}Li^{+}$ to $^{238}U^{+}$) every 30 μs. For our analyses, 5000 full TOF-mass spectra were averaged and reported in one output data point (integration time 0.151 s). Data reduction of the TOF spectrum includes gas black correction, baseline correction based on two anchor points on both sides of the mass peak, as well as signal integration over a mass window (0.25 atomic mass unit), centred at the mass peak, in order to obtain the peak intensity (unit in mV) [52]. Further TOF- and SF-MS instrument parameters are listed in Supplementary Table S1.

For LA-ICP-SF-MS, we used the zircon standard GJ-1 (601.86 ± 0.37 Ma, from $^{206}Pb/^{238}U$ age results) [53,54] as a primary dating standard, with additional zircon secondary standards Plešovice (337.13 ± 0.37 Ma, $^{206}Pb/^{238}U$ age) [55], Temora 1 (416.8 ± 1.1 Ma, $^{206}Pb/^{238}U$ age) [56], Zircon 91500 (1063.51 ± 0.39 Ma, $^{206}Pb/^{238}U$) [57]. The secondary reference zircon ages from measurements in this study are in accordance with published results. Data processing was carried out using software packages Iolite (version 2.5) [58] and VizualAge (version 2013-02) [59] for age calculation and uncertainty propagation. Laser induced element fractionation (LIEF) was corrected based on the primary reference material analysed with identical conditions as the sample. Unless otherwise mentioned, calculated ages are quoted as two times the standard error (2SE) and propagation of uncertainty is by quadratic addition. Reproducibility and age uncertainty of reference material are propagated where appropriate.

For TOF-MS, we employed an in-house zircon single crystal SSEF87244 with an age of 523 ± 5 Ma, as primary standard for U–Pb dating. This crystal was characterized by ICP-SF-MS to have a high uranium concentration (approximately U 2000 mg/kg) and having no detectable amount of common lead (^{204}Pb). The ages obtained from secondary standard Zircon 91500 (1059.62 ± 8.70 Ma, $^{206}Pb/^{238}U$ age) are in accordance with the published result within the uncertainty. For dating analyses of zirconolite, we applied the same set of zircon standards, due to the lack of a matrix-matched zirconolite standards. For data processing, we used an in-house developed calculation script based on MATLAB (R2018b, The MathWorks, Inc., Natick, MA, USA). Benefiting from the full mass spectrum analysis in TOF-MS, it was possible to recheck if the whole measurements were done on targeted inclusions or if the ablation moved into host materials (ruby, spinel, etc.) by monitoring transient signals of various isotopes in TofDaq software (V1.2.97, TOFWERK AG, Thun, Switzerland). Furthermore, raw transient signals were carefully checked for common Pb interference based on isotope ^{204}Pb. In most of our samples, the ^{204}Pb signal was below the detection limit. Therefore, no correction for common Pb was applied. However, there may be still a small amount of common Pb present in our samples due to the low natural abundance of ^{204}Pb isotope, as shown in some discordant results in Wetherill and Tera-Wasserburg plots, and these discordant ages are omitted in the reported ages. The intensity ratios of U–Pb isotopes were monitored with TofDaq software to select appropriate sections in the transient signal that could be used for dating calculation. The radiometric ages based on $^{206}Pb/^{238}U$ and $^{207}Pb/^{235}U$ ratios are calculated using gas blank corrected signals. The radiometric ages are calculated as a ratio of the mean method and errors (2SE) are propagated following the procedures suggested by [60].

The age uncertainty of reference materials is propagated afterwards and other uncertainties are not included. All dating results are plotted using *IsoplotR* (V3.0) [61]. Analysis is accepted only when its age with 2SE overlaps the concordia curve [62]. For TOF-MS, the REEs were quantified using a non-matrix matched NIST610 SRM glass as external standard, due to its nearly full coverage of the periodic table of the elements. Additionally, Ca was used as internal standard and a total mass normalization procedure was applied [63].

4. Results

In total, we analysed 109 zircon grains and 14 zirconolite grains for this study. For the detailed U–Pb data and trace element data of all samples we refer the interested reader to the Supplementary Data.

4.1. U–Pb Dating of Zircons

For zircon inclusions in gem-quality ruby and spinel showing growth zoning patterns in VPSE imaging (e.g., Figure 3c), we analysed both core and rim separately. Individual spots from zircon core reveal a wide range of ages, from ~94 Ma to ~26 Ma (Figure 4a). The zircon rim displays ages between ~30 Ma to ~22 Ma (Figure 4a). The youngest ages are concordant at 22.26 ± 0.36 Ma for zircon in ruby (Figure 5a) and 22.88 ± 0.72 Ma for zircon in spinel (Figure 5b).

In the ruby- and spinel-bearing Bawlongyi marble, we observed both zircon and zirconolite as accessory mineral phases. The U–Pb ages of zircon grains cover a wide range from Cretaceous (~95 Ma) to Early Miocene (~17 Ma, Figure 4b). This is similar to the zircon inclusions in gemstones. The youngest age of zircon is concordant at 17.11 ± 0.22 Ma (Figure 5c) which is younger than that from our investigated rubies and spinels. The core ages from accessory zircon intergrown with zirconolite reveal ages ranging from ~65 Ma to ~55 Ma. These ages are within the range as the age of zircon cores found in gemstones (ruby and spinel).

Analysed zircon core results in garnet-orthopyroxene gneiss from Aunglan Taung yield groups of Jurassic (~180 Ma to ~140 Ma) and Cretaceous (~110 Ma to ~70 Ma) ages (Figure 6). The U–Pb ages of zircon rims were found to be concordant at 31.98 ± 0.30 Ma, as is shown in Figure 5d.

U–Pb ages of accessory zircon in biotite-garnet gneiss from Kinn reveal mainly discordant ages (Supplementary Figure S2). Two zircon rim measurements give a concordant age of 26.13 ± 1.24 Ma (Figure 5e). These rim ages are also consistent with ages of zircon rims in gemstones from the present study. The discordia may correspond to a mixture with different overgrowth ages, or Pb loss. The oldest zircon core age is ~985 Ma, and a VPSE image of this zircon is shown in Supplementary Figure S3.

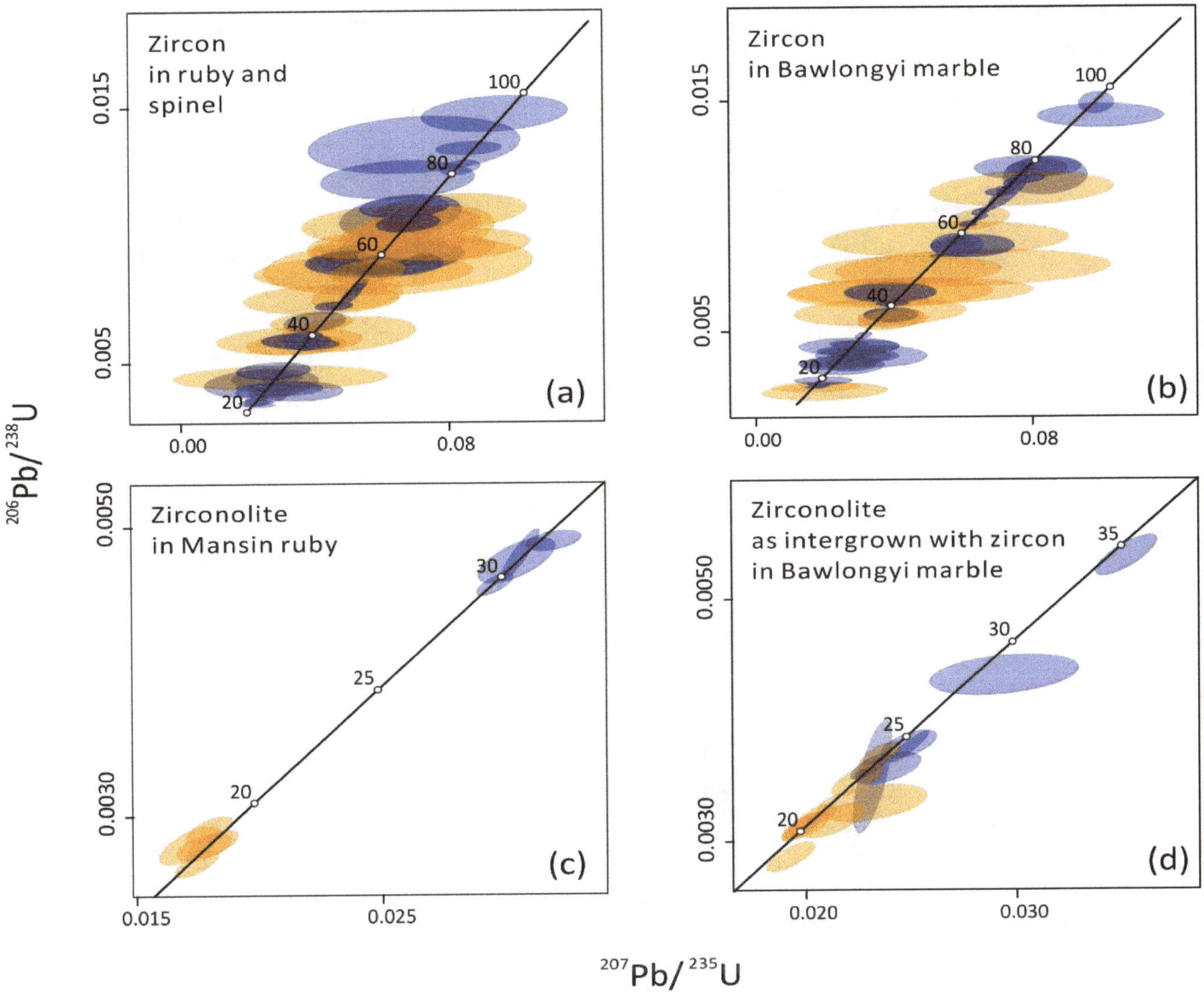

Figure 4. Wetherill diagrams showing wide range of U–Pb age of (**a**) zircon inclusion in gem-quality ruby and spinel, (**b**) accessory zircon in Bawlongyi marble, (**c**) zirconolite inclusion in ruby from Mansin and (**d**) zirconolite intergrown with zircon in Bawlongyi marble. Ellipses indicate two standard error (2SE) uncertainty of measurements. Blue is by sector-field (SF)-MS and orange is by time-of-flight (TOF)-MS.

4.2. U–Pb Dating of Zirconolite

Zirconolite grains from Mansin ruby and from ruby- and spinel-bearing marble (Bawlongyi) were analysed by TOF-MS and SF-MS, as shown in Figure 4c,d. Zirconolite from Mansin ruby reveals a concordia age of 29.78 ± 0.38 Ma (Figure 5f) analysed by SF-MS, whereas TOF-MS analysis adjacent to the SF-MS measurements reveals an age of 18.06 ± 0.36 Ma (Figure 5g). Furthermore, zirconolites observed as intergrowth of zircons in ruby- and spinel-bearing marble yield apparent concordia ages at 23.61 ± 0.36 Ma using the SF-MS setup (Figure 5h) and 20.30 ± 0.30 Ma using the TOF-MS (Figure 5i). These different results for zirconolite may be due to different growth zonings in the inclusions, which could not be identified by VPSE imaging. The discrepancy may also be due to the application of a non-matrix matched standard (zircon), which induces various fractionation effects with different instrument operating parameters and corrections [53,64]. Since both measurements have revealed secondary reference zircon (Zircon91500) in accordance with the published ages within uncertainty, the two instrumental setups were operated normally. Therefore, the reported zirconolite ages may lack of accuracy and 2SE of uncertainty is related to measurement precision. Due to lack of a zirconolite reference material, it is important to mention that the zirconolite ages obtained here may be only indicative and can be improved further.

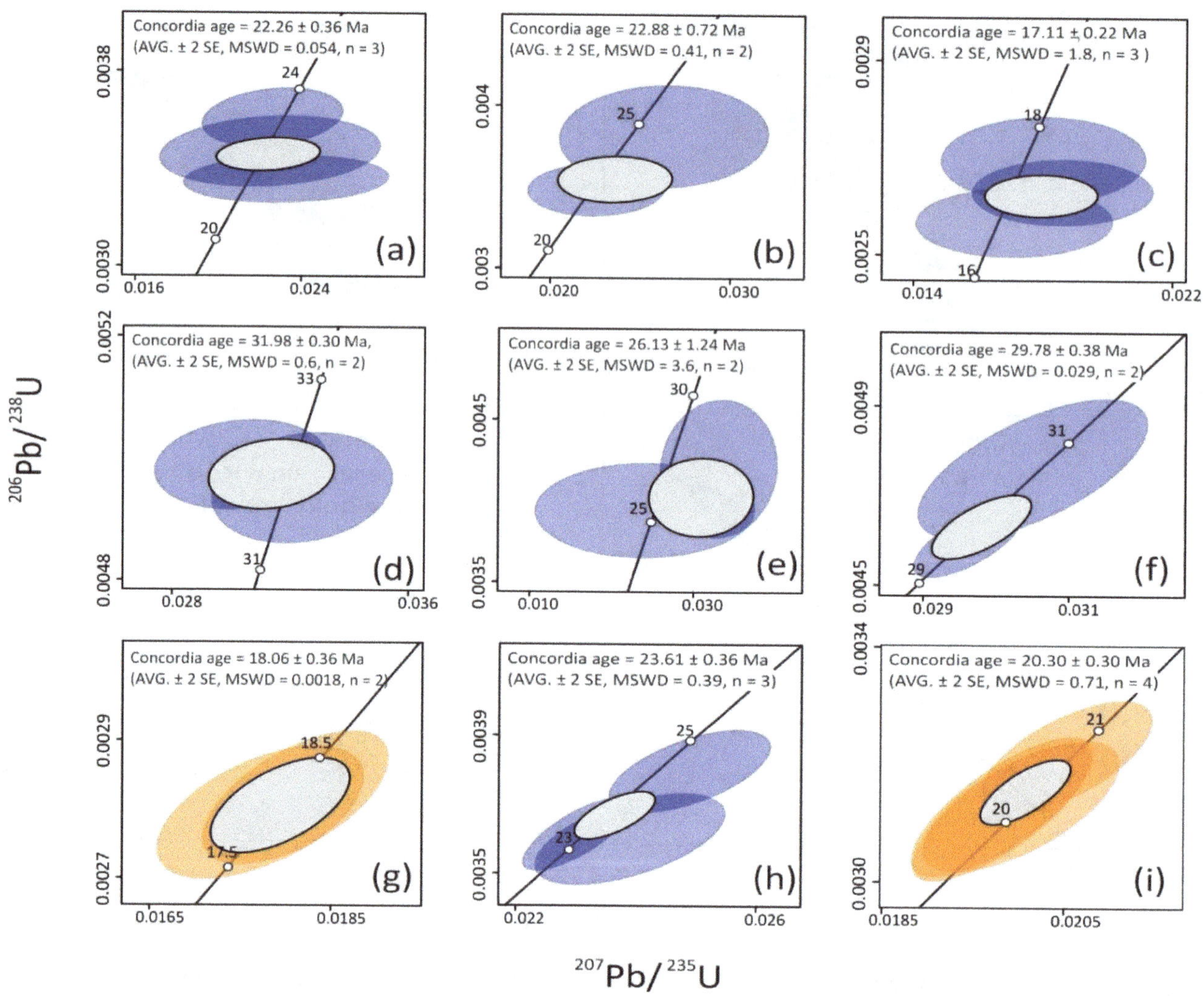

Figure 5. Concordia ages analysed on the rim of zircon inclusion in (**a**) Mansin and Kyatpyin ruby (22.26 ± 0.36 Ma), (**b**) Kyauksin spinel (22.88 ± 0.72 Ma) and accessory zircon in (**c**) Bawlongyi marble (17.11 ± 0.22 Ma), (**d**) Aunglan Taung gneiss (31.98 ± 0.30 Ma) and (**e**) Kinn gneiss (26.13 ± 1.24 Ma). Concordia age of zirconolite inclusion from Mansin ruby showing (**f**) 29.78 ± 0.38 Ma by SF-MS, and (**g**) 18.06 ± 0.36 Ma by TOF-MS as well as of zirconolite intergrown with zircon in Bawlongyi marble showing (**h**) 23.61 ± 0.36 Ma by SF-MS, (**i**) 20.30 ± 0.30 Ma by TOF-MS. Reported ages are concordia ages with two standard error (2SE) of uncertainty of measurements. Blue ellipses are by SF-MS and orange ellipses are by TOF-MS with 2SE of uncertainty. Gray ellipses indicate concordia age with 1SE of uncertainty.

4.3. REE Composition of Zircon and Zirconolite

Using TOF-MS, we collected multi-element data, specifically of the rare earth elements (REE), simultaneously to the U–Pb measurements. For a detailed list of the REE concentrations of selected zircon and zirconolite samples present as inclusions in ruby and spinel and as accessory minerals in host rock (Bawlongyi marble) and adjacent gneisses (Aunglan Taung gneiss and Kinn gneiss) the reader is referred to Supplementary Data. Most of the zircon REE data are from the core of the inclusions/accessory minerals, whereas in the zirconolite no zoning was observed by VPSE imaging.

The chondrite-normalized Coryell-Masuda diagrams of our samples reveal for zircon a distinct enrichment in heavy REE (HREE) in all samples [50]. Zircon inclusions from gem-quality ruby

(Figure 7a) vary quite largely in their REE concentrations, thus resulting in a rather broad REE pattern in the chondrite-normalized plot. Still, all investigated zircon inclusions in ruby show a very similar trend with part of light REE (LREE) at or even below detection limit. In addition, they show only a weak negative Eu anomaly and no positive Ce anomaly. The REE pattern of zircon in spinel (blue traces in Figure 7a) follows the same trend, but in addition these inclusions contain even less LREE, mostly below detection limit.

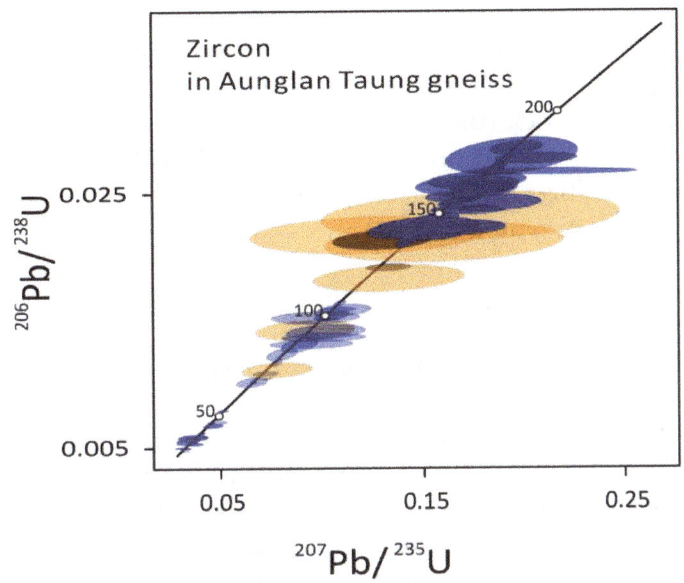

Figure 6. Wetherill diagram of accessory zircon in Aunglan Taung gneiss. Ages of accessory zircon in Kinn gneiss show discordia ages and are reported in Supplementary Figure S2. Ellipses indicate two standard error (2SE) uncertainty of measurements. Blue is by SF-MS and orange is by TOF-MS.

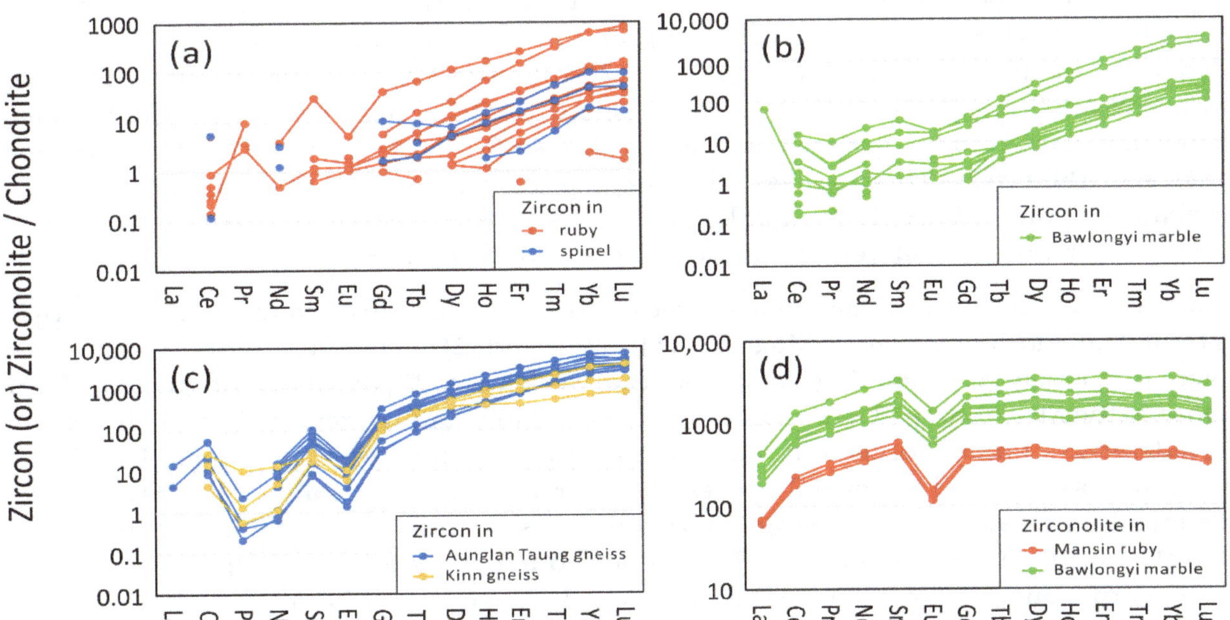

Figure 7. Chondrite normalized Coryell-Masuda rare earth element (REE) diagrams of (**a**) zircon inclusion in gem-quality ruby and spinel (**b**) marble (**c**) gneiss and (**d**) zirconolite from ruby and marble. (Chondrite REE values from [65]).

Zircon as accessory mineral in Bawlongyi marble (Figure 7b) shows a very similar pattern as the zircon inclusions in ruby, again with only a weak negative Eu anomaly, but in contrast to the zircon inclusions in ruby with a (weak) positive Ce anomaly. In general, the concentrations of REE in zircons from the marble are higher than from zircon inclusions in ruby (and spinel), as can also be seen in the chondrite-normalized plot. The most pronounced enrichment in HREE is observed in zircons from the gneiss samples (Aunglan Taung and Kinn) (Figure 7c). In addition, these zircons show also distinct negative Eu and positive Ce anomalies.

Compared to zircon, the analysed zirconolites show a distinctly different REE pattern in their chondrite-normalized plot (Figure 7d). Both, zirconolite as inclusions in ruby (from Mansin) and as accessory minerals in marble host rock (Bawlongyi) reveal a rather flat REE pattern, dominated by a distinct negative Eu anomaly and a general relative depletion in LREE. Similar to the zircon inclusions in ruby (and spinel), the concentration of REE is lower in the zirconolite inclusions in gem-quality ruby than in the host-rock (Bawlongyi marble).

5. Discussion and Conclusions

The zircons from gemstone samples and ruby- and spinel-bearing marble from Mogok area reveal consistent radiometric U–Pb ages over a wide time range from Cretaceous to Early Miocene. The youngest ages (Late Oligocene to Early Miocene) are found in the rim of accessory zircon in the Bawlongyi marble. They are in good agreement with previously reported U–Pb ages on zircon inclusions [16], as well as on a titanite inclusion in ruby from Thurein Taung (western Mogok) [14]. Similar U–Pb ages are also known from analyses carried out by commercial gemmological laboratories on faceted gem-quality ruby and sapphire from Mogok [17–19].

The U–Pb ages calculated in the rim of zircon indicate a late regional metamorphic event and are interpreted to be related to the formation of the marble-hosted ruby and spinel in the Mogok area. The youngest concordia ages recorded from the rim of zircon inclusions in our studied gemstones are 22.26 ± 0.36 Ma (ruby) and 22.88 ± 0.72 Ma (spinel). Based on the evident similarity of ages of the zircon rims in both, ruby and spinel, we presume that ruby and spinel both formed during Oligocene to Early Miocene and under similar granulite-facies regional metamorphic conditions as recorded in the surrounding Mogok area [4,35,66,67]. Interestingly, the youngest age of zircon grains in Bawlongyi marble is even younger with 17.11 ± 0.22 Ma. This is interpreted to be related to post-regional metamorphic contact metamorphism [2,26,45,68] caused by the late intrusion of Kabaing granite nearby (Figure 2). This youngest zircon age is in accordance with dating results on zircons from Kabaing granite (16.8 ± 0.5 Ma) [26]. A comparison of geochronological data from this study with previously reported data from literature is presented in Figure 8.

Moreover, we observe two age groups in the core of accessory zircon in the garnet-orthopyroxene gneiss from Aunglan Taung (central Mogok): (1) Jurassic and (2) Cretaceous (Figure 6). Jurassic zircon ages were previously recorded in granite from the Mogok area [2] and in orthogneiss of Kyanigan Hills close to the Mogok area [39]. According to Mitchell [69] and references therein, the Mogok belt was considered as a former northward continuation of the Sumatra-western Myanmar Late Cretaceous magmatic arc, as suggested by several age results on granodioritic to tonalitic plutons and batholiths from Banmauk area. The youngest rim ages of zircon grains in Aunglan Taung gneiss reveal a concordia age at 31.98 ± 0.30 Ma, similar ages have been reported in other studies [16,70]. Accessory zircon in the biotite-garnet gneiss from Kinn area shows discordant ages. The oldest U–Pb ages (Precambrian) are found in the core of a zircon grain, which we interpreted as a detrital zircon based on its growth zoning pattern (see Supplementary Figure S3).

These old age results are in accordance with former studies which attribute the metasediments of Mogok a Precambrian age [27,28,41]. The Late Oligocene ages (e.g., 26.39 ± 1.24 Ma) from zircon rims in Kinn gneiss are interpreted to represent the most recent regional metamorphic event.

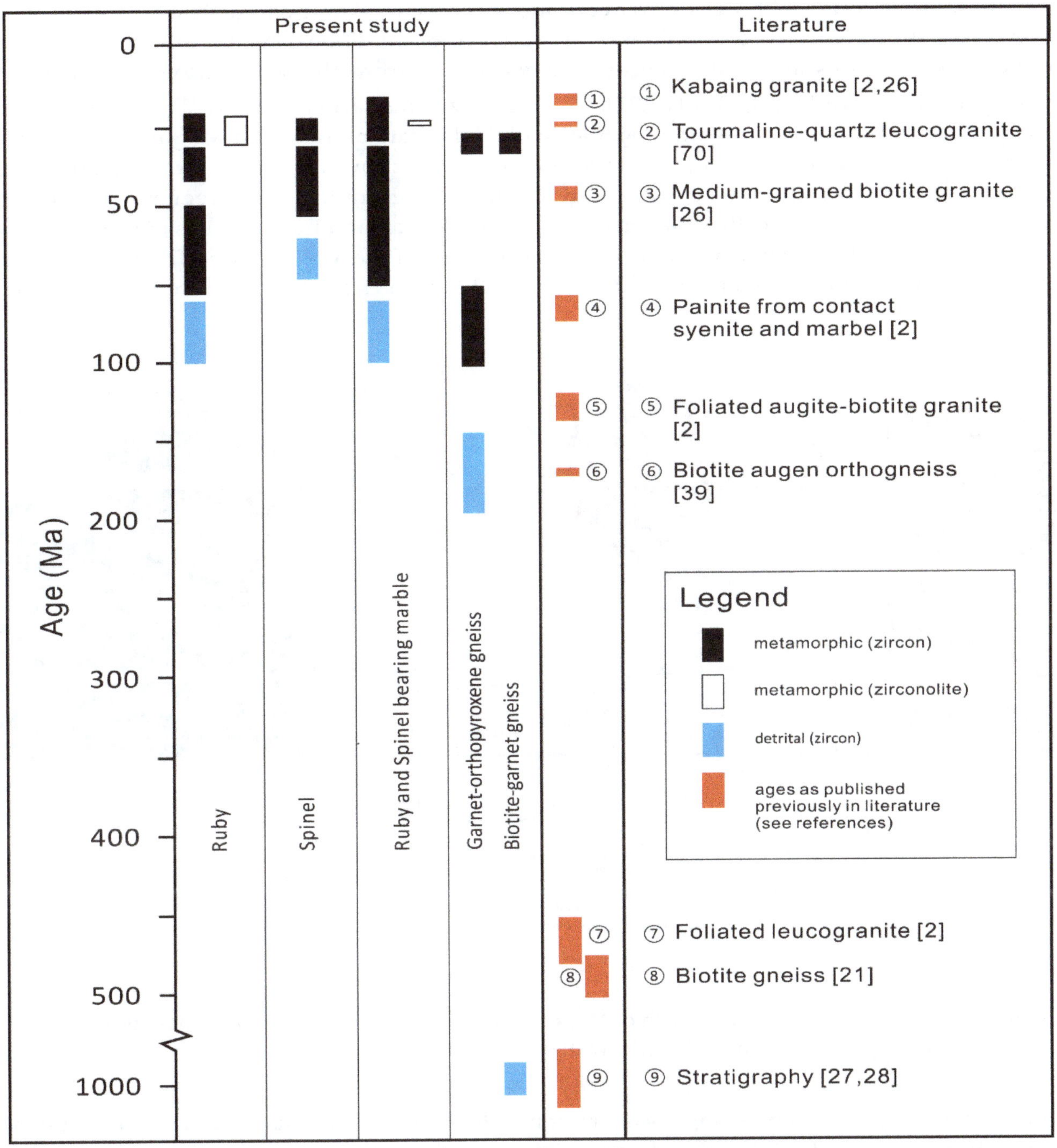

Figure 8. U–Pb ages reported in this study compared to previously published ages from the Mogok Stone Tract.

Zirconolite in ruby from Mansin reveals a U–Pb age of 18.06 ± 0.36 Ma using TOF-MS and 29.78 ± 0.38 Ma using SF-MS. A difference is also observed with zirconolite intergrown with zircon in ruby- and spinel-bearing marble (20.30 ± 0.30 Ma with TOF-MS and 23.61 ± 0.36 Ma with SF-MS). This discrepancy is discussed in previous section. Despite these differences, the U–Pb ages are still within the range of those previously reported in literature [2,20,21,28]. Further research is ongoing to understand the cause of this discrepancy and to find a method to eliminate it in order to obtain more accurate U–Pb ages on zirconolite.

To summarize, our U–Pb ages on zircon and zirconolite found as inclusions in gem-quality ruby and spinel and as accessory minerals in ruby- and spinel-bearing marble and in adjacent granulite-facies gneisses are well in agreement with data from the wider Mogok area reported previously in literature [14,21,38–40,68,71]. To visualize this larger context, Figure 9 presents a summary of U–Pb ages from this study together with U–Pb data of the wider Mogok area [72]. Based on the results of this study and in accordance with previous investigations, we conclude that the Mogok area and its surroundings experienced a Paleogene granulite-facies regional metamorphic overprint, influenced by post-collisional tectonics of the Eurasian and Indian plates [70,73]. We further assume that the marble-hosted gem-quality ruby and spinel of the Mogok Stone Tract formed during this regional metamorphic event between 30 Ma and 22 Ma, i.e., from Oligocene to Early Miocene.

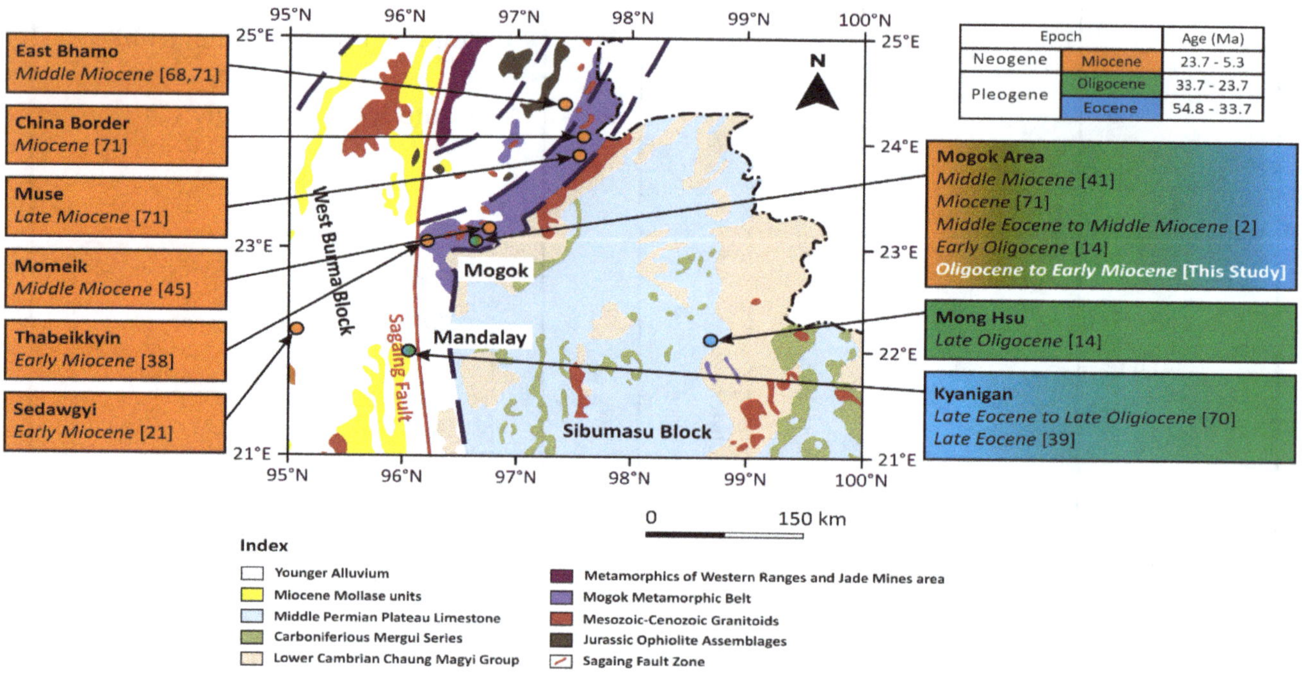

Figure 9. Map showing current research of Mogok combined with previous geochronological studies in its surrounding area (modified after [74] with permission).

Simultaneously to U–Pb dating measurements, REE concentrations in zircon and zirconolite were measured by LA-ICP-TOF-MS and chondrite-normalized REE diagrams were plotted in Figure 7a–d. A positive Ce anomaly and a negative Eu anomaly such as observed in the investigated accessory zircons in Aunglan Taung and Kinn gneisses (Figure 7c) are common in zircon [75,76] and generally related to magmatic fractionation processes (e.g., by feldspar crystallizations and/or a function of the oxygen fugacity and the Eu anomaly of the melt) [77].

Our focus is to characterize the studied zircon and zirconolite inclusions and accessory minerals from Mogok (Myanmar). Specifically, as to our knowledge, it is the first time that the full range of REEs in zirconolite from Mogok is reported. The zircon and zirconolite REE concentrations (see Supplementary Data) and REE plots may aid in country of origin determination for gem-quality ruby and spinels in commercial gemmological laboratories, similar to a study on apatite inclusions in gem-quality corundum and spinel from Mogok and other deposits [78].

The present study adds considerable new geochronological (U–Pb) and geochemical (REE) data to previous research about the Mogok area in Myanmar. With a special focus on zircon and zirconolite present as inclusions in gem-quality ruby and spinel and as accessory minerals in their marble host rock and adjacent gneisses, our data provides a better understanding of ruby and spinel formation within the Mogok Stone Tract, one of the finest ruby and spinel sources in the world.

Author Contributions: All authors were involved in designing the experiments. M.M.P., L.F., W.A.B., M.S.K. collected and prepared the samples. A.B., M.M.P. conducted VPSE imaging. H.A.O.W., M.M.P. conducted measurements using TOF-MS. M.G., M.M.P., H.A.O.W. conducted measurements using SF-MS. M.M.P., M.G., H.A.O.W. performed data analysis. M.M.P. prepared manuscript. All authors contributed in figures and in the manuscript. M.S.K. and H.A.O.W. conceived of and supervised the project. All authors have read and agreed to the published version of the manuscript.

Acknowledgments: The authors would like to thank Ma Mie (Silken East Co. Ltd., Thailand), Ko Chu (Kyauksaung mine, Mogok) and local mine owners in Mogok area for their generous support and for the donation of gem-quality samples for this study. We appreciate the kind help of Ko Kyaw Swar, Ko Nay, Sebastian Hänsel, U Aung Kyaw Htun, Ko Ja Mu, Ah Ba and Anty Phyu during the field trip to Mogok in 2016. We thank Pascal Tschudin at University of Basel for the preparation of thin sections and Evi Bieler from the Nano Imaging Lab of the University Basel for numerous SEM-EDS and BSE images. Thanks also to the academic editors of this special issue, four anonymous reviewers and Laurent E. Cartier for detailed suggestions and comments on the manuscript.

References

1. Gordon, R. On the Ruby Mines near Mogok, Burma. *Proc. R. Geogr. Soc. Mon. Rec. Geogr.* **1888**, *10*, 261–275. [CrossRef]
2. Thu, K. The Igneous Rocks of the Mogok Stone Tract: Their Distribution, Petrography, Petrochemistry, Sequence, Geochronology and Economic Geology. Ph.D. Thesis, University of Yangon, Yangon, Myanmar, 2007.
3. Themelis, T. *Gems and Mines of Mogok*; A&T Pub.: Los Angeles, CA, USA, 2008.
4. Phyo, M.M. Mineralogical, Gemmological and Petrological Study of the Mogok Stone Tract in Myanmar with a Special Focus on Gem-Quality Ruby and Spinel. Ph.D. Thesis, Universität Basel, Basel, Switzerland, 2019.
5. Harlow, G.E.; Bender, W. A study of ruby (corundum) compositions from the Mogok Belt, Myanmar: Searching for chemical fingerprints. *Am. Mineral.* **2013**, *98*, 1120–1132. [CrossRef]
6. Sutherland, F.L.; Zaw, K.; Meffre, S.; Yui, T.-F.F.; Thu, K. Advances in trace element "fingerprinting" of gem corundum, ruby and sapphire, Mogok area, Myanmar. *Minerals* **2014**, *5*, 61–79. [CrossRef]
7. Giuliani, G.; Fallick, A.E.; Garnier, V.; France-Lanord, C.; Ohnenstetter, D.; Schwarz, D. Oxygen isotope composition as a tracer for the origins of rubies and sapphires. *Geology* **2005**, *33*, 249–252. [CrossRef]
8. Yui, T.F.; Zaw, K.; Wu, C.M. A preliminary stable isotope study on Mogok Ruby, Myanmar. *Ore Geol. Rev.* **2008**, *34*, 192–199. [CrossRef]
9. Zaw, K.; Sutherland, L.; Yui, T.-F.; Meffre, S.; Thu, K. Vanadium-rich ruby and sapphire within Mogok Gemfield, Myanmar: Implications for gem color and genesis. *Miner. Depos.* **2015**, *50*, 25–39. [CrossRef]
10. Giuliani, G.; Fallick, A.E.; Boyce, A.J.; Pardieu, V.; Pham, V.L. Pink and red spinels in marble: Trace elements, oxygen isotopes, and sources. *Can. Mineral.* **2017**, *55*, 743–761. [CrossRef]
11. Phyo, M.M.; Franz, L.; Bieler, E.; Balmer, W.; Krzemnicki, M.S. Spinel from Mogok, Myanmar–A detailed inclusion study by raman microspectroscopy and scanning electron microscopy. *J. Gemmol.* **2019**, *36*, 418–435. [CrossRef]
12. Giuliani, G.; Dubessy, J.; Banks, D.A.; Lhomme, T.; Ohnenstetter, D. Fluid inclusions in ruby from Asian marble deposits: Genetic implications. *Eur. J. Mineral.* **2015**, *27*, 393–404. [CrossRef]
13. Sutherland, F.L.; Piilonen, P.C.; Zaw, K.; Meffre, S.; Thompson, J. Sapphire within zircon-rich gem deposits, Bo Loei, Ratanakiri Province, Cambodia: Trace elements, inclusions, U–Pb dating and genesis. *Aust. J. Earth Sci.* **2015**, *62*, 761–773.
14. Sutherland, F.; Zaw, K.; Meffre, S.; Thompson, J.; Goemann, K.; Thu, K.; Nu, T.; Zin, M.; Harris, S. Diversity in ruby geochemistry and its inclusions: Intra- and inter-continental comparisons from Myanmar and Eastern Australia. *Minerals* **2019**, *9*, 28. [CrossRef]
15. Coenraads, R.R.; Lin Sutherland, F.; Kinny, P.D. The origin of sapphires: U–Pb dating of zircon inclusions sheds new light. *Mineral. Mag.* **1990**, *54*, 113–122. [CrossRef]

16. Zaw, K.; Sutherland, F.L.; Graham, I.; Meffre, S.; Thu, K. Dating zircon inclusions in gem corundum deposits and genetic implications. In Proceedings of the 13th Quardrennial IAGOD Symposium, Adelaide, Australia, 6–9 April 2010.
17. Elmaleh, E.; Schmidt, S.T.; Karampelas, S.; Link, K.; Kiefert, L.; Süssenberger, A.; Paul, A. U–Pb ages of zircon inclusions in sapphires from Ratnapura and Balangoda (Sri Lanka) and implications for geographic origin. *GEMS Gemol.* **2019**, *55*, 18–28.
18. Krzemnicki, M.S.; Wang, H.A.O.; Phyo, M.M. Age dating applied as a testing procedure to gemstones and biogenic gem materials. In Proceedings of the 36th IGC Conference, Nante, France, 27–31 August 2019; pp. 48–50.
19. Link, K. Age Determination of zircon inclusions in faceted sapphires. *J. Gemmol.* **2016**, *34*, 692–700. [CrossRef]
20. Garnier, V.; Maluski, H.; Giuliani, G.; Ohnenstetter, D.; Schwarz, D. Ar–Ar and U–Pb ages of marble-hosted ruby deposits from central and southeast Asia. *Can. J. Earth Sci.* **2006**, *43*, 509–532. [CrossRef]
21. Mitchell, A.; Chung, S.L.; Oo, T.; Lin, T.H.; Hung, C.H. Zircon U–Pb ages in Myanmar: Magmatic-metamorphic events and the closure of a neo-Tethys ocean? *J. Asian Earth Sci.* **2012**, *56*, 1–23. [CrossRef]
22. Sutherland, F.L.; Duroc-Danner, J.M.; Meffre, S. Age and origin of gem corundum and zircon megacrysts from the Mercaderes-Rio Mayo area, South-west Colombia, South America. *Ore Geol. Rev.* **2008**, *34*, 155–168. [CrossRef]
23. Sorokina, E.S.; Rösel, D.; Häger, T.; Mertz-Kraus, R.; Saul, J.M. LA-ICP-MS U–Pb dating of rutile inclusions within corundum (ruby and sapphire): New constraints on the formation of corundum deposits along the Mozambique belt. *Miner. Depos.* **2017**, *52*, 641–649. [CrossRef]
24. Link, K. New age data for blue sapphire from Mogok, Myanmar. *J. Gemmol.* **2016**, *35*, 107–109.
25. Balmer, W.A.; Hauzenberger, C.A.; Fritz, H.; Sutthirat, C. Marble-hosted ruby deposits of the Morogoro Region, Tanzania. *J. Afr. Earth Sci.* **2017**, *134*, 626–643. [CrossRef]
26. Gardiner, N.J.; Robb, L.J.; Morley, C.K.; Searle, M.P.; Cawood, P.A.; Whitehouse, M.J.; Kirkland, C.L.; Roberts, N.M.W.; Myint, T.A. The tectonic and metallogenic framework of Myanmar: A Tethyan mineral system. *Ore Geol. Rev.* **2016**, *79*, 26–45. [CrossRef]
27. Chhibber, H.L. *The Geology of Burma*; Macmillan and Co. Limited: London, UK, 1934.
28. Iyer, L.A.N. *The Geology and Gem-Stones of the Mogok Stone Tract, Burma. Memoirs of the Geological Survey of India*; Calcutta: Delhi, India, 1953; Volume 82.
29. Mitchell, A.H.G.; Htay, M.T.; Htun, K.M.; Win, M.N.; Oo, T.; Hlaing, T. Rock relationships in the Mogok metamorphic belt, Tatkon to Mandalay, central Myanmar. *J. Asian Earth Sci.* **2007**, *29*, 891–910. [CrossRef]
30. Bertrand, G.; Rangin, C.; Maluski, H.; Han, T.A.; Thein, M.; Myint, O.; Maw, W.; Lwin, S. Cenozoic metamorphism along the Shan scarp (Myanmar): Evidences for ductile shear along the Sagaing fault or the northward migration of the eastern Himalayan syntaxis? *Geophys. Res. Lett.* **1999**, *26*, 915–918. [CrossRef]
31. Bender, F. *Geology of Burma*; Schweizerbart Science Publishers: Stuttgart, Germany, 1983.
32. Zaw, K. Geological, petrological and geochemical characteristics of granitoid rocks in Burma: With special reference to the associated W-Sn mineralization and their tectonic setting. *J. Southeast Asian Earth Sci.* **1990**, *4*, 293–335. [CrossRef]
33. Gardiner, N.J.; Searle, M.P.; Morley, C.K.; Whitehouse, M.P.; Spencer, C.J.; Robb, L.J. The closure of Palaeo-Tethys in Eastern Myanmar and Northern Thailand: New insights from zircon U–Pb and Hf isotope data. *Gondwana Res.* **2016**, *39*, 401–422. [CrossRef]
34. Lee, H.Y.; Chung, S.L.; Yang, H.M. Late Cenozoic volcanism in central Myanmar: Geochemical characteristics and geodynamic significance. *Lithos* **2016**, *245*, 174–190. [CrossRef]
35. Thu, Y.K.; Win, M.M.; Enami, M.; Tsuboi, M. Ti–rich biotite in spinel and quartz–bearing paragneiss and related rocks from the Mogok metamorphic belt, central Myanmar. *J. Mineral. Petrol. Sci.* **2016**, *111*, 270–282.
36. La Touche, T.H.D. *Geology of the Northern Shan State*; Office of the Geological Survey of India: Calcutta, India, 1913.
37. Thein, M. Modes of occurrence and origin of precious gemstone deposits of the Mogok Stone Tract. *J. Myanmar Geosci. Soc.* **2008**, *1*, 75–84.
38. Brook, M.; Snelling, N.J. K/Ar and Rb/Sr age determinations on rocks and minerals from Burma. In *Report of the Isotope Geology Unit 76/12*; Institute of Geological Science: Keyworth, Nottingham, UK, 1976.

39. Barley, M.E.; Pickard, A.L.; Zaw, K.; Rak, P.; Doyle, M.G. Jurassic to Miocene magmatism and metamorphism in the Mogok metamorphic belt and the India-Eurasia collision in Myanmar. *Tectonics* **2003**, *22*, 1–11. [CrossRef]
40. Searle, M.P.; Waters, D.J.; Morley, C.K.; Gardiner, N.J.; Htun, U.K.; Nu, T.T.; Robb, L.J. Chapter 12 Tectonic evolution of the Mogok metamorphic and Jade mines belts and ophiolitic terranes of Burma (Myanmar). In *Myanmar: Geology, Resources and Tectonics*; Barber, A.J., Zaw, K., Crow, M.J., Eds.; Geological Society: London, UK, 2017; Volume 48, pp. 261–293.
41. Searle, D.L.; Haq, B.T. The Mogok belt of Burma and its relationship to the Himalayan orogeny. In Proceedings of the 22nd International Geological Congress, New Delhi, India, 14–22 December 1964; Volume 22, pp. 132–161.
42. Clegg, E.L.G. The Cretaceous and associated rocks of Burma. *Mem. Geol. Surv. India* **1941**, *74*, 1–102.
43. Sone, M.; Metcalfe, I. Parallel Tethian sutures in mainland Sout-East Asia: New insights for Palaeo-Tethys closure and implications for the Indosinian Orogeny. *C. R. Geosci.* **2008**, *340*, 166–179. [CrossRef]
44. Zaw, K.; Swe, K.; Barber, A.J.; Crow, M.J.; Nwe, Y.Y. Chapter 1 Introduction to the geology of Myanmar. In *Myanmar: Geology, Resources and Tectonics*; Barber, A.J., Zaw, K., Crow, M.J., Eds.; Geological Society: London, UK, 2017; Volume 48.
45. Damon, P.E. *Granities from the Kabaing Batholithic Complex, Mogok Area, Burma*; Correlation and Chronology of ore Deposits and Volcanic Rocks; Annual Progress Report No. C00-689-60; United States Atomic Energy Commission: Washington, DC, USA, 1966.
46. Phyo, M.M.; Franz, L.; De Capatini, C.; Balmer, A.W.; Krzemnicki, M.S. Petrology and PT-conditions of quartz- and nepheline-bearing gneisses from Mogok Stone Tract, Myanmar. In Proceedings of the 15th Swiss Geoscience Meeting, Davos, Switzerland, 17–18 November 2017.
47. Bayliss, P.; Mazzi, F.; Munno, R.; White, T.J. Mineral nomenclature: Zirconolite. *Mineral. Mag.* **1989**, *53*, 565–569. [CrossRef]
48. Hughes, R.W. *Ruby & Sapphire*; RWH Publishing: Bangkok, Thailand, 1997.
49. Schaltegger, U.; Schmitt, A.K.; Horstwood, M.S.A. U–Th–Pb zircon geochronology by ID-TIMS, SIMS, and laser ablation ICP-MS: Recipes, interpretations, and opportunities. *Chem. Geol.* **2015**, *402*, 89–110. [CrossRef]
50. Hoskin, P.W.O.; Schaltegger, U. The Composition of Zircon and Igneous and Metamorphic Petrogenesis. *Rev. Mineral. Geochem.* **2003**, *53*, 27–62. [CrossRef]
51. Hendriks, L.; Gundlach-Graham, A.; Hattendorf, B.; Günther, D. Characterization of a new ICP-TOFMS instrument with continuous and discrete introduction of solutions. *J. Anal. At. Spectrom.* **2017**, *32*, 548–561. [CrossRef]
52. Wang, H.A.O.; Krzemncki, M.S. Multi-Element Analysis of Minerals using Laser Ablation Inductively Coupled Plasma Time Of Flight Mass Spectrometry. Manuscript near submission.
53. Jackson, S.E.; Pearson, N.J.; Griffin, W.L.; Belousova, E.A. The application of laser ablation-inductively coupled plasma-mass spectrometry to in situ U–Pb zircon geochronology. *Chem. Geol.* **2004**, *211*, 47–69. [CrossRef]
54. Horstwood, M.S.A.; Košler, J.; Gehrels, G.; Jackson, S.E.; McLean, N.M.; Paton, C.; Pearson, N.J.; Sircombe, K.; Sylvester, P.; Vermeesch, P.; et al. Community-Derived Standards for LA-ICP-MS U–(Th–)Pb Geochronology–Uncertainty Propagation, Age Interpretation and Data Reporting. *Geostand. Geoanal. Res.* **2016**, *40*, 311–332. [CrossRef]
55. Sláma, J.; Košler, J.; Condon, D.J.; Crowley, J.L.; Gerdes, A.; Hanchar, J.M.; Horstwood, M.S.A.; Morris, G.A.; Nasdala, L.; Norberg, N.; et al. Plešovice zircon—A new natural reference material for U–Pb and Hf isotopic microanalysis. *Chem. Geol.* **2008**, *249*, 1–35. [CrossRef]
56. Black, L.P.; Kamo, S.L.; Allen, C.M.; Aleinikoff, J.N.; Davis, D.W.; Korsch, R.J.; Foudoulis, C. TEMORA 1: A new zircon standard for phanerozoic U–Pb geochronology. *Chem. Geol.* **2003**, *200*, 155–170. [CrossRef]
57. Wiedenbeck, M.; Alle, P.; Corfu, F.; Griffin, W.L.; Meier, M.F.O.; von Quadt, A.; Roddick, J.C.; Spiegel, W. Three natural zircon standards for U–Th–Pb, Lu–Hf, trace element and REE analyses. *Geostand. Newsl.* **1995**, *19*, 1–23. [CrossRef]
58. Paton, C.; Hellstrom, J.; Paul, B.; Woodhead, J.; Hergt, J. Iolite: Freeware for the visualisation and processing of mass spectrometric data. *J. Anal. At. Spectrom.* **2011**, *26*, 2508–2518. [CrossRef]
59. Petrus, J.A.; Kamber, B.S. VizualAge: A Novel Approach to Laser Ablation ICP-MS U–Pb Geochronology Data Reduction. *Geostand. Geoanal. Res.* **2012**, *36*, 247–270. [CrossRef]

60. Ulianov, A.; Müntener, O.; Schaltegger, U.; Bussy, F. The data treatment dependent variability of U–Pb zircon ages obtained using mono-collector, sector field, laser ablation ICPMS. *J. Anal. At. Spectrom.* **2012**, *27*, 663–676. [CrossRef]
61. Vermeesch, P. IsoplotR: A free and open toolbox for geochronology. *Geosci. Front.* **2018**, *9*, 1479–1493. [CrossRef]
62. Spencer, C.J.; Kirkland, C.L.; Taylor, R.J.M. Strategies towards statistically robust interpretations of in situ U–Pb zircon geochronology. *Geosci. Front.* **2016**, *7*, 581–589. [CrossRef]
63. Guillong, M.; Hametner, K.; Reusser, E.; Wilson, S.A.; Günther, D. Preliminary Characterisation of New Glass Reference Materials (GSA-1G, GSC-1G, GSD-1G and GSE-1G) by Laser Ablation-Inductively Coupled Plasma-Mass Spectrometry Using 193 nm, 213 nm and 266 nm Wavelengths. *Geostand. Geoanal. Res.* **2005**, *29*, 315–331. [CrossRef]
64. Sylvester, P.J.; Ghaderi, M. Trace element analysis of scheelite by excimer laser ablation-inductively coupled plasma-mass spectrometry (ELA-ICP-MS) using a synthetic silicate glass standard. *Chem. Geol.* **1997**, *141*, 49–65. [CrossRef]
65. McDonougha, W.F.; Sun, S.-S. The composition of the Earth. *Chem. Geol.* **1995**, *120*, 223–253. [CrossRef]
66. Thu, Y.K.; Enami, M.; Kato, T.; Tsuboi, M. Granulite facies paragneisses from the middle segment of the Mogok metamorphic belt, central Myanmar. *J. Mineral. Petrol. Sci.* **2017**, *112*, 1–19.
67. Win, M.M.; Enami, M.; Kato, T. Metamorphic conditions and CHIME monazite ages of Late Eocene to Late Oligocene high-temperature Mogok metamorphic rocks in central Myanmar. *J. Asian Earth Sci.* **2016**, *117*, 304–316. [CrossRef]
68. Bertrand, G.; Rangin, C.; Maluski, H.; Bellon, H. Diachronous cooling along the Mogok Metamorphic Belt (Shan scarp, Myanmar): The trace of the northward migration of the Indian syntaxis. *J. Asian Earth Sci.* **2001**, *19*, 649–659. [CrossRef]
69. Mitchell, A.H.G. Cretaceous-Cenozoic tectonic events in the western Myanmar (Burma)-Assam region. *J. Geol. Soc. Lond.* **1993**, *150*, 1089–1102. [CrossRef]
70. Searle, M.P.; Noble, S.R.; Cottle, J.M.; Waters, D.J.; Mitchell, A.H.G.; Hlaing, T.; Horstwood, M.S.A. Tectonic evolution of the Mogok metamorphic belt, Burma (Myanmar) constrained by U–Th–Pb dating of metamorphic and magmatic rocks. *Tectonics* **2007**, *26*. [CrossRef]
71. Min, M. Thermochronology Applied to Strike-Slip Zones Central America and Myanmar. Ph.D. Thesis, Technische Universität Bergakademie Freiberg, Freiberg, Germany, 2007.
72. Crow, M.J.; Zaw, K. Appendix Geochronology in Myanmar (1964–2017). In *Myanmar: Geology, Resources and Tectonics*; Barber, A.J., Zaw, K., Crow, M.J., Eds.; Geological Society: London, UK, 2017; Volume 48, pp. 713–759.
73. Dewey, J.F.; Cande, S.; Pitman, W.C. Tectonic evolution of the India-Eurasia collision zone. *Eclogae Geol. Helv.* **1989**, *82*, 717–734.
74. Jiang, H.; Li, W.Q.; Jiang, S.Y.; Wang, H.; Wei, X.P. Geochronological, geochemical and Sr–Nd–Hf isotopic constraints on the petrogenesis of Late Cretaceous A-type granites from the Sibumasu Block, Southern Myanmar, SE Asia. *Lithos* **2017**, *268–271*, 32–47. [CrossRef]
75. Hoskin, P.W.O.; Ireland, T.R. Rare earth element chemistry of zircon and its use as a provenance indicator. *Geology* **2000**, *28*, 627–630. [CrossRef]
76. Gieré, R. Zirconolite, allanite and hoegbomite in a marble skarn from the Bergell contact aureole: Implications for mobility of Ti, Zr and REE. *Contrib. Mineral. Petrol.* **1986**, *93*, 459–470. [CrossRef]
77. Trail, D.; Bruce Watson, E.; Tailby, N.D. Ce and Eu anomalies in zircon as proxies for the oxidation state of magmas. *Geochim. Cosmochim. Acta* **2012**, *97*, 70–87. [CrossRef]
78. Bieri, W.; Grobety, B.; Peretti, A.; Hametner, K.; Günther, D. Chemical composition of apatite inclusions in corundum and spinel determined by LA-ICP-MS and its potential for authentication and provenance determination. In Proceedings of the Goldschmidt 2010: Earth, Energy, and the Environment, Knoxville, TN, USA, 13–18 June 2010; p. 89.

8

Ruby Deposits: A Review and Geological Classification

Gaston Giuliani [1,*], Lee A. Groat [2], Anthony E. Fallick [3], Isabella Pignatelli [4] and Vincent Pardieu [5]

1 Géosciences Environnement, Université Paul Sabatier, GET/IRD et Université de Lorraine, C.R.P.G./C.N.R.S., 15 rue Notre-Dame-des-Pauvres, BP 20, 54501 Vandœuvre, France

2 Earth, Ocean and Atmospheric Sciences, University of British Columbia, Vancouver, BC V6T 1Z4, Canada; groat@mail.ubc.ca

3 Isotope Geosciences Unit, Scottish Universities Environmental Research Centre (S.U.E.R.C.), Rankine Avenue, East Kilbride, Glasgow G75 0QF, UK; anthony.fallick@glasgow.ac.uk

4 Georessources, Université de Lorraine, UMR 7539 CNRS-UL, BP 70239 Vandœuvre, France; isabella.pignatelli@univ-lorraine.fr

5 Field gemmologist & Consultant at VP Consulting, Manama, Bahrain; vince@fieldgemology.org

* Correspondence: giuliani@crpg.cnrs-nancy.fr

Abstract: Corundum is not uncommon on Earth but the gem varieties of ruby and sapphire are relatively rare. Gem corundum deposits are classified as primary and secondary deposits. Primary deposits contain corundum either in the rocks where it crystallized or as xenocrysts and xenoliths carried by magmas to the Earth's surface. Classification systems for corundum deposits are based on different mineralogical and geological features. An up-to-date classification scheme for ruby deposits is described in the present paper. Ruby forms in mafic or felsic geological environments, or in metamorphosed carbonate platforms but it is always associated with rocks depleted in silica and enriched in alumina. Two major geological environments are favorable for the presence of ruby: (1) amphibolite to medium pressure granulite facies metamorphic belts and (2) alkaline basaltic volcanism in continental rifting environments. Primary ruby deposits formed from the Archean (2.71 Ga) in Greenland to the Pliocene (5 Ma) in Nepal. Secondary ruby deposits have formed at various times from the erosion of metamorphic belts (since the Precambrian) and alkali basalts (from the Cenozoic to the Quaternary). Primary ruby deposits are subdivided into two types based on their geological environment of formation: (Type I) magmatic-related and (Type II) metamorphic-related. Type I is characterized by two sub-types, specifically Type IA where xenocrysts or xenoliths of gem ruby of metamorphic (sometimes magmatic) origin are hosted by alkali basalts (Madagascar and others), and Type IB corresponding to xenocrysts of ruby in kimberlite (Democratic Republic of Congo). Type II also has two sub-types; metamorphic deposits sensu stricto (Type IIA) that formed in amphibolite to granulite facies environments, and metamorphic-metasomatic deposits (Type IIB) formed via high fluid–rock interaction and metasomatism. Secondary ruby deposits, i.e., placers are termed sedimentary-related (Type III). These placers are hosted in sedimentary rocks (soil, rudite, arenite, and silt) that formed via erosion, gravity effect, mechanical transport, and sedimentation along slopes or basins related to neotectonic motions and deformation.

Keywords: ruby deposits; classification; typology; magmatism; metamorphism; sedimentary; metasomatism; fluids; stable and radiogenic isotopes; genetic models; exploration

---

1. Introduction

Ruby is the red gem variety of the mineral corundum. The name corundum was introduced in the early 19th century by French mineralogist De Bournon when gemmological terminology was transformed from a rather confusing system mainly based on color and origin into a system based on a new science:

mineralogy. The term "corundum" likely originated from the Sanskrit kurunvinda meaning "hard stone", which became kurund in vernacular. It was the name used by the native people met during the field expedition sent by De Bournon to describe the stones they were mining in Southern India.

If the name "corundum" originates from Sanskrit, in Sanskrit rubies were called ratnaraj, queen of precious stones and symbol of permanent internal fire. In Greco-Roman times, the reference to fire was also common; in 400 before Christ Theophrastus [1] related ruby to anthrax (coal). Two thousand years ago Pliny [2] called all red stones carbunculus (diminutive of carbo), which became escarboucle in old French. In the 11th century, the Arab scholar Al Biruni separated different types of rubies based on specific gravity and hardness, but his work remained unknown in Europe for centuries except possibly for a few scholars like French bishop Marbode [3] who, a few years later, also separated carbuncles into three different types.

The term "ruby" appeared in European languages during the Middles Ages. How and why it was introduced is not clear but it was used by [4]. It is derived from the Latin ruber, meaning red. It became popular and then, until the introduction of the word corundum in the 19th century, most red stones were just called rubies. Experts and authors commonly separated different types based on their origin: "Oriental Ruby" was used for stones mined from Burma or Ceylon (ruby), "Balas Ruby" for gems mined from the modern-day Tajikistan (spinel), and "Bohemian ruby" was used for gems coming traditionally from the modern-day Czech Republic ("red garnet", either pyrope, almandine, and/or spessartite). However, as rubies, spinels, and garnets are often found in association and as new deposits were discovered, the system became very confusing. Ironically, the majority of the most famous rubies (like the "Black Prince Ruby" and the "Timur Ruby" of the British Crown Jewels or the "Cote de Bretagne Ruby" of the French Crown Jewels) that contributed significantly to the public recognition of the word "ruby" would actually, according to the new terminology based on mineralogy, be spinels. As gemmology is not simply about science, but also about history, art, and trade, these gems are still called by their historic "ruby" name, maintaining the taste for either curiosity, controversy, or confusion among many authors, particularly those with a non-scientific and/or historical background.

Currently we know that ruby is the chromium (Cr)-rich variety of corundum, but other chromophores such as vanadium (V) and iron (Fe) may be present in the structure of the mineral. The ideal formula of corundum is Al_2O_3 and it crystallizes in the trigonal system, with a hardness of 9 on the Mohs scale (while spinel and garnets crystallize in the cubic system and are softer with a hardness of 8). Unlike spinel and garnet, ruby is a dichroic mineral, which will show a different color (either purplish- or orangey-red) depending on which direction light travels inside the gem. In gemmology this is of critical importance because lapidaries, wishing to optimize the color from a rough stone by cutting it, can produce a red, a pinkish red, a purplish red, or an orangey-red stone, depending on how they decide to orientate and cut the gem.

The ruby industry is very lucrative [5] but it is far more complex than many ore industries such as iron, silver, or gold because the values of ruby do not depend only on weight or purity. The value depends on a much more complex quality, with grading systems including many artistic, subjective, and cultural aspects. In addition to the well-known and rather comprehensive system established by the diamond trade based on the 4Cs (i.e., color, clarity, cut, and carat weight of the gem) in the middle of the 20th century, with colored stones such as rubies additional factors have a critical importance for market value.

First, it has to be established that the gem was actually mined and was not produced from a factory. Synthetic rubies created either by flux, flame fusion, or some hydrothermal process appeared at the end of the 19th century. If the arrival of synthetics created some initial scandals and panic, the trade answered by establishing gemmological laboratories that were able to identify natural gems versus synthetics and thus re-establish the necessary confidence for the trade to grow.

Second, enhancements and treatments are also of importance to the final evaluation. To satisfy the strong demand for large clean rubies, some very efficient heat treatment techniques were developed in the 1990s in Thailand. These were different from the traditional blow-pipe technique used in Sri Lanka for centuries and described by many travellers beginning with Al Burini in the 10–11th centuries.

Whilst the blow pipe was used to remove the purplish aspect in some rubies, the new technique went further: using borax as a flux additive it could heal fractures in rubies and thus enable the cutting of larger stones. Due to this new treatment technique and the ones that followed, new deposits, such as Mong Hsu in eastern Myanmar, started to dominate the ruby trade in the early 1990s and Thailand increased its importance as the World's main ruby trading centre [6]. In the early 1990s, in Bangkok, it was very rare to find gem corundum that had not been heat-treated [7]. Drucker in [8], at the end of 1990, explained that the average price of ruby fell after controversy over its treatments. However, if the average price for rubies fell, prices for fine untreated rubies have exploded, particularly since the early 2000s (as observed in the reports from auction houses, many available online).

As the auction results show, not only did the prices for unheated rubies increase, but also the prices for unheated Myanmar rubies. Geographic origin is the last important factor to consider in establishing the value of a fine ruby (Figure 1). At this point it is critical to understand that, as fine rubies are both very rare but also very durable products, they usually survive for centuries after being mined, passing from one owner to another. As a consequence, the few very fine gems coming from existing mines are actually in competition with all the fine gems mined in the past, which are mainly from Myanmar (formerly known as Burma). Due to this, modern ruby traders are actually acting more like antique dealers than exploration geologists or Indiana Jones [9]. Public recognition and romance are very important factors contributing to demand for fine gemstones [10].

The assignment of a commercial value to ruby and other gems is carried out by gem merchants with the help of a few specialized gemmological laboratories providing independent third-party opinions. These labs provide answers for the following key questions [11]: (1) Is the gem the ruby variety of corundum? (2) Is it natural or synthetic ruby? (3) Has it been enhanced? (4) Where is it from (geographic origin)? If for most of these laboratories the answer to the first question is more or less routine work, the question about origin determination remains a great challenge [11]. The geographic origin is often controversial as in many cases it is based on expert opinion more than on provable facts [12–15]. Origin determination was introduced by Eduard Gübelin who worked on inclusions in gems using mainly microscopy [16,17], but also Raman and infrared spectroscopy. Such techniques are usually suitable for high value large rubies (that are usually never really free of mineral inclusions) but find their limits in the case of small clean stones or for heated stones (with inclusions that have been altered or even melted by heat treatment). Currently, these techniques are combined with the analysis of trace elements using energy dispersive X-ray fluorescence (EDXRF) and more especially since the early 2000s laser ablation-inductively coupled plasma-mass spectroscopy (LA-ICP-MS). To be able to produce origin reports, gemmological laboratories require sophisticated instrumentation, an experienced and qualified team, and a solid and very reliable reference collection.

The pricing of natural ruby is unique in terms of color and weight in carats. Currently if factory-grown synthetic rubies are cheap, if carving- or bead-grade treated rubies of low transparency and either dark or light color can be worth only a few dollars per kilo, a natural untreated ruby is still perhaps the world's most expensive gemstone. The best Myanmar rubies are more highly valued than an equivalent-sized flawless colorless diamond. The world record price for a single ruby sold at auction currently belongs to the Sunrise ruby, sold in 2015 for US$32.42 million (US$1.27 million per carat at 25.59 ct) [18]. In December 2015, the 15.04 ct Crimson Flame ruby was sold for UU$18.3 million (US$1.21 million per ct). Another notable ruby is the 10.05 ct Ratnaraj, sold for US$10.2 million (just over US$1 million per ct) in November 2017. Fine-quality unheated rubies from the recently discovered Didy mine in Madagascar also command high prices [19]. A set of eight faceted rubies, ranging from seven to more than 14 cts in weight, had an estimated market value of US$10 million [20]. The discovery of the Montepuez ruby deposits in Mozambique in May 2009 [21–23] and their exploitation by various companies including Gemfields Group Ltd. [24,25] changed the international ruby market [6,26]. A recent Competent Persons Report by SRK Consulting forecast production of 432 million carats from Montepuez over 21 years [27]. The resource at Montepuez is divided into primary reserves of 253 million cts (projection of 115 cts per ton) and secondary (mainly colluvial) reserves of 179 million cts (projection of 7.07 cts per ton). Twelve Montepuez auctions held since June 2014 have generated US$512 million in aggregate revenue (see Gemfields website). The auction by Gemfields Group Ltd. in June 2018 yielded a record sale of US$71.8 million with an average price of US$122 per ct. The success of the Gemfields ruby auctions are due to product from the Mugloto colluvial deposit, which provided clean and very high-quality red stones with typical sizes between one and 10 cts, and occasionally larger crystals [24].

Figure 1. Rubies worldwide. (**A**) Ruby crystal (1.7 cm × 1.7 cm × 0.5 cm) in marble from Bawpadan, Mogok, Myanmar. Photo: Louis-Dominique Bayle © le Règne Minéral; (**B**) rubies (from left to right: 2.1 cm × 1.3 cm × 1.3 cm and 1.7 cm × 1 cm × 1 cm) from Ambohimandrosoa, Antsirabe area, Madagascar. Collection: L. Thomas; Photo: Louis-Dominique Bayle © le Règne Minéral; (**C**) rubies from the Montepuez mine, Mozambique. Vincent Pardieu © Gemmological Institute of America; (**D**) Barrington Tops alluvial ruby concentrate, crystals up to 9 mm in size, New South Wales, Australia. Photo: Australian Museum Collection, courtesy Gayle Webb; (**E**) ruby crystal (6 cm × 4.8 cm × 1 cm) extracted from amphibolite rocks at Ampanihy, Southern Madagascar. Collection: J. Béhier. Photo: Louis-Dominique Bayle © le Règne Minéral; (**F**) rubies in phlogopite matrix with plagioclase (white) and minor kyanite (blue) hosted in between leucogabbro and meta-ultramafic rocks affected by fluid circulation and metamorphic metasomatism, Aappaluttoq deposit, southwest Greenland. Photo: Courtesy True North Gems Inc. (Vancouver, BC, Canada).

Over the last decade, knowledge on the formation of ruby deposits has improved significantly as their origin determination using a variety of new methods and analytical procedures. The present article synthesizes the state of knowledge on the geology and genesis of ruby deposits. The different issues focused on their location on the planet, their physical and chemical properties, pressure–temperature conditions of crystallization, the source of the constituent elements (viz. aluminum, chromium, vanadium, and iron), their age of formation, and finally the different proposed genetic models.

In this review we propose a geological classification scheme for ruby based on scientific works and own experience. The classification typology should open a new framework for mineral exploration guidelines and geographic determination for gemmological laboratories.

2. Worldwide Ruby Deposits

The distribution of major and minor commercial, industrial, scientific, and historical world sources of ruby is shown in Figure 2.

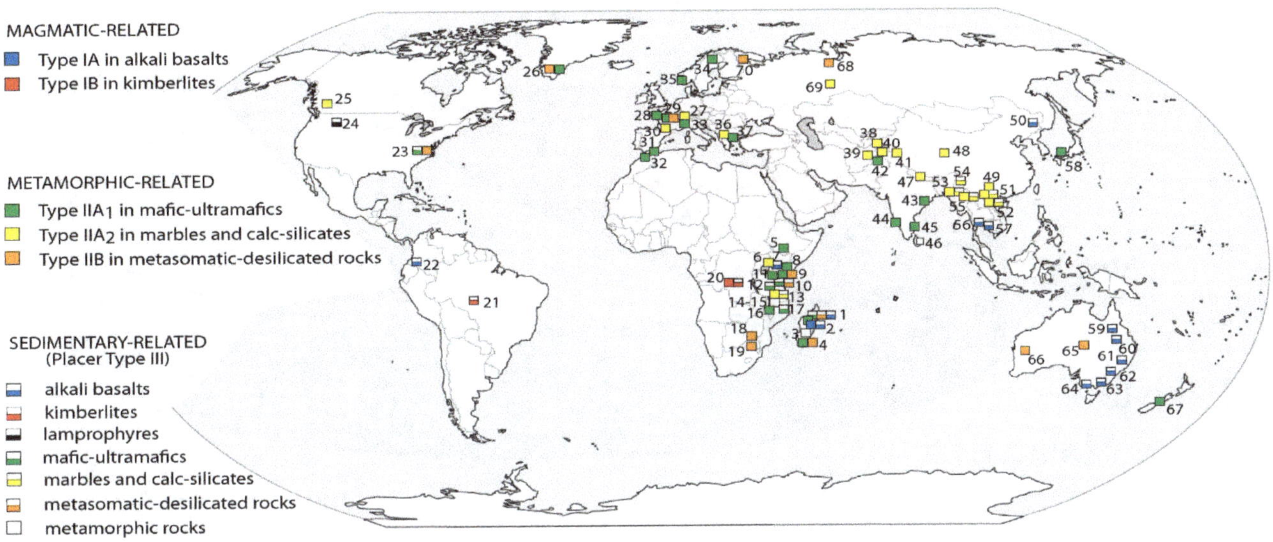

Figure 2. Main major and minor world sources of ruby plotted according to their geological types. Madagascar: (1) Andilamena, Ambodivoangy-Zahamena, Didy, Vatomandry; (2) Andriba; Ankaratra (Antsirabe-Antanifotsy region, Soamiakatra-Ambohimandroso; Ambohibary); (3) Zazafotsy; Ilakaka-Sakaraha; (4) Bekily-Vohibory area (Ambatomena; Ianapera; Fotadrevo; Anavoha; Maniry; Gogogogo; Vohitany; Ejeda). Ethiopia: (5) Kibre mengist, Dilla. Kenya: (6) West Pokot; (7) Baringo; (8) Kitui; (9) Mangare (Rockland-formerly John Saul mine; Penny Lane; Aqua, Hard Rock). Tanzania: (10) Umba valley; (11) Longido, Lossogonoi; (12) Winza; Loolera (Kibuko); (13) Morogoro (Mwalazi; Visakazi; Nyama, Kibuko, etc.); Mahenge (Lukande; Mayote; Kitwaro; Ipanko; (14) Songea; (15) Tunduru. Malawi: (16) Chimwadzulu. Mozambique: (17) Montepuez (Namahumbire/Namahaca), M'sawize, Ruambeze. Zimbabwe: (18) O'Briens (verdites). South Africa: (19) Barberton (verdites). Democratic Republic of Congo: (20) Mbuji-Mayi. Brazil: (21) São Luis-Juina River. Colombia: (22) Mercaderes-Rio Mayo. United States: (23) North Carolina (Corundum Hill; Cowee Valley; Buck Creek); (24) Montana (Rock Creek). Canada: (25) Revelstoke. Greenland: (26) South of Nuuk region (Fiskenæsset district-Aappaluttoq); North of Nuuk (Storø; Kangerdluarssuk). Switzerland: (27) Campo Lungo. France: (28) Brittany (Champtoceaux); (29) French Massif Central (Haut-Allier-Chantel; Peygerolles; Lozère-Aveyron-Marvejols; Vialat-du-Tarn); (30) Pyrenées (Arignac). Spain: (31) Alboran sea; Marrocos: (32) Beni Bousera. It aly: (33) Piedmont. Norway: (34) Froland. Finland: (35) Kittilä. Macedonia: (36) Prilep. Greece: (37) Gorgona-Xanthi; Paranesti-Drama. Tajikistan: (38) Snezhnoe, Turakuloma, Badakhshan. Afghanistan: (39) Jegdalek, Kash. Pakistan: (40) Hunza valley; (41) Batakundi, Nangimali; (42) Dir. India: (43) Orissa, Kalahandi; (44) Karnakata (Mysore); (45) Andhra Pradesh (Salem district). Sri Lanka: (46) Ratnapura, Polonnaruwa; Elahera. Nepal: (47) Chumar, Ruyil. China: (48) Qinghai; (49) Yuan Jiang. (50) Muling. Vietnam: (51) Luc Yen-Yen Bai; (52) Quy Chau. Myanmar: (53) Mogok, (54) Namya; (55) Mong Hsu. Thailand: (56) Chanthaburi-Trat (Bo Waen, Bo Na Wong, Wat Tok Phrom, Bo Rai, Nong Bon). Cambodia: (57) Pailin, Samlaut. Japan: (58) Ida. Australia: (59) Lava Plains; (60) Anakie fields-Rubyvale; (61) New England fields (Inverell); (62) Macquarie-Cudgegong, Barrington Tops-Yarrowitch; (63) Tumbarumba; (64) Western Melbourne fields; (65) Harts range; (66) Poona. New Zealand: (67) Westland (Hokitika). Russia: (68) Polar Urals (Hit Island); (69) Ural mountains (Kuchin-Chuksin-Kootchinskoye); (70) Karelia.

High-value rubies are important mineral resources that contribute significantly to the gross domestic product of many countries. In 2005, Africa was officially (statistics are rare on the production of most ruby deposits) responsible for around 90% of world ruby production (in weight, not in quality; 8000 tons [28]; Figure 3) and this was before 2009 when the Montepuez ruby deposits were discovered in Mozambique [21–23].

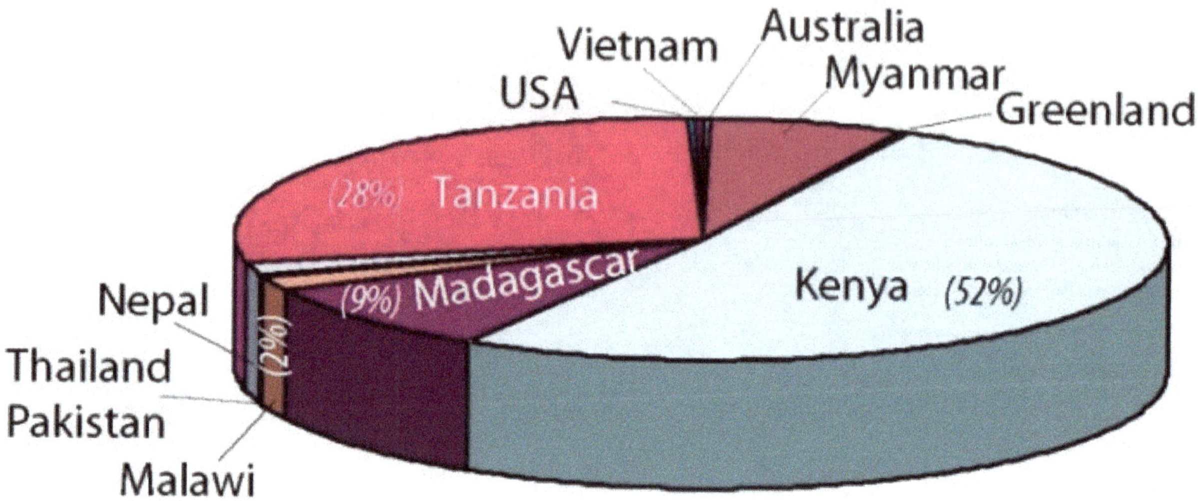

Figure 3. Ruby production worldwide in 2005 showing the production (in %) of some African countries, and others; modified from [28].

The highest-quality ruby crystals come from Central and South-East Asia and Eastern Africa ([27]; see Figure 1C). Myanmar, with the Mogok Stone Tract (see Figure 1A), is famous for having produced rubies of the finest quality since 600 CE [5,29,30]. These are often called "pigeon blood", a term of Myanmar origin popularized by English traders in the 19th century [31]. Other Asian producers including Thailand (Chanthaburi-Trat mines) and Cambodia (Pailin and Samlaut) emerged as serious players in the 1970s, but their production has declined dramatically over the past 20 years [6]. Ruby deposits in Vietnam (Luc Yen, Yen Bai, and Quy Chau areas), Afghanistan (Jegdalek), Pakistan (Hunza valley), Azad-Kashmir (Batakundi and Nangimali), and Tajikistan (Kukurt, Turakoluma, and Badakshan) have become significant producers since the end of the 1980s, but their production remains limited. In the early 1990s Mong Hsu in Eastern Myanmar emerged as a major deposit but its production has declined since the early 2000s [6].

Today the high production and quality of rubies from Mozambique (and to a lesser extend Madagascar) has changed the way rubies are traded and with the arrival of these new players, the production of many other deposits collapsed [6]. Production from the Winza deposit diminished to the point of essentially disappearing, and the Rockland (former John Saul) mine in Kenya stopped producing. In Africa (Figure 4) gem rubies are mined in Kenya (Mangare and Baringo), Tanzania (Umba-Kalalani, Morogoro, Mahenge, and Loolera), Malawi (Chimwadzulu), Madagascar (Vatomandry, Andilamena, Didy, Zahamena, and Ilakaka), and Mozambique (Montepuez, Ruambeze, and M'sawize). The Montepuez mine is the most highly mechanized ruby mine worldwide [6,26]. In Southern Greenland True North Gems Inc. started working on the Aappaluttoq deposit near Fiskenæsset in 2005 (see Figure 1F) [32–37] and a mine (operated by "Greenland Ruby") entered into production in 2017.

To explain these morphological differences, Sorokina et al. [49] introduced the concept of crystal "ontogeny" that encompasses nucleation, growth, and alteration of single crystals. According to their data, the variations of crystal habits in rubies from deposits in Africa and South-East Asia can be directly correlated with Cr and Fe contents and changes in P–T conditions during crystal growth.

3.2. The Trapiche Texture of Ruby

The conditions under which the ruby grows can give rise to a particular texture called 'trapiche' (Figure 10A). This texture was first described by [50] for an unusual Colombian emerald that resembles the spokes of a milling wheel used to process sugar cane. The term "trapiche" was used exclusively for emeralds until the 1990s, when trapiche rubies appeared in the Thai and Burmese gem markets [47,51,52].

The trapiche texture is composed of growth sectors separated by more or less sharp boundaries of inclusions [53,54], known as "sector boundaries". These boundaries separate a central "core" (corresponding to the pinacoidal growth sectors) from the surrounding growth sectors. The boundaries are generally inclined with respect to the c axis. This texture should not be confused with that seen in purple-pink and blue sapphires (Figure 10B) and resulting from the distribution of different inclusions in alternating portions of the crystal and called "trapiche-like" [55,56].

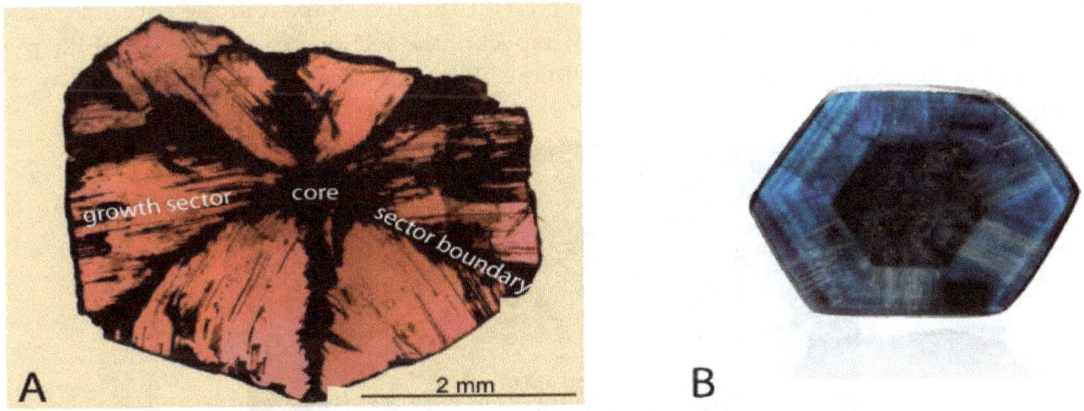

Figure 10. The trapiche texture of corundum. (**A**) Photomicrograph in transmitted light of a trapiche ruby from Mong Hsu, Myanmar (Photo: Virginie Garnier) and (**B**) aspect of a "trapiche-like" sapphire from southern Africa (2 cm × 1.8 cm × 0.2 cm, 13.46 cts). Photo: Louis-Dominique Bayle.

The trapiche texture is best seen in cross-sections perpendicular to the *c* axis. The presence and size of the core depends on the orientation of the cross-section in the crystal [57] (Figure 11A). Thus, two cross-section appearances are possible: (1) the core is present and the sector boundaries surround it and develop from its edges to the rim of the crystal, separating six trapezoidal growth sectors (Figure 11B) and (2) the core is absent and the sector boundaries intersect at a central point, giving rise to triangular growth sectors (Figure 11C).

Variable contents of chromophores, in particular Cr but also Ti and V have been detected in different rubies as well as within the same samples [52,58,59]. This can explain why the core can be red (Cr > Ti) or black (Ti > Cr) [60], although the core is sometimes yellowish/white-to-brown like the sector boundaries. For both core and sector boundaries, the yellowish/white-to brown color is due to the presence of calcite, dolomite, corundum, unidentified silicate inclusions containing K-Al-Fe-Ti, and Fe-bearing minerals. Iron oxides/hydroxides formed during late weathering that was favored by the high porosity of the sector boundaries (Figure 12) [52,57]. The sector boundaries develop along the $\langle 1\bar{1}0 \rangle$ directions and can contain other solid inclusions such as plagioclase, rutile, titanite, margarite, zircon, etc. [57–59], but they do not contain matrix material from the host rock. This absence is an important feature differentiating trapiche rubies from trapiche emeralds [61].

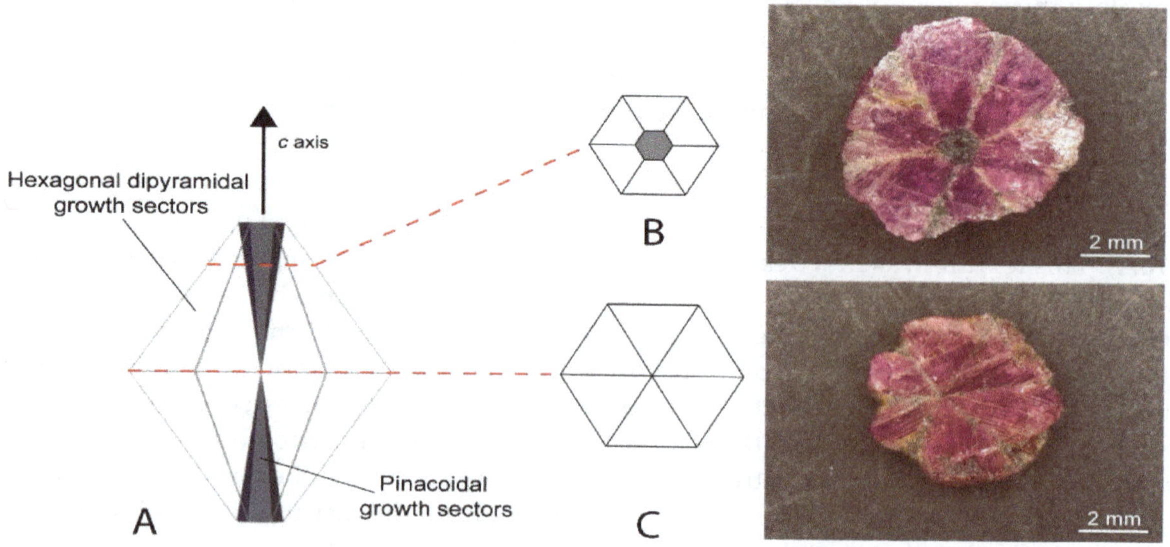

Figure 11. Variations of the ruby trapiche texture. (**A**) Variation of the core, which corresponds to the pinacoidal growth sectors (in grey). These sectors have a tapered shape and the core size varies as a function of the orientation of the cross-section; (**B**) aspect of the section near the vertex of the pyramid; and (**C**) aspect of the section if cut in the vertex plane. Photos by Isabella Pignatelli of two trapiche rubies from Mong Hsu, Myanmar.

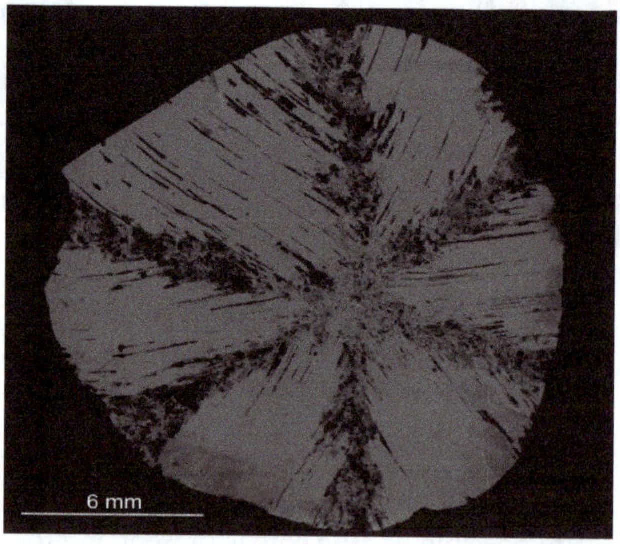

Figure 12. X-ray computed tomography image showing the high porosity (in black) of a trapiche ruby from Luc Yen, North Vietnam [57]. The porosity corresponds to 0.6% of the ruby volume and is developed in the sector boundaries. Photo: Christophe Morlot.

Tube-like voids filled with carbonates, liquid, and/or gas may be observed in the growth sectors (Figure 13) [51,57,62]. These can develop from the core or sector boundaries and extend into the growth sectors, running perpendicular to the sectors with an inclination of 5° relative to the {0001} ruby faces. Sometimes the tube-like voids can be parallel to the c axis.

Garnier et al. [58,59] found single-phase (liquid), two-phase (liquid + gas), and three-phase (liquid + gas + solid) inclusions between the tube-like voids and/or their extensions into the sector boundaries. Microthermometry and Raman spectrometry indicate that these inclusions correspond to the trapping of two immiscible fluids during the ruby formation: a carbonic fluid in the CO_2-H_2S-COS-S_8-AlO(OH) system and molten salts [62–64]. These fluids are considered the product of metamorphism of evaporites during the devolatization of carbonates and thermochemical-sulfate reduction.

Trapiche rubies are found in Myanmar, Vietnam, Tajikistan, Pakistan, and Nepal, but have never been described by geologists in situ in primary deposits. The trace element contents of the Myanmar and Vietnamese trapiche rubies prove that they originated from marble-type deposits [57–59]. Moreover, similar mineralogical and chemical features have been observed for trapiche and non-trapiche rubies from Mong Hsu in Myanmar and Luc Yen in Vietnam, confirming that they formed in the same geological context, i.e., marble-hosted ruby deposits [57,58,62,65]. The rubies formed in the amphibolite facies, during retrograde metamorphism at $620 < T < 670$ °C and $2.6 < P < 3.3$ kbar [65,66].

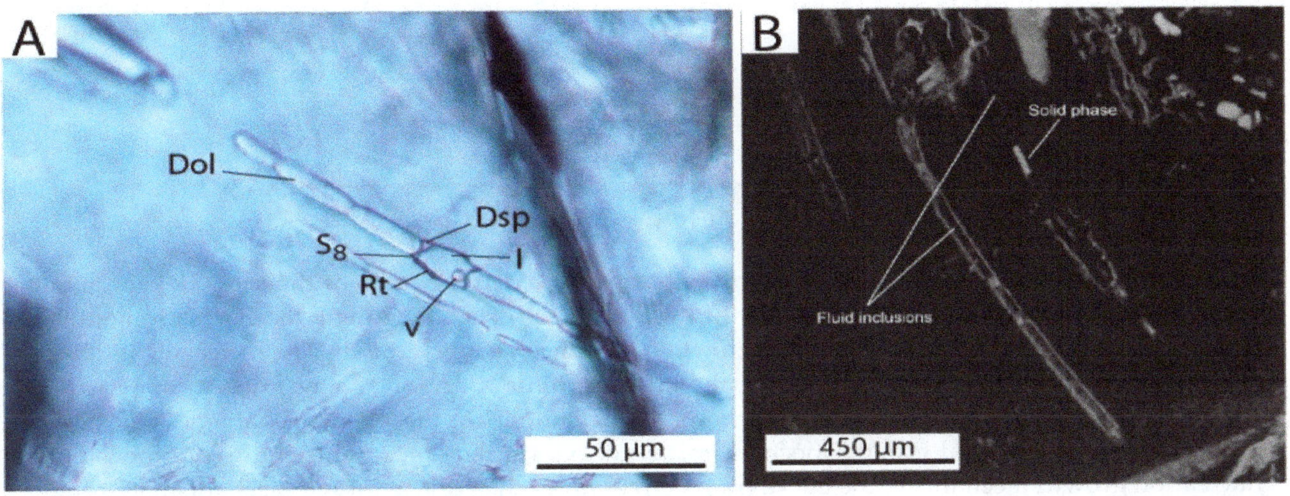

Figure 13. Solid and fluid inclusions trapped by trapiche rubies. (**A**) Primary fluid inclusions trapped along the sector boundary of a trapiche ruby from Mong Hsu, Myanmar. The fluid inclusion was trapped within tube-like structures formed by dolomite (Dol). It is a multi-phase fluid inclusion containing a two-phase (liquid + vapor) CO_2-H_2S-S_8-bearing fluid and minerals including diaspore (Dsp) and rutile (Rt). V = vapor phase, l = liquid phase; from [62]. Photo: Gaston Giuliani; (**B**) X-ray computed tomography image of a trapiche ruby from Luc Yen, north Vietnam, showing tube-like voids that contain solid and fluid inclusion cavities; from [57]. Photo: Christophe Morlot.

In this geological context, the trapiche texture developed very quickly, as witnessed by the trapping of several solid and fluid inclusions and the formation of tube-like voids. The growth episodes in minerals are due to changes in the driving-force conditions [47]. In solution growth, mass transfer plays a key role. The driving force is defined by the difference in concentration between the bulk and subsurface supersaturation, i.e., the concentration gradient within the boundary layer [46]. Three growth episodes due to changes in driving force conditions were identified by [47] (Figure 14). Firstly, the core formed under low driving-force conditions via layer-by-layer growth (Step 1). It s development was hindered by the rapid formation of sector boundaries, which enveloped it and constitute the "skeleton" of the trapiche ruby (Step 2). These boundaries formed when the driving-force conditions increased and adhesive-type growth took place. Finally, a decrease in the driving-force conditions allowed the formation of growth sectors and the corresponding crystal faces through ordinary layer-by-layer growth (Step 3). The growth sectors filled the interstices left by the sector boundary, completing the trapiche texture.

3.3. Chemical Composition and Geographic Origin of Ruby

Currently, the accurate chemical composition of rubies can be determined by several analytical techniques, such as X-ray fluorescence (XRF), electron microprobe analysis (EMPA), laser ablation-inductively coupled plasma-mass spectrometry (LA-ICP-MS), and secondary ion mass spectrometry (SIMS). Chemical data facilitate the differentiation between natural and synthetic rubies [67] (Figure 15) and help to determine their geological origin worldwide [11,68–72].

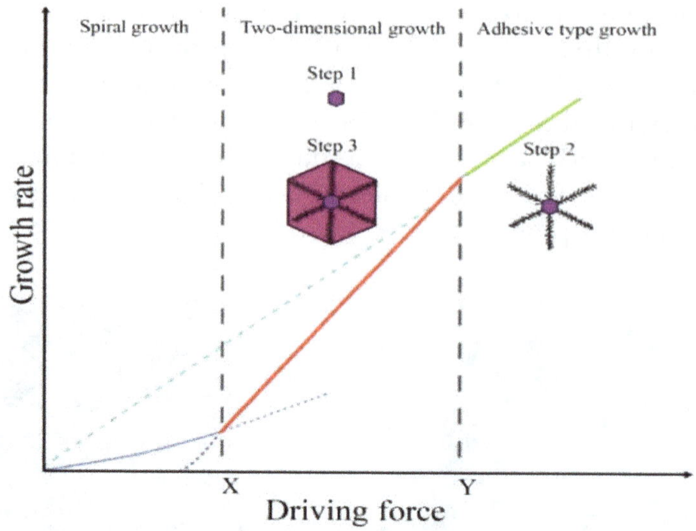

Figure 14. Schematic diagram showing growth rate versus driving force of the three development stages of trapiche ruby. The diagram shows the different types of crystal growth mechanisms, modified from [47]: spiral growth, two-dimensional growth, and adhesive-type growth. The increasing intensity of the driving force is presented between the critical points X and Y, where nucleation occurred. The curves represent growth rate versus driving force relations for the three mechanisms. The curve (in blue) shows the spiral growth mechanism; the curve (in red) represents the two-dimensional nucleation growth mechanism; the curve (in green) denotes the adhesive-type mechanism. A rough interface is expected above point Y and it corresponds to the dendritic stage of trapiche-ruby formation. The nucleation occurred between X and Y domains where the growth sectors formed in trapiche texture. Finally, a smooth interface is expected below X where nucleation by dislocation occurred either for non-trapiche ruby or in overgrowth zone in trapiche ruby.

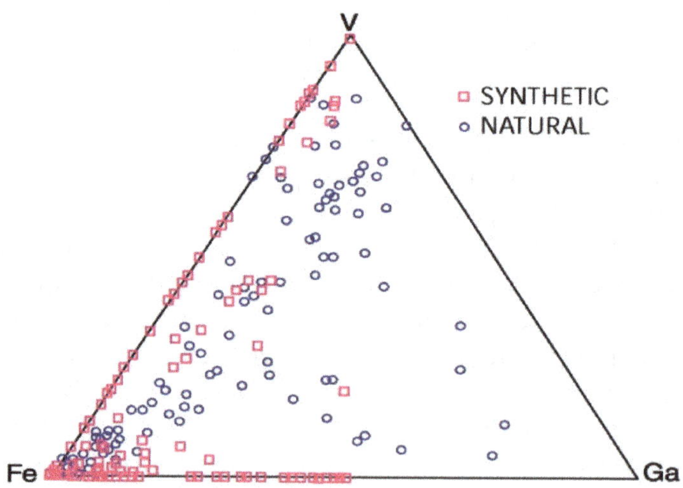

Figure 15. Gallium (Ga)–vanadium (V)–iron (Fe) ternary showing the distinction between natural and synthetic rubies, modified from [67].

Sutherland et al. [73–75] used a Fe_2O_3/TiO_2 versus Cr_2O_3/Ga_2O_3 diagram to decipher the metamorphic versus magmatic origin of corundum. This diagram is commonly used for the separation of blue sapphires from several deposits worldwide [76]. The corundum crystals termed "metamorphic" have pastel (blue, pink, and orange) to red colors, and are rich in chromium and poor in gallium (Cr_2O_3/Ga_2O_3 ratio > 3). The corundum crystals defined as "magmatic" are xenocrysts of blue-green-yellow (BGY) sapphires with a Cr_2O_3/Ga_2O_3 ratio < 1. The diagram was applied to rubies and sapphires related to alkali basalt from the provinces of Antananarivo (the Soamiakatra, Kianjanakanga, and Ambatomainty deposits) and

Toamasina (the Vatomandry and Andilamena deposits) in Madagascar [77,78] (Figure 16). Note that all of the points related to rubies plot in the metamorphic field. They overlap data for rubies from Andilamena, which is located in the north of the Vatomandry area, and where the corundum is hosted in different types of rocks (viz. metamorphosed mafic and ultramafic rocks (M-UMR) and plumasites in M-UMR). The majority of the sapphires plot in the magmatic domain. The distribution of sapphires from the Vatomandry area is heterogeneous, and while most are located in the magmatic domain yet others are in the metamorphic one. The latter corresponds to colorless, light to dark green, and light to purplish blue sapphires with significant Cr_2O_3 contents (150 ppm < Cr_2O_3 < 2100 ppm).

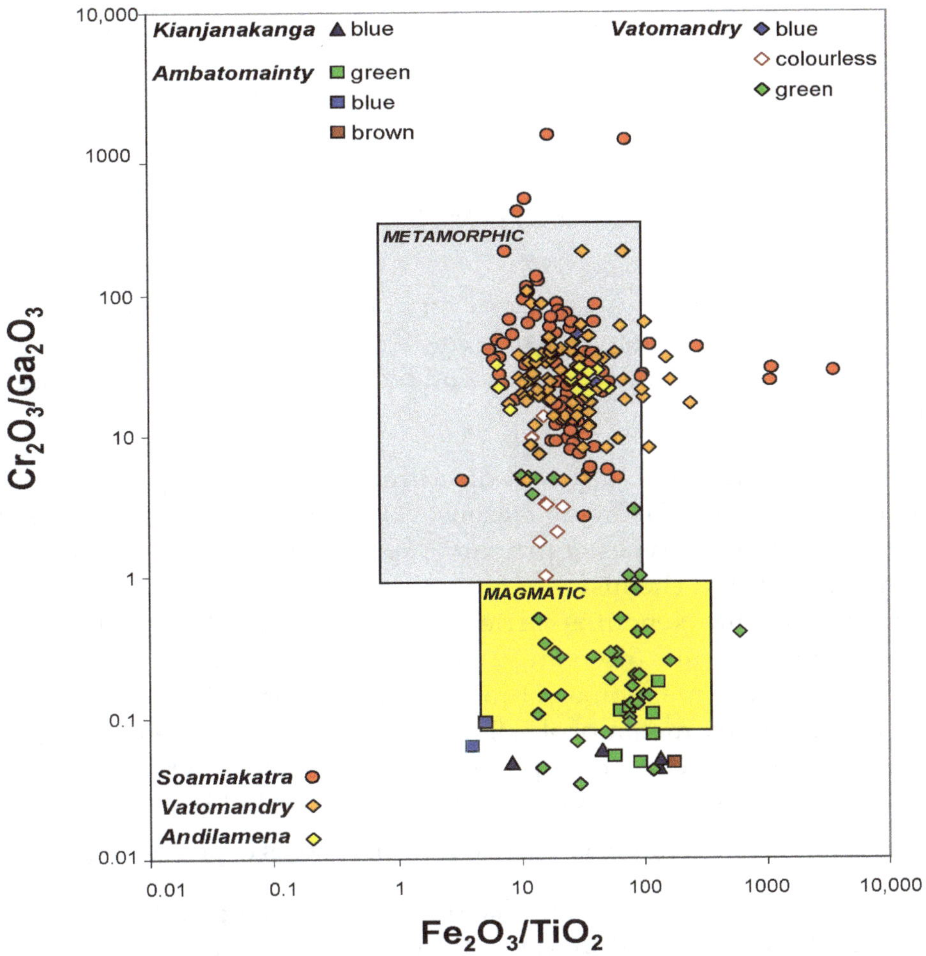

Figure 16. Cr_2O_3/Ga_2O_3 versus Fe_2O_3/TiO_2 diagram for rubies and sapphires associated with alkali basalts from the Antananarivo and Toamasina Provinces in Madagascar. The metamorphic and magmatic boxes defined for Australian sapphires are from [68,75].

Giuliani et al. [72] and Uher et al. [79] proposed a new chemical classification diagram for gem corundum based on the different types of primary corundum deposits worldwide (see [40,80]). It used a database of 2000 EMPA analyses of ruby and sapphire that were obtained under the same analytical conditions. The $FeO + TiO_2 + Ga_2O_3$ versus $FeO–Cr_2O_3–MgO–V_2O_3$ diagram in Figure 17 uses: (1) the FeO content to discriminate between iron-poor rubies in marbles and iron-rich rubies in M-UMR and (2) the second discriminator between ruby and sapphire is the addition to (Y-axis parameter) or subtraction from (X-axis parameter) FeO of trace elements associated preferentially with sapphire (TiO_2 and Ga_2O_3) or ruby (Cr_2O_3, V_2O_3, and MgO). The fields of the different gem corundum deposits are: for rubies, (R1) marble; (R2) John Saul mine-type; (R3) M-UMR; (R4) metasomatites; for sapphires, (S1) syenitic rocks; (S2) metasomatites; and (S3) alkali-basalt and lamprophyre. Finally, the chemical diagram permits a clear separation with limited overlap of the main types of deposit found worldwide.

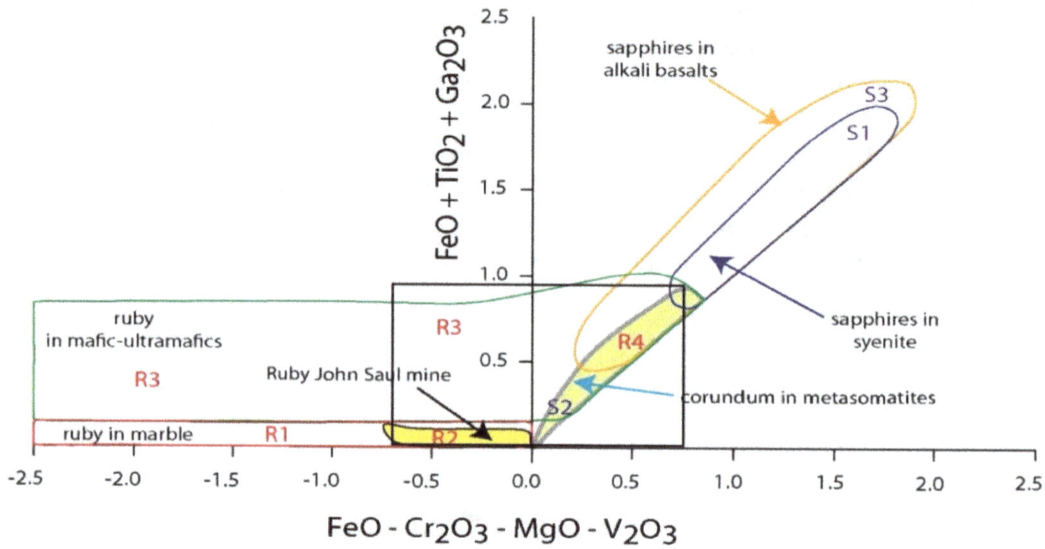

Figure 17. FeO-Cr_2O_3-MgO-V_2O_3 versus $FeO + TiO_2 + Ga_2O_3$ diagram based on a database of 2000 electron microprobe analysis (EMPA) analyses (in wt % oxides) showing the main chemical fields defined for different types of corundum deposits worldwide [72]. Different geochemical domains are defined (see text).

The chemical diagram provides sufficient discrimination to distinguish between the different types of corundum deposits for the following reasons: (1) the database contains more than 2000 EMPA analysis representative of all of the types of primary deposits and (2) the magmatic sapphire domain (S1) is defined by the geology and chemistry of corundum. The sapphires are hosted in syenitic rocks or occur as megacrysts in syenitic xenoliths carried by alkali basalts. Most of the sapphires in placers linked to the erosion of alkali basalts (domain S2) overlap the domain S1. The main superposition of geological domains is for ruby of metasomatic (-metamorphic) origin that overlaps the domain of ruby associated with M-UMR (strictly metamorphic); (3) the diagram uses Ga and Mg despite their high detection limits, and the two elements are opposed: Ga is added to Ti while Mg is added to Cr and V. This choice corresponds with the fact that Ga and Ti are concentrated in magmatic sapphires (syenitic rocks) while Mg, Cr, and V are concentrated in metamorphic rubies hosted in Mg-Cr-(V)-rich protoliths.

A statistical classification by discriminant factors analysis using oxide concentrations (Ti, Mg, V, Cr, Fe, and Ga) of more than 900 corundum crystals from the main primary deposits worldwide was proposed by [72]. The deposits were classified into different types: (1) for ruby as marble, M-UMR, and metasomatites and (2) for sapphires as syenitic rocks, metasomatites, plumasites, and skarns.

The discriminating factors F1 and F2 calculated for the different types of ruby deposits (known classes) with their respective centroid for each class are as follows:

$$\text{Factor } 1 = 0.271691 \times (Cr_2O_3 - 0.294456)/0.376245 + 0.796762 \times (FeO - 0.26294)/0.224015 + 0.0206904 \times (Ga_2O_3 - 0.0112047)/0.0124095 + 0.136135 \times (MgO - 0.00545581)/0.00526035 - 0.124367 \times (TiO_2 - 0.0274651)/0.0539087 - 0.108332 \times (V_2O_3 - 0.0102233)/0.0111099$$

$$\text{Factor } 2 = 0.124957 \times (Cr_2O_3 - 0.294456)/0.376245 - 0.121871 \times (FeO - 0.26294)/0.224015 - 0.287007 \times (Ga_2O_3 - 0.0112047)/0.0124095 - 0.0527783 \times (MgO - 0.00545581)/0.00526035 + 0.0584802 \times (TiO_2 - 0.0274651)/0.0539087 - 0.439517 \times (V_2O_3 - 0.0102233)/0.0111099$$

The discriminating factors F1 and F2 calculated for the different types of sapphire deposits (known classes) with their respective centroid for each class are as follows:

$$\text{Factor 1} = 0.23095 \times (Cr_2O_3 - 0.0613814)/0.0596296 - 0.723839 \times (FeO - 0.489593)/0.372367 - 0.0403934 \times (Ga_2O_3 - 0.0159195)/0.00946033 + 0.0767757 \times (MgO - 0.00613983)/0.0040002 - 0.00901195 \times (TiO_2 - 0.0236271)/0.0567399 + 0.0791335 \times (V_2O_3 - 0.00402119)/0.00329669$$

$$\text{Factor 2} = -0.530426 \times (Cr_2O_3 - 0.0613814)/0.0596296 - 0.638704 \times (FeO - 0.489593)/0.372367 - 0.00893226 \times (Ga_2O_3 - 0.0159195)/0.00946033 - 0.200622 \times (MgO - 0.00613983)/0.0040002 + 0.296868 \times (TiO_2 - 0.0236271)/0.0567399 - 0.165072 \times (V_2O_3 - 0.00402119)/0.00329669$$

Finally, prediction diagrams were proposed for the different types of deposits and applied to placers of alkali basalt-related deposits [72]. Figure 18 shows how such a diagram can be used to determine the most probable deposit type for ruby from the Soamiakatra deposit in Madagascar.

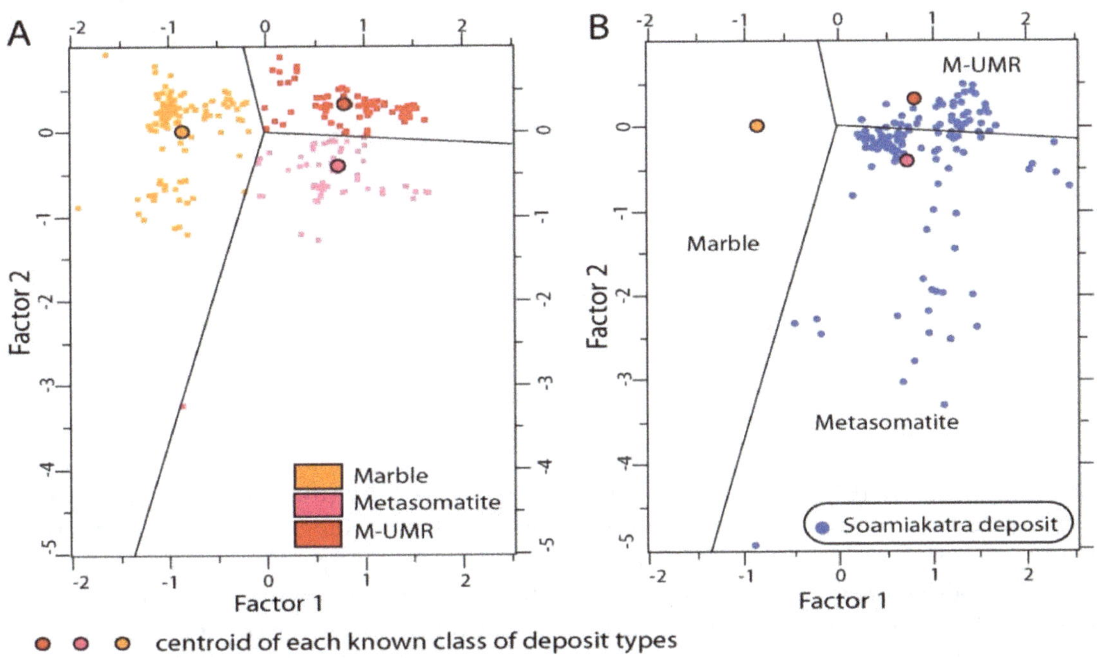

Figure 18. Discriminant factors analysis using oxide concentrations in ruby; modified from [72]. (**A**) Diagram used to distinguish between the different types of ruby deposit (marble, metamorphosed mafic and ultramafic rocks (M-UMR), and metasomatite) with the centroids of the known deposit classes and (**B**) prediction of the geologic origin of rubies from the Soamiakatra deposit in the Antanarivo province of Madagascar.

Peucat et al. [70] obtained LA-ICP-MS analyses of corundum from different geological settings and suggested using their Fe, Ti, Cr, Ga, and Mg contents and chemical ratios such as Ga/Mg, Fe/Ti, Fe/Mg, and Cr/Ga to determine their geological origins. The Ga/Mg ratio versus the Fe content has proven to be an efficient tool for discriminating between some metamorphic and magmatic blue sapphires, even if some localities overlap and are difficult to discriminate. Natural rubies have normally limited Ga (<200 ppm) and low Ga/Mg ratios (<10) and in some cases their trace-element contents are unusual and they plot in the field for sapphire [81].

The use of Fe versus Ga/Mg for ruby is presented here for different types of deposits (Figure 19). The main types of deposits are discriminated as a function of the Fe content of the ruby: (1) range 10,000–1000 (ppm) for ruby in M-UMR (metamorphism sensu stricto and metamorphic metasomatic) with the example of different ruby deposits in Southern Madagascar (Vohitany, Anavoha, and Andriba), Greenland (Aappaluttoq), Mozambique (Maninge Nice at Montepuez), ruby as xenocrysts in alkali basalts (examples include Pailin, Macquarie River, New England, and Chanthaburi-Trat) and kimberlite (Mbuji-Mayi) or alluvial from kimberlites (Juina, Brazil); (2) range 1000–10 ppm for ruby in marbles (Lu Yen-Yen Bai and Quy Chau in Vietnam, Mogok and Mong Hsu in Myanmar); and (3) less than 10 ppm for metasomatites as plumasites (John Saul mine, Kenya).

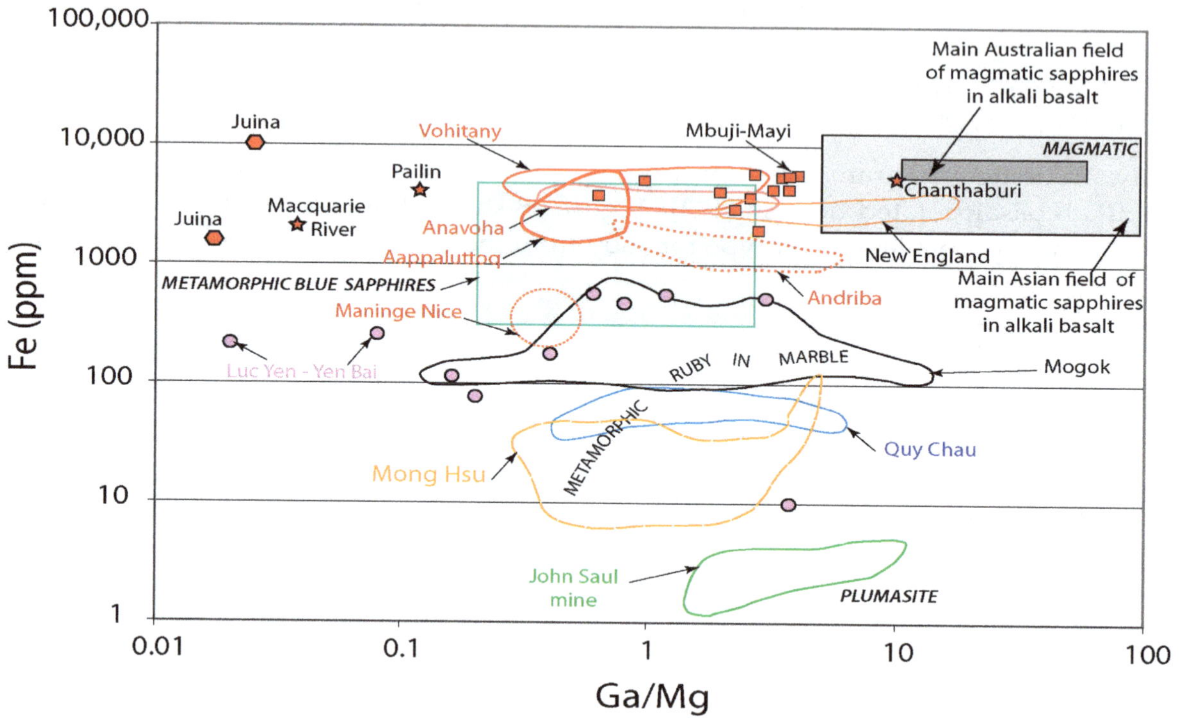

Figure 19. Fe (ppm) versus Ga/Mg discrimination diagram for ruby (this work; modified from [69]), showing the variation of Fe content for some rubies relative to their geological environments. The boxes for magmatic sapphire related to alkali basalts are reported for comparison with the plot of Pailin sapphire in Cambodia (data from [72,82]) as well as the box of metamorphic sapphires (drawn with the range of values defined in the tables of [70]). The following different types of ruby and/or their chemical fields are shown: (1) the chemical domain of rubies associated with M-UMR (red domains) from (i) Madagascar [72] with the occurrences of Vohitany (analysis, n = 25), Anavoha (n = 6), and Andriba (n = 6); (ii) Greenland and the chemical domain of Aappaluttoq ruby deposit (n = 24, data from [83]); (iii) Mozambique (Montepuez mining district) with the colluvial deposit of Maninge Nice from (Pardieu, unpublished data); (2) inclusions of ruby in diamonds from Juina in Brazil (red hexagon) [43]; (3) xenocrysts of ruby from the Mbuji-Mayi kimberlite in the Democratic Republic of Congo (red square) [84]; (4) xenocrysts of rubies in alluvial of alkali basalts such as Pailin, Chantaburi-Trat, and Macquarie River and New England in Australia (red star) [82,85,86]; (5) rubies in marble from Luc Yen-Yen Bai represented by single measurements (violet circle) and Quy Chau (blue domain) in Vietnam [57,66], Mogok (black domain) [87], and Mong Hsu (orange domain) [88]; and (6) rubies from the former John Saul (now Rockland) mine in Kenya (green domain) from desilicated pegmatites (Giuliani et al., unpublished data). These rubies show high variations in Fe, which permits discrimination of their geological type; this is characterized by a decrease of Fe from rubies in M-UMR to marble and metasomatic rocks (plumasite and shear-zones). The Ga/Mg ratio of rubies from M-UMR is different from magmatic sapphires but is within the range of values of metamorphic sapphires as defined by [69]. The domains represent the distribution of several spot analyses. The others analyses are the average values of punctual measurements.

The Ga/Mg ratio is highly variable, between 0.1 and 10, and some rubies such as those occurring as xenocrysts in alkali basalt (Chanthaburi-Trat and New England) and M-UMR (Madagascar) plot in the magmatic field of sapphire (right side of Figure 19). Such an overlap opens the question of the significance of such geochemical signatures and will be discussed elsewhere.

Geographic origin determination of ruby has become a great challenge during recent decades [72]. The sampling of rubies from reliable sources worldwide [20,36,89] has facilitated the assembly of reliable reference collections in the gem laboratories [90]. The mineral assemblages and inclusions combined with trace element chemistry sometimes make it possible to decipher the geographic origin of rubies. Trace elements are used by most of the laboratories and the data are collected with several analytical techniques [91] mainly LA-ICP-MS.

Representative analyses of rubies originating from deposits of different geological types are presented in Tables 1 and 2. The quantification of the various trace elements is done using commercially available calibration standards such as those provided by the National Institutes of Standards and Technology (NIST) or the United States Geological Survey (USGS). The difference in composition between the standards and the analyzed corundum has led to some uncertainties in the data generated by the different laboratories. The Gemmological Institute of America laboratory developed matrix-matched corundum standards of ruby and sapphire and the standard deviation of the analyzed ruby became negligible and therefore can be compared with a database of stones with known provenance [92]. The application of this analytical protocol to ruby made it possible to confirm the geological classification of gem corundum (ruby and pink corundum) for (1) Fe-poor ruby hosted in marble and (2) Fe-rich ruby in M-UMR (Figure 20). The geographic determination is not based on the Cr content of ruby because it appears by experience that it is not useful for origin certification [11]. The discrimination for Fe-rich rubies in M-UMR is based on the contents of V-Mg-Fe as shown in Figure 20 for deposits in Mozambique, Madagascar, and Thailand/Cambodia.

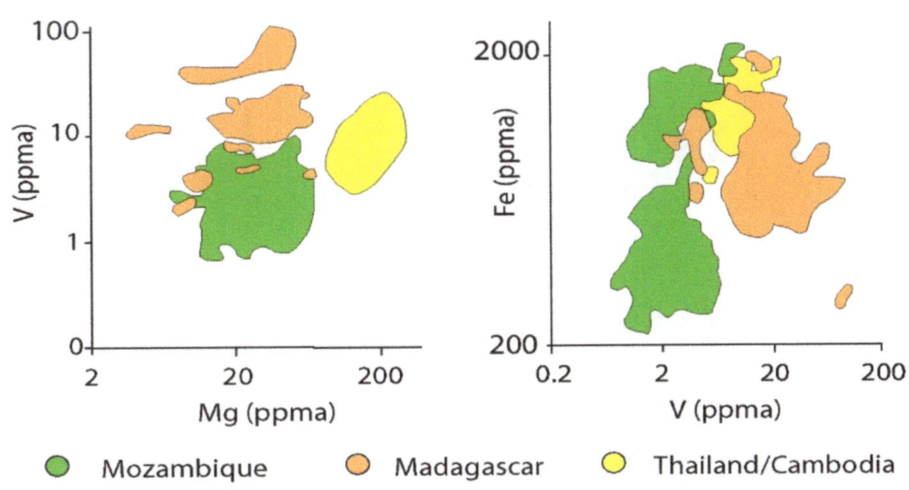

Figure 20. The geographic origin of rubies from Mozambique, Madagascar, and Thailand/Cambodia using V versus Mg and Fe versus V diagrams, modified from [11]. The limits of detection of the analyses by laser ablation-inductively coupled plasma-mass spectroscopy (LA-ICP-MS) are for Mg, V, and Fe respectively, 0.1–0.3, 0.03–0.2, and 5–20 ppm.

The geographic origin of ruby can usually be achieved based on inclusions in association with trace element chemistry despite some overlaps [11–14]. The oxygen isotopic compositions of these rubies overlap strongly and are not useful for geographic origin determination [93].

Table 1. Representative oxide (wt %) EMPA analyses of rubies originating from deposits of different geological types relative to the new enhanced geological classification for ruby.

Type of Deposit	Deposit	Color Corundum	Country	Al_2O_3	MgO	TiO_2	V_2O_3 (in wt %)	Cr_2O_3	FeO	Ga_2O_3	Total	References
Type IA	Somiakatra	deep red	Madagascar	98.16	na	0.03	0.060	0.858	0.52	0.01	99.64	[94]
		pink		98.75	na	0.01	0.000	0.243	0.26	0.01	99.28	
Type IB	Mbuji Mayi	fuchsia	RDC	93.76	0.01	0.01	0.029	5.641	0.47	0.01	99.94	[84]
	Paranesti	red	Greece	97.58	bdl	0.01	0.010	2.650	0.39	0.07	100.71	[95]
Type IIA$_1$	Montepuez	red	Mozambique	99.38	bdl	0.01	na	0.080	0.45	bdl	99.92	[96]
	Vohitany	red	Madagascar	99.06	0.01	0.01	0.000	0.219	0.59	0.01	99.90	[72]
	Ronda	red	Spain	99.33	0.02	0.01	na	0.120	0.19	na	99.67	[97]
	Luc Yen	red to pink	Vietnam *	99.42	0.02	0.02	0.010	0.370	0.02	0.01	99.87	[98]
Type IIA$_2$	Mogok	red	Myanmar	99.01	0.00	0.01	0.047	0.794	0.01	0.01	99.88	[72]
	Mong Hsu	red		99.20	0.01	0.08	0.041	0.521	0.00	0.00	99.85	
	Nangimali	pink	Pakistan	99.85	0.01	0.02	0.005	0.055	0.00	0.01	99.94	
	Anavoha	pink	Madagascar	98.76	0.00	0.00	0.002	0.023	0.45	0.00	99.25	
Type IIB$_1$	Naxos	pink	Greece	100.70	0.35	0.27	0.090	0.180	0.40	0.15	102.14	[95]
	Polar Urals	red	Russia	95.16	bdl	bdl	na	3.860	0.44	na	99.46	[99]
	John Saul (Kimbo)	pink	Kenya	100.09	0.03	0.04	0.016	0.595	0.00	0.04	100.80	[72]
Type IIB$_2$	Zazafotsy	red	Madagascar	98.63	0.01	0.03	0.005	0.188	0.30	0.01	99.17	[100]
	Hokitika	red	New Zealand	90.65	na	bdl	0.050	9.050	0.18	na	99.93	[42]
Type IIIA	Vatomandry	red	Madagascar	98.43	0.02	0.04	0.040	0.761	0.52	0.02	99.83	[94]
		pink		99.29	0.03	0.04	0.005	0.161	0.44	0.02	99.99	
		red brown		97.68	0.03	0.04	0.053	0.561	0.63	0.03	99.02	
	Antsabotraka	red brown		99.48	0.02	0.02	0.010	0.130	0.39	0.02	100.06	
	Ambatomainty	red brown		99.50	0.03	0.02	0.012	0.127	0.21	0.01	99.91	
Type IIIB	Macquarie	red	Australia	99.07	0.09	bdl	na	0.450	0.22	na	99.83	[75]
	Ilakaka	pink	Madagascar	99.28	0.01	0.01	0.001	0.178	0.29	bdl	99.77	[72]
	Quy Chau	red to pink	Vietnam ▫	99.28	0.00	0.03	0.020	0.580	0.06	0.01	99.98	[98]
	Luc Yen	red to pink	Vietnam ▫	99.76	0.00	0.01	0.070	0.540	0.01	0.01	100.40	

na = not analyzed; *, ▫ = number of analyses for calculating the presented average composition of ruby or pink corundum with respectively; * n = 29, ▫ n = 40 and n = 16; bdl = below detection limit.

Table 2. Representative analyses (in ppm) obtained by LA-ICP-MS of rubies originating from deposits of different geological types relative to the new enhanced geological classification for ruby.

Type of Deposit	Deposit or Mining District	Country	Mg	Ti	V (in ppma)	Fe	Ga	Cr	Number of Analyses	References
Type IIA$_1$	Longido	Tanzania	(7–42)	np	(3–25)	(62–544)	(6–10)	(1604–5059)	16	[101]
in M-UMR	Winza		(0–118)	np	(0–1)	(405–1596)	(4–11)	(161–1094)	14	
	Chimwadzulu	Malawi	(10–39)	np	(1–10)	(953–2760)	(5–15)	(269–1816)	18	
Type IIB$_1$	Aappaluttoq	Greenland	(2–81)	(15–210)	(0.1–14)	(655–2516)	(0.7–25)	(50–2871)	17	[102]
Type IIA$_2$	Mogok (Mo)	Myanmar	(5–75)	(7–80)	(48–1089)	(bdl-301)	(2–51)	np	65 = Mo + MH	[11]
in marble	Dattaw (Mo)		(bdl)	(105–702)	(66–97)	(bdl-237) *	(32–42)	(105–702)	4	[103]
	Kadoke Tat (Mo)		(bdl-25)	(bdl-580)	(150–1472)	(bld-644) *	(11–140)	(61–2963)	13	
	Baw-Padan (Mo)		(33–40)	(70–74)	(206–347)	(270–442)	(106–166)	(29–2339)	2	[87]
	Namya		(bdl-66)	(79–739)	(184–489)	(bdl-234) *	(23–107)	(479–6864)	12	[103]
	Mong Hsu (MH**)		(48–115)	(778–1815)	(324–417)	(bdl-29)	(66–89)	(2026–7015)	65 = Mo + MH	[88]
	not precise	Vietnam	(1–130)	(bdl-1214)	(2–214)	(bdl-28)	(1–122)	np	93	[11]
	Jegdalek	Afghanistan	(7–380)	(9–486)	(4–135)	(bdl-1128)	(5–35)	np	74	
	Snezhnoe	Tajikistan	(6–58)	(bdl-579)	(24–112)	(bdl-155)	(14–28)	np	48	
			(4–72)	(6–75)	(51–122)	Bdl *	(61–83)	(1696–4204)	6	[104]
Type IIIA	Pailin	Cambodia	(97–258)	(32–128)	(5–22)	(818–1935)	(5–11)	np	34	[11]
in alkali			(118–226)	(95–210)	(16–33)	(2454–3620)	(18–34)	(450–7761)	14	[105]
basalt	Chantaburi-	Thailand	(126–181)	(32–138)	(4–14)	(756–1442)	((5–10)	np	7	[11]
	Trat		(102–163)	(81–219)	(9–30)	(2175–3794)	(16–32)	(955–2856)	10	[105]
	New England	Australia	(9–55)	(6–203)	(4–14)	(2560–3150)	(170–310)	(870–3370)	?	[81]
Type IIIB	Mugloto	Mozambique	(11–50)	np	(2–7)	(851–2034)	(5–12)	(506–1737)	14	[101]
in M-UMR	Maninge Nice		(19–47)	np	(1–6)	(261–402)	(6–9)	(1886–4941)	5	
	Montepuez		(8–65)	(bdl-59)	(bdl-10)	(231–2154)	(5–14)	np	85	[11]
Type IIIB	Zahamena	Madagascar	(13–61)	np	(10–100)	(284–1994)	(10–30)	(135–3922)	14	[101]
in MetamR	Didy		(0–67)	np	(2–17)	(533–1274)	(3–17)	(32–2698)	21	
	Andilamena		(19–53)	np	(4–29)	559–1559)	(7–21)	(53–1569)	14	

np = not precise because Cr is no longer used in the GIA laboratory as a discriminant for geographic origin determination; * = iron was measured by EMPA; MH** = range of averages composition of core and rim for two samples (n = 33 analyses) from Mong Hsu (MH); MetamR = metamorphic rocks environment; M-UMR = mafic-ultramafic rocks; Type IIA2 refers to the proposed new geological classification for ruby; bdl = below detection limit.

4. Geological Environment and Age of Primary Ruby Deposits

The global distribution of ruby deposits is closely linked to the Wilson cycle, i.e., collision, rift and subduction geodynamics [80,106,107]. Ruby forms in mafic, felsic, and metamorphosed carbonate platform geological environments, but it is always associated with rocks depleted in silica and enriched in alumina. In the presence of silica, Al is preferentially incorporated into Al-Si-bearing minerals such as feldspars and micas. The mineral association of corundum commonly consists of plagioclase, sapphirine, biotite, phlogopite, amphibole, pyroxene, and carbonates [108].

Two major geological environments are favorable for the presence of ruby, specifically amphibolite to medium-pressure granulite facies metamorphic belts, and alkaline basaltic volcanism in continental rifting environments. Three questions arise concerning ruby in metamorphic environments, which formed primarily via changes in temperature and pressure and fluid–rock interaction such as diffusion or percolation (metasomatism): (1) the origin and role of the parental fluids; (2) the nature and importance of the protolith; and (3) the characteristics of the paleogeography of the depositional sedimentary environment (in the case of carbonate platforms).

Gem ruby is rare because it requires the existence of several conditions: (1) an environment containing aluminum or the circulation of an Al-bearing metasomatic fluid; (2) a parental fluid issued either from the gem host-rock environment or exotic fluid circulations at medium to high temperatures (T > 500 °C); (3) a seed surface and sufficient space for the growth of the crystal; (4) the incorporation of trace elements especially Cr, Fe, and V from the parental fluid in the unit cell of the mineral; and (5) the absence of internal crystalline deformation during and after growth.

In the amphibolite to medium-pressure granulite facies metamorphic belts, the gem corundum pressure-temperature field corresponds to a pressure domain above 2 kbar, and temperatures between 500 and 850 °C (Figure 21).

The host lithologies are alumina-rich and/or silica-poor rocks such as marble, aluminous gneiss, metamorphosed M-UMR, or juxtaposed meta-felsic and silica-poor rocks affected by the circulation of fluids at their contact and altered by metasomatism (metasomatites such as desilicated pegmatite called plumasite, skarn, etc.). Marble is a silica- and alumina-depleted rock, and under these conditions the transport of alumina by a fluid phase seems necessary for the crystallization of corundum; however, Al is usually considered to be an immobile element but investigations on different ruby deposits associated with fluid circulation and metasomatism have shown that Al was mobile [83].

Ruby-bearing skarns have not been clearly identified but the presence of silica enrichments in certain rubies from Mogok [87,103] led some authors to suspect the influence of thermal anomalies accompanying the different Si-rich intrusions affecting the mining district. This opens debate on the accuracy of determination of Si content in ruby, and the use of accurate ruby standards is necessary to decrease the detection limit of Si as done by [87,92]. Stone-Sundberg et al. [92] measured by LA-ICP-MS analysis Si ranging from 70 to 305 ppm in rubies from Mogok as well as other elements such as B (11–82 ppm), Ga (280–800 ppm), and Sn (2–35 ppm) that recorded potential skarn trace-element enrichments [88]. Emmett et al. [109] focused on the low mass resolution of quadrupole ICP-MS as the cause of that technique inability to measure accurately Si in corundum. Another problem is that ICP-MS occurs in a silica glass torch, which causes large variations in the Si background leading to high detection limits [109]. The results published up to now for different rubies [91,104] remain questionable. Measurements of Si by SIMS could be a more accurate method as explained by [109].

In alkali basalt volcanism, rubies are xenocrysts carried to the surface by the magmas. The corundum is found either as xenocrysts or crystal within xenoliths in lava flows and plugs of subalkaline olivine basalts, high alumina alkali basalt, and basanite. These magmas occur in crustal extensional environments impacted by the rise of upwelling mantle plumes. Ruby-bearing xenoliths are very rare because the mother rocks generally melt or disintegrate during storage in the magma chamber or during the magma transport. In Madagascar, xenoliths of ruby-bearing clinopyroxenite and metagabbro have been found in the Soamiakatra deposit. The metamorphic ruby formed in the lower crust at a temperature around 1100 °C and a pressure of about 20 kbar [77,94].

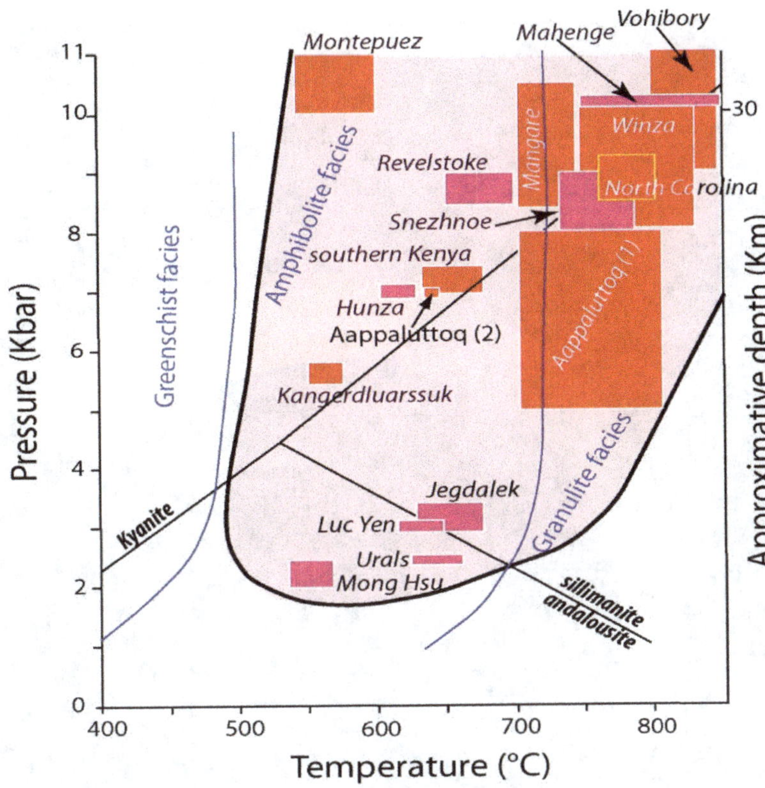

Figure 21. Pressure–temperature (P–T) formation conditions of metamorphic gem ruby (modified from [40,110]). The P–T fields of: (1) the marble type deposits from Mong Hsu [60], Urals [111], Luc Yen and Jegdalek [65], Hunza [112], Revelstoke [113], Mahenge [114], and Snezhnoe [115] are represented by violet rectangles and (2) the M-UMR deposits, metamorphic sensu stricto or associated with metasomatic metamorphism are represented by red rectangles: Greenland-Kangerdluarssuk [116], Aappaluttoq 1 [34] and 2 [83]; Southern Kenya [117,118]; Montepuez [96]; Mangare [119]; Winza [120]; North Carolina [121]; and Vohibory [122]. The P–T fields of ruby and pink corundum in clinopyroxenites and metagabbro xenoliths associated with alkali basalts are not shown. The schematic fields of Greenschist, Amphibolite, and Granulite facies are reported.

The ages of corundum formation are defined by indirect dating of a series of minerals (zircon, monazite, rutile, titanite, and micas) present either in the host rocks or as syngenetic inclusions in the corundum. These minerals have different blocking temperatures for closure to isotopic migration that make it possible to establish a cooling history for the corundum. Four main periods of gem corundum formation that include ruby and sapphire and encompass the main economic production worldwide, are presented in Figure 22 [80,106,107]. The others primary ruby occurrences known worldwide as in Australia, Norway, Finland, United-States, and other countries (see Figure 2) were not considered in the main periods of gem ruby formation. They are either uneconomical or without determination of ages.

The oldest deposits were formed in the Archean (2.97–2.6 Ga). An example is the Aappaluttoq ruby deposit in the 2.97 Ga Fiskenæsset anorthosite complex, Greenland (Figure 2). The deposit occurs at the contact between two types of rocks, i.e., ultramafic rocks (meta-peridotite or dunite) and an anorthite-rich rock (anorthosite and leucogabbro) [34,83]. Ruby is hosted in phlogopite-calcic amphibole metasomatic rocks formed by fluid–rock interaction at the contact of the two lithologies. The metasomatic fluids were triggered by the intrusion of pegmatites dated between 2.72 and 2.70 Ga with U–Pb zircon geochronology by LA-ICP-MS [83]. U/Pb dating of monazite gave age at 2.64 Ga [34] and a direct Pb-Pb isochron age of 2.68 Ga was obtained on ruby by Thermal Ionisation Mass Spectrometry [103]. The age of ruby is correlated to the age of the pegmatites intrusion; consequently an average age at 2.71 Ga is considered for ruby formation [83].

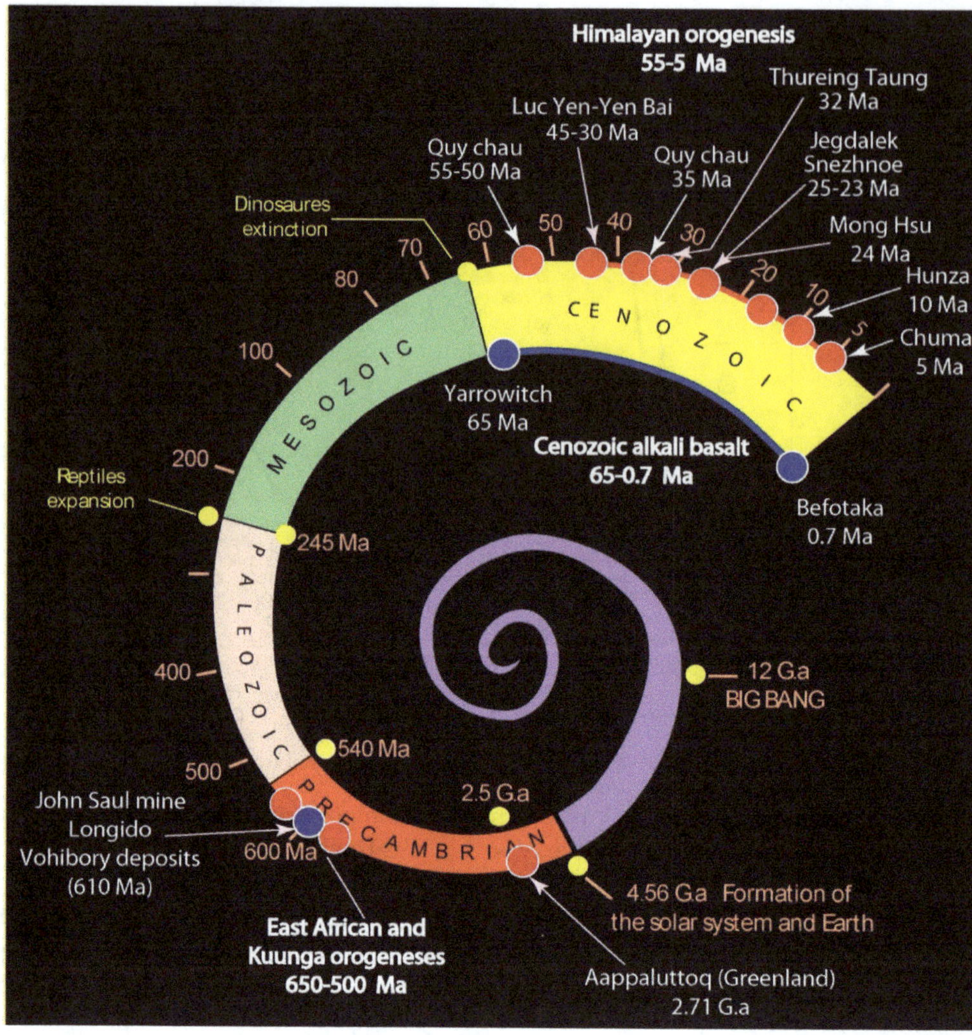

Figure 22. The spiral time diagram of corundum formation that includes economic ruby (and sapphire) deposits. Four main economic corundum periods have been defined worldwide by different methods of indirect dating of corundum. The oldest deposit is Aappaluttoq in Greenland where the ruby formed during the Archean at 2.71 Ga. The main period of gem rubies formation was the Pan-African orogeny (750–450 Ma). This includes primary ruby deposits in the Pan-African tectonic-metamorphic belt such as the John Saul mine in Kenya, Longido in Tanzania, and the deposits of Southern Madagascar, all formed at around 610 Ma. The third period corresponds to the Cenozoic Himalayan orogeny (55 Ma to the Quaternary) with the famous marble-hosted ruby deposits in Central and South-East Asia as in the deposits of Thureing Taung at Mogok and Mong Hsu in Myanmar [88]. The fourth period is dominated by the extrusion of alkali basalts in the Cenozoic (65 Ma to Quaternary) where ruby (and sapphire) were transported as xenocrysts or in xenoliths by the magmas. The age of the parental rocks of the ruby is unknown but a metamorphic origin is preferentially evidenced. Most of the ages are minimum cooling ages except those obtained on zircon by the U/Pb method.

The second period of corundum formation was the Pan-African orogeny (750–450 Ma). This includes primary ruby and sapphire deposits in the gemstone belt of eastern Africa, Madagascar, India, and Sri Lanka that are linked to collisional processes between Eastern and Western Gondwana during Pan-African tectonic-metamorphic events [123]. The metamorphic ruby and sapphire deposits in Southern Madagascar have numerous geological similarities with those in Eastern Africa, Sri Lanka, and Southern India (see [40]). U/Pb dating of zircon from the John Saul mine in Kenya (612 ± 6 Ma; [118]), Longido in Tanzania (610 ± 6 Ma; [124]), and the Vohibory deposits in Madagascar (612 ± 5 Ma; [125]) reveal similar periods of formation related to the East African orogeny. The precise age of formation for the ruby deposit of Montepuez in Mozambique is unknown but a Pan-African age is probable [126]. Moreover, the U/Pb ages of zircon coeval with sapphire in the Andronandambo skarn deposit in Southern Madagascar range between 523 and 510 Ma [127], and the $^{40}Ar/^{39}Ar$ ages of biotite-phlogopite syngenetic with sapphire and ruby from the Ambatomena, Sahambano, and Zazafotsy deposits in Madagascar are between 494 and 487 Ma [128]. LA-ICP-MS dating of rutile inclusions in ruby from the John Saul and Aqua mines in the Mangare area gave ages of 533 and 526 Ma, respectively [129]. These cooling ages, lower that those obtained by the U/Pb method, must be considered as minimum ages but as confirming the existence of a metamorphic corundum episode during the late Pan-African orogenic cycle (Cambrian) and corresponding to the Kuunga orogeny [130].

The third period corresponds to the Cenozoic Alpine orogeny with tectonics events registered in the Alps; Rhodopes and Himalayas (55 Ma to the Quaternary). Examples include the marble-hosted ruby deposits in Campo Lungo (Switzerland) and Xanthi (Greece), and in Central and South-East Asia. In the Himalayas, the ruby deposits in marble occur in metamorphic blocks that were affected by major tectonic events during the collision of the Indian and Asian plates [131–134]. The ruby has been indirectly dated by $^{40}Ar/^{39}Ar$ stepwise heating experiments performed on single grains of coeval phlogopite, and by ion-probe U–Pb analyses of zircon included in the corundum [40,88,106,135]. Other indirect dating for the Snezhnoe deposits in Central Pamir was realized on bulk rock (ruby-bearing phlogopite-plagioclase rocks) by the Rb-Sr method [134]. The Rb–Sr errorchron age yielded 23 ± 1.6 Ma, which had to be considered as a minimum age for ruby formation. All of the Eocene to Pliocene ages (55–5 Ma) are consistent with extensional tectonic events that were active from Afghanistan to Vietnam in the ruby-bearing metamorphic belt.

The fourth main period of gem economic ruby is dominated by the extrusion of alkali basalts in the Cenozoic (65 Ma to Quaternary). Gem corundum occurs worldwide as xenocrysts or megacrysts in xenoliths or enclaves incorporated in basaltic magmas during their ascent. Such sapphire and ruby deposits occur from Tasmania through Eastern Australia, South-East Asia, and Eastern China to far Eastern Russia [68,85,88,106]. They are also found in Nigeria and Cameroon in the Aïr and Hoggar regions, the French Massif Central in the Limagne rift, and in Northern, Central, and Eastern Madagascar [39]. In the Australian West Pacific margins intraplate basaltic fields, the U/Pb ages on zircons are widespread between 60 and 1 Ma [69,73,85,86,106]. Nevertheless, U/Pb dating of zircons xenocrysts in the Barrington Tops deposits in New South Wales indicated multiple episodes of corundum formation and eruptive discharges between 60 and 3 Ma [136]. Few U/Pb data gave ages of 150 and 180 Ma for zircons from the deposit of Cudgegong [136], and 400 Ma for those from Tumbarumba [137]. These oldest ages signify that the Australian West Pacific margins has been essentially a passive margin in a couple hundred million years until the Cenozic basalts related to continental rifting brought corundum to the surface.

5. Classification of Ruby Deposits

5.1. Genetic Classifications

Gem corundum deposits are classified as either primary or secondary deposits. The primary deposits contain corundum either in the rock where it crystallized or as xenocrysts and in xenoliths in the rock that carried it from the zone of crystallization in the crust or mantle to the Earth's surface [39]. The secondary deposits, i.e., present-day placers, are of three types [138]: (1) eluvial concentrations derived by in situ weathering, or weathering plus gravitational movement, or accumulation; (2) colluvial deposits corresponding to decomposed primary deposits which have moved vertically and laterally down-slope as the hillside eroded; and (3) alluvial deposits resulting from erosion of the host rock and transport of corundum by streams and rivers. Concentration occurs where water velocity drops at a slope change in the hydrographical profile of the river, e.g., at the base of a waterfall, broad gullies, debris cones, meanders, and inflowing streams. Sometimes the corundum placers are mined in beach sands [139,140].

Classification systems for corundum deposits have evolved over time and are based on different mineralogical and geological features including: (1) the morphology of the corundum [141]; (2) the geological context of the deposits [5]; (3) the lithology of the host rocks [142]; (4) the genetic processes responsible for corundum formation [110]; (5) the genetic type of the deposit [143], (6) the geological environment and nature of the corundum host-rock [40,80,108,140]; (7) the oxygen isotopic composition of the corundum [144], and (8) discriminant factors developed using oxide concentrations in 900 corundum samples from the main primary deposits [72]. Classification systems for ruby deposits have not been previously proposed.

5.2. An Enhanced Classification for Ruby Deposits

In this work we proposed classifying ruby from primary and secondary deposits into three main types (Table 3).

Primary ruby deposits were subdivided into two types based on their geological environment of formation: (Type I) magmatic and (Type II) metamorphic.

(Type I) Magmatic-related with two sub-types:

(1) Sub-Type IA: Rubies either as xenocrysts or in xenoliths hosted by magmatic rocks such as alkali basalts (Madagascar, and others; see Figure 2);

(2) Sub-Type IB: Xenocrysts of ruby in kimberlite (Democratic Republic of Congo).

(Type II) Metamorphic-related with two sub-types:

(1) Sub-Type IIA: Metamorphic deposits sensu stricto in metamorphosed mafic and ultramafic rocks (M-UMR; Sub-Type IIA_1) and marble (Sub-Type IIA_2). Examples of Sub-Type IIA_1 include the ruby deposits in M-UMR from Montepuez (Mozambique), Bekily-Vohibory region (Madagascar), and others, examples of Sub-Type IIA_2 concern the ruby deposits in marble from the Mogok Stone Track, and others.

(2) Sub-Type IIB: Metamorphic-metasomatic deposits characterized by high fluid–rock interaction and metasomatism (Sub-Type IIB_1), i.e., plumasite or desilicated pegmatites as at the former John Saul mine in Kenya, and metasomatites in M-UMR as at Aappaluttoq (Greenland), and marble, and shear zone-related or fold-controlled metasomatic-metamorphic deposits in different substrata, corundum-bearing Mg-Cr-biotite schist and gneiss or marble (Sub-Type IIB_2). Examples are respectively the ruby occurrences at Sahambano, Zazafotsy, and Ambatomena in Madagascar, and Mahenge and the Uluguru Mountains in Tanzania.

Secondary ruby deposits are defined as Type III, i.e., sedimentary-related. They are divided into three sub-types hosted in sedimentary rocks:

(1) Sub-Type IIIA: Gem placers in alkali basalt environments (Eastern Australia, Central Madagascar, South-East Asia, and others).

(2) Sub-Type IIIB: Gem placers in metamorphic environments (Montepuez in Mozambique and the Mogok Stone Track in Myanmar, and others).

(3) Sub-Type IIIC: Gem placers with ruby originating from multiple and unknown sources, such as Ilakaka in Madagascar, Tunduru and Songea in Tanzania, and others.

Table 3. Typological classification of ruby occurrences and deposits worldwide.

Type of Deposit	Magmatic-Related		Metamorphic-Related								Sedimentary-Related *	
Geological Environment	Magmatic		Metamorphic								Sedimentary	
Metamorphic Conditions	Amphibolite to Granulite or Eclogite Facies		Greenschist–Amphibolite–Granulite Facies								No Metamorphism	
Host-Rocks	Magmatic and Metamorphic Rocks		Metamorphic Rocks								Detritic Rocks	
Types of Deposit	TYPE I Magmatic-Metamorphic		Type IIA Metamorphic *sensu-stricto*					Type IIB Metamorphic-Metasomatism (Fluid-Rock Interaction)			TYPE III (Placers)	
Sub-Types	TYPE IA Alkali Basalt	TYPE IB Kimberlite *	TYPE IIA_1 Mafic-Ultramafic Rocks (M-UMR) *				TYPE IIA_2 Marble	Type IIB_1 Plumasite * or Metasomatite		Type IIB_2 Shear-Fault-fold in Gneiss, Schist, M-UMR	Type IIIA Rudites–Arenites–Lutites	Type IIIB
Host-rock of Ruby	Xenocryst-Xenolith	Xenocryst	"Verdite"	"Anyolite"	Metamorphosed M-UMR	Mafic Gneiss	Calc-Silicate Rocks	M-UMR	Marble		Alkali Basalt-Kimberlite Fields	Metamorphic Environement
Corundum	Metamorphic (and magmatic?) ruby	Metamorphicruby	Ruby**	Ruby	Ruby	Ruby	Ruby*	Ruby**	Ruby	Ruby	Ruby**	
Deposits Worldwide	All in alkali basalt environment with the presence of either disseminated *ruby xenocrysts* in volcanoclastics, epiclastics, diatremes, or in *ruby-bearing xenoliths* of pyroxenite and metagabbro (Antanifotsy, Soamiakatra*, Madagascar)	DRC (Mbuji-Mayi) *xenocrysts*	*Zimbabwe* (O'Briens) *South Africa* (Barberton)	*Tanzania* (*Longido*)	*Tanzania* (Winza). *Madagascar* (Andriba; Bekily-Vohibory (Ejeda; Ianapera; Fotadrevo; Anavoha, Gogogogo, Maniry; Vohitany; *Ethiopia* (Kibre mengist). *Kenya (Kitui; Mangare).* *Malawi (Chimwadzulu).* *Mozambique* (Montepuez; M'sawize; Ruambeze). *France* (Brittany; Lozère; French Massif Central, Haut-Allier, Peygerolles). *Italy* (Piemont). *Norway* (Froland). *Finland* (Kittilä). *Greece* (Paranesti). *Pakistan* (Dir). *India* (Orissa; Kerala). *New Zealand* (Hokitika) *Spain* (Alboran Sea) *New Zealand* (Hokitika).	*India* (Mysore).	*Afghanistan* (Jegdalek). *Nepal* (Chumar; Ruyil). *Pakistan* (Hunza valley; Batakundi; Nangimali). *Tajikistan* (Turakuloma**; Badakhshan). *Myanmar* (Mogok; *Mong Hsu).* *China* (Qinghai; Yuan Jiang). *Vietnam* (Luc Yen; Yen Bai; Quy Chau). *Tanzania* (Morogoro; Mahenge). *Kenya* (West Pokot). *Canada* (Revelstoke ***). *Macedonia* (Prilep). *Switzerland* *(Campo Lungo).* *France* (Pyrenees). *Russia* (Urals). Greece (Xanthi).	*South Africa* (Transvaal). *Zimbabwe* (O'Briens). *Kenya* (Mangare area, Aqua; Penny lane; Rockland-John Saul mine; Hard Rock). *Tanzania* (Umba-Kalalani). *Madagascar* (Vohibory area - Bekily, Anavoha***; Vohitany; Andilamena). *Russia* (Polar Urals, Hit Island; Karelia). *Australia* (Poona ****; Harts Range$^{\square}$). *United-States* (Corundum Hill). *France* (Haut Allier, Chantel) Greenland (Aappaluttoq; Nuuk-Stovø); *Kangerdluarssuk).* *India* (Kerala) *United-States* (North Carolina).	*Tanzania* (Mahenge $^{\square}$, Kitwalo and Greyson mines).	*Madagascar* (Sahambano *; Zazafotsy *; Ionaivo *, Ambatomena **). *India* (Kerala). *Kenya* (Mangare area). *Tanzania* (Morogoro, Mahenge; Uluguru Mountains ***).	*Australia* (Lava Plains; Anakie fields; New England fields; Macquarie-Cudgegong; Barrington Tops; Tumbarumba; Western Melbourne fields). *China* (Muling). *Thailand* (Chantaburi-Trat); *Cambodia* (Pailin). *Madagascar* (Ankaratra massif; Vatomandry). *Kenya* (Baringo). *Democratic Republic of Congo* (Mbuji-Mayi). *Brazil* (São Luis).***	*Sri Lanka* (Ratnapura; Elahera, Polonnaruwa). *Myanmar* (Mogok; Mong Hsu). *Madagascar* (Ilakaka****; Andilamena; Didy; Zahamena). *Vietnam* (Luc Yen; Yen Bai; Quy Chau). *Tanzania* (Tunduru; Songea; Winza; Umba valley,. Morogoro; Mahenge). *Mozambique* (Montepuez; M'sawize; Ruambeze). *Malawi* (Chimwadzulu). *United-States* (North Carolina-Cowee Creek)

Table 3. *Cont.*

Type of Deposit	Magmatic-Related		Metamorphic-Related								Sedimentary-Related *	
Geological Environment	Magmatic		Metamorphic								Sedimentary	
Metamorphic Conditions	Amphibolite to Granulite or Eclogite Facies		Greenschist–Amphibolite–Granulite Facies								No Metamorphism	
Host-Rocks	Magmatic and Metamorphic Rocks		Metamorphic Rocks								Detritic Rocks	
Types of Deposit	TYPE I Magmatic-Metamorphic		Type IIA Metamorphic *sensu-stricto*					Type IIB Metamorphic-Metasomatism (Fluid-Rock Interaction)			TYPE III (Placers)	
Sub-Types	TYPE IA Alkali Basalt	TYPE IB Kimberlite *	TYPE IIA$_1$ Mafic-Ultramafic Rocks (M-UMR) *				TYPE IIA$_2$ Marble	Type IIB$_1$ Plumasite * or Metasomatite		Type IIB$_2$ Shear-Fault-fold in Gneiss, Schist, M-UMR	Type IIIA Rudites–Arenites–Lutites	Type IIIB
Host-rock of Ruby	Xenocryst-Xenolith	Xenocryst	"Verdite"	"Anyolite"	Metamorphosed M-UMR	Mafic Gneiss	Calc-Silicate Rocks	M-UMR	Marble		Alkali Basalt-Kimberlite Fields	Metamorphic Environement
Remarks	*very rare economic deposit	*very rare occurrence	*including metamorphosed mafic-ultramafic rocks, mafic gneisses, and special cases of retromorphosed UMR as verdites and anyolite. ** most of the time "corundum".				* include pink corundum. ** Al-rich metapelite or metabauxite intercalated in marble *** metapelite intercalated in marble	* desilicated -pegmatite or -gneiss and other -felsic metamorphic rocks. ** sometimes corundumite. *** rocks called sakenites by Lacroix (1922).	***** ruby associated with emerald. ¤ in amphibolite. ¤¤metasomatism of mafic dyke in marble.	* in biotitites with sapphires ** circulation of fluids in MR **** circulation of fluids in folds hinge.	* netotectonics and basins. ** include pink sapphire. *** inclusion of ruby in diamond.	**** placers are hosted by detritic rocks of the Triassic Isalo sedimentary serie, unknown origin for ruby.
Economy	non-economic		carving, and gems	economic for some deposits		non-economic	economic for some deposits	economic for some deposits	±economic	economic for some deposits	economic for most of the deposits	

6. The Different Types of Ruby Deposits

We will now examine the main ruby deposits worldwide using the above classification.

6.1. Magmatic-Related (Type I)

6.1.1. Sub-Type IA: Metamorphic Rubies either as Xenocrysts or Xenoliths in Magmatic Rocks such as Alkali Basalts (See Table 3)

During recent decades, studies of sapphire and ruby xenocrysts in alkali basalts focused on the nature of solid and fluid inclusions [11,75,85,145–150], trace element chemistry [70,72–74,81,150], oxygen isotopes [40,144,151–153], and the nature of their parental xenoliths [75,77,81,145,154] to show that gem corundum may have different magmatic and metamorphic origins. Whatever their origin, these rubies are classified in the Type I category because the classification is based on their geological environment and not on their genetic origin.

The origin of ruby has long been debated but its metamorphic origin was accepted based on the work done by Australian researchers on different deposits from eastern Australia [68,106], work that focused on trace element diagrams, mineral inclusions, and oxygen isotopes [155]. A metamorphic origin has been questioned once more by recent studies showing melt inclusions in ruby from the Chanthaburi-Trat region of Thailand and Pailin in Cambodia [105].

The metamorphic origin of xenocrystic rubies as in Eastern Australia, the Kanchanaburi-Bo Rai and Nam Yuen deposits in Thailand, and Pailin in Cambodia, and Central and Eastern Madagascar is characterized by their trace element distributions and inclusions. The classical Fe/Ti versus Cr/Ga chemical variation diagram proposed by [74] permits characterization of "metamorphic" ruby and pink corundum, which are rich in Cr and poor in Ga (Cr_2O_3/Ga_2O_3 ratio > 3), and contain inclusions of chromiferous spinel, pleonaste, Al-rich diopside, and sapphirine. Recently, [81] characterized rubies (up to 3370 ppm Cr) and pink corundum (up to 1520 ppm Cr) from New England placers, which have distinctive trace element features compared to those found elsewhere in Eastern Australia. In particular they exhibit very high contents of Ga (up to 310 ppm) and Si (up to 1820 ppm). High Ga and Ga/Mg values are unusual in ruby, and the trace elements suggest a magmatic–metasomatic influence for the genesis of New England rubies. The oxygen isotopes ($\delta^{18}O = 4.4 \pm 0.4‰$, $n = 22$) fit within the range of $\delta^{18}O$ values defined for ruby hosted in M-UMR. Zoned ruby and pink corundum (Figure 23A) have high Mg, Ga, and Cr and the Fe versus Ga/Mg diagram suggests magmatic affinity (Figure 23B).

The geological origin of New England rubies before their transport by alkali basalts remains a great mystery. Sutherland et al. [81] proposed for their formation either the interaction of intrusive fractionated I-type granitoid magmas, during the New England orogeny, with Cr-bearing ophiolites, or ruby-bearing eclogites and -cliopyroxenites in metamorphosed M-UMR. It is important to note that the different chemical parameters (such as Ga/Mg, Fe/Ti, Fe/Mg, and Cr/Ga) proposed by [70] were designed only for origin determination of blue sapphires. In some cases, the use of these chemical diagrams is not useful for discriminating geologic or geographic origin of corundum. In the case of Australian rubies, the majority of the data are scattered in the field of metamorphic sapphires, but those of New England overlap the field of magmatic ones. The New England case, and other as for Chanthaburi-Trat and Pailin rubies illustrate the complexity and ambiguity of the use of trace element chemistry [11].

These chemical domains have no genetic significance for some cases and they illustrate only the variation of metals in the parental fluids before their incorporation in the structure during crystal growth, i.e., a kind of chemical color signature. On the contrary, the Fe (ppm) versus Ga/Mg discrimination diagram applied to ruby (see Figure 19) shows that the high variation of Fe in ruby correlates with the geological classification of ruby deposits characterized by decreasing Fe in ruby from M-UMR to marble and plumasite. The overlap is not extensive, as it is for sapphires and the diagram can apparently be used for deciphering the geological origins of ruby.

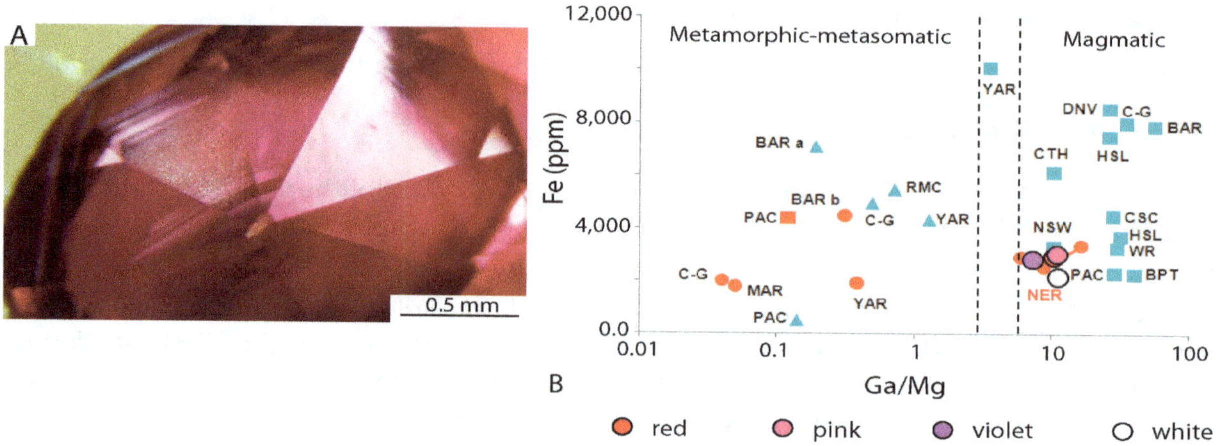

Figure 23. Rubies from the New England locality of Eastern Australia. (**A**) Apex culet view of an oval-cut 2.31 carats zoned ruby showing alternating growth zones passing through violet to white and pink bands and to ruby on the extreme margin of the top table, Collection Australian Museum. Photo: Gayle Webb; (**B**) Fe (ppm) versus Ga/Mg diagram for ruby and pink, violet, and white corundum from New England ruby (NER) and other Eastern Australian localities; chemical data from [86]. The red symbols indicate ruby, the blue triangles metamorphic sapphire suites, and the blue squares magmatic-transitional sapphire suites. The different localities are: NSW = New South Wales; BAR = Barrington Tops; YAR = Yarrowitch; C-G = Cudgegong-Gulgong; MAR = Macquarie River; WR = Weld River, NE Tasmania; PAC = Pailin, Cambodia; BPT = Bo Phloi, Thailand; HSL = Huai Sai, Laos; CSC = Changle, Shandong, China; DNV = Dak Nong, Vietnam; CTH = Chanthaburi-Trat, Thailand; RMC = Rio Mayo, Colombia.

Mineral inclusions are also key features in the debate about metamorphic versus magmatic origin of rubies. The most common solids are clinopyroxene (Al-rich diopside), plagioclase (bytownite to andesine), K-feldspar-spinel association [148,149], garnet (pyrope, ± almandine), sapphirine, scapolite, Cr-spinel, meionite, phlogopite, anatase, apatite, sillimanite, and sulphides [5,68,77,105,145,146,154,156]. All of these minerals are described in rubies found in metamorphic environment, although some of them are also found in magmatic assemblages. The discovery of melt inclusions in ruby reopened the debate about their possible magmatic origin [105,148,149]. Glassy melt inclusions are described frequently for sapphires in alkali basalts [157] and lamprophyres [147,158] but not commonly in rubies. Melt inclusions in rubies from Thailand and Cambodia were reported by [16], recognized by [156], and confirmed by [148,149]. Two types of melt inclusions were observed in the Thai and Cambodian rubies [106]: (1) glassy melt inclusions with negative crystal morphology, generally 25–100 μm in size, and with a contraction bubble; (2) polycrystalline aggregates, 200–500 μm across, considered to be recrystallized melt inclusions. The EMPA analyses of two melt inclusions in ruby from Thailand and Cambodia gave similar chemical compositions in wt % oxides: 53.4. $< SiO_2 < 55$; $23.6 < Al_2O_3 < 24$; $4.9 < CaO < 6.2$; $4.4 < Na_2O < 5.2$; $0.1 < K_2O < 3.1$; $1.2 < FeO < 1.6$; $1.4 < MgO < 1.9$; $0.1 < TiO_2 < 0.2$, and $6.3 <$ volatiles < 7.5. These chemical data are comparable to those found for melt inclusions in Yogo Gulch sapphires from Montana. Palke et al. [105] suggested that corundum formed during partial melting of hydrated and carbonated lower crustal anorthosites or troctolites by local intrusion of the lamprophyres or alkali basalts (for the Thai and Cambodian deposits) that eventually carried the corundum to the surface. The genetic model proposed by [105] for the Thai and Cambodian rubies is an anorthosite that was converted at higher pressure to a garnet-clinopyroxenite. In this scenario, Palke et al. [105] argued that ruby is not an "accidental xenocryst" transported by the alkali basalt but rather an "in situ" xenocryst enclave in the alkali basalt itself before its transport to the surface. To the contrary and despite the presence of melt inclusions in ruby from Bo Rai, Promwongan et al. [148,149] and Sutthirat et al. [154] considered that the assemblage Al-rich pyroxene, plagioclase, pyrope, and spinel closely resembles the mineralogy of mafic granulite xenoliths in alkali basalt associated with the Bo Rai gem field.

The main discussion that arises from this genetic scenario is that the proposed anorthosite protolith for ruby formation has been affected by deformation and consequent metamorphism at high temperature. Under these conditions, how can we define the exact nomenclature and the genesis of ruby, i.e., magmatic origin (anorthosite) or metamorphic (clinopyroxenite or metagabbro in the eclogite facies), or both? The following studies are examples of the ongoing debate:

(1) Sutthirat et al. [145] reported ruby (0.27 wt % Cr_2O_3) included in a clinopyroxene (0.22 wt % Cr_2O_3; 16.7 wt % Al_2O_3; and 1.8 wt % Na_2O) xenocryst in the Nong Bon alkali basalt, one of the Cenozoic Chantaburi-Trat basalts of Eastern Thailand. Ruby from the alluvial deposits contains inclusions of garnet (pyrope), sapphirine, and Al-Na-clinopyroxene. Thermodynamic calculation constrained the temperatures of ruby + clinopyroxene crystallization to between 800 and 1150 ± 100 °C and pressures of between 10 and 25 kbar, corresponding to depths of 35–90 km. The pyrope + ruby assemblages formed at depths greater than 60 km and the source rock proposed was a mafic ruby-pyrope-pyroclasite or ruby-pyrope-clinopyroxenite. Sutthirat et al. [145] rejected a magmatic crystallization for the ruby and this was confirmed by later studies of Thai ruby [148,149,154].

(2) Sutherland et al. [75] studied a suite of alluvial-sourced pink corundum and ruby (up to 0.6 wt % Cr_2O_3) from the paleodrainage deposits along the Cudgegong-Macquarie River in New South Wales, Australia. The ruby contained inclusions of clinopyroxene (0.1 wt % Cr_2O_3; 20.3 wt % Al_2O_3; and 1.0 wt % Na_2O), meionite, and anatase. Primary fluid inclusions are either two-phase (liquid + gas) or three-phase (liquid + gas + daughter phases), but melt inclusions were not reported. The comparison of clinopyroxene-ruby thermometry at Cudgegong-Macquarie suggested ruby formation at temperatures between 1000 and 1300 °C. The P–T diagram showing thermometric paths for Al-rich diopside-ruby pairs from different ruby deposits in Thailand and Australia represented in Figure 24 [75,81]. Saminpanya et al. [146] confirmed the presence of inclusions of garnet (pyrope), Al-rich diopside, and sapphirine in Thai rubies. They proposed that the rubies crystallized in high-pressure metamorphic rocks of UMR composition in the upper mantle.

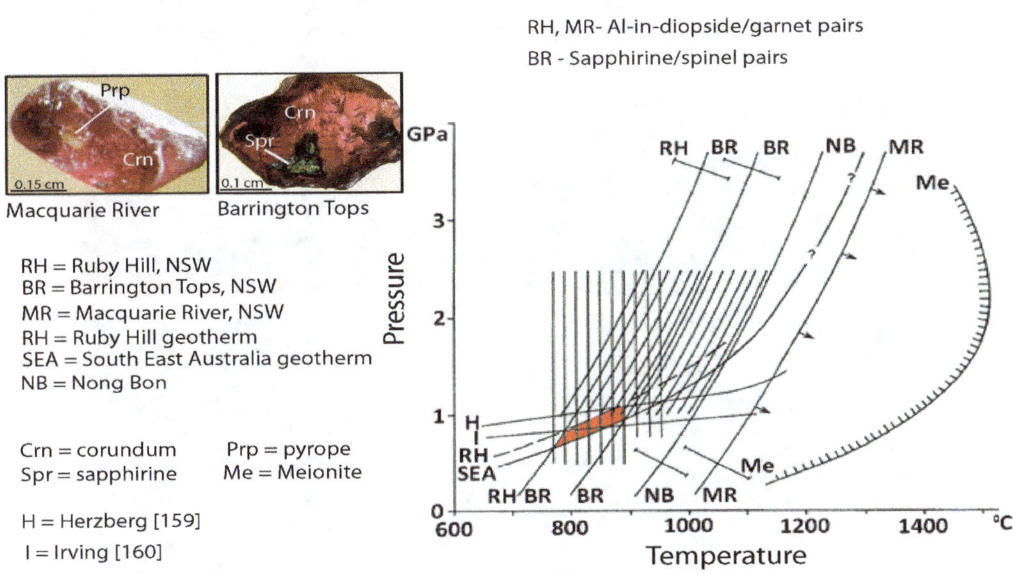

Figure 24. Thermometric traces for mineral inclusion pairs for rubies from Australia (Ruby Hill, Barrington Tops, and Macquarie River) and Thailand (Nong Bon), modified from [81]. The P (GPa)–T (°C) diagram is based on Al- in diopside/garnet and sapphirine/spinel pairs, after [75,136]. The traces intersect the T range for Barrington Tops rubies (vertical lines), the East Australian ruby-bearing xenolith-derived Mesozoic garnet granulite (RH) and late Cenozoic (SEA) geotherms, the garnet to spinel pyroxenite transition of [159] (H), and the garnet-clinopyroxene-plagioclase transition of [160] (I). The red area represents P 6–8 kbar for the Barrington Tops ruby suite. The extreme Al-rich diopside and meionite-bearing Macquarie River ruby suggests a P–T crystallization between thermometric trace (MR) and the meionite stability boundary (Me).

(3) Sutthirat et al. [154] studied different types of xenoliths contained in basalts from Bo Rai viz. (1) peridotites; (2) clinopyroxenites; and (3) ruby-bearing mafic granulites with plagioclase-clinopyroxene-spinel-garnet. Peridotitic xenoliths have characteristics of depleted mantle (lherzolite to harzburgite and dunite) and the P–T estimates yielded approximately 1000–1300 °C at 1.6–1.9 GPa. Clinopyroxenite and mafic granulites xenoliths are mantelic cumulates formed at P-T, respectively, between 950 and 1200 °C and 1.6 and 2.2 GPa, and 700 and 1000 °C and 0.95 and 1.3 GPa. The xenoliths suffered a few metamorphic events and the authors argued that ruby formation is clearly related to mafic granulite xenoliths (which appear to have formed at about 30–50 km depths before transport by basaltic eruption).

(4) The Soamiakatra ruby deposit in the Antsirabe area, Antananarivo province, was an economic in situ mining site such as at Pailin, Cambodia, where ruby was found either as xenocrysts in alkali basalt or in xenoliths brought to the surface by the Cenozoic Ankaratra volcanism [77]. The deposit at Morarano known as Soamiakatra [95,161–163] was discovered in 1997. Ruby has been extracted by panning from alluvial and soils, but prospecting and mechanization of operations exposed the weathered primary deposit at the top of the basaltic plug (Figure 25A). The ruby-bearing clay and phyllitic material of whitish color (20 m thick) was exploited for two years before bedrock was reached (Figure 25B). The secondary deposit assured an easy and effective ruby production of high quality.

Figure 25. The Soamiakatra ruby deposit (Madagascar). (**A**) View of the quarry showing the contact of the alkali basalt (β) plug with the Neoproterozoic metamorphic formations (Nmf). The economic weathered upper zone of the alkali basalt (β) was mined in the past up to the non-weathered basalt; (**B**) aspect of the ruby-bearing weathered basaltic zone, around 20 m thick, mined in benches; (**C**) ruby-bearing metagabbro xenolith (mg) carried by the alkali basalt (β); (**D**) macroscopic aspect of a ruby (Crn)-bearing metagabbro xenolith (mg); (**E**) xenocryst of ruby with an olivine (Ol) crystal in the alkali basalt (β); and (**F**) ruby-garnet (Grt)-plagioclase (Pl)-bearing metagabbro crosscut by the alkali basalt (β) flow. Photos: Gaston Giuliani and Saholy Rakotosamizanany.

The alkali basalt dyke intruded graphitic gneisses, khondalites, and migmatites of the Neoproterozoic Ambatolampy and Tolongoina series. The alkali basalt hosts numerous enclaves and xenoliths (Figure 25C), 2–20 cm across, of crustal (gneiss, migmatite, and granitoid) and mafic (peridotite, clinopyroxenite, gabbro, and garnetite) origin. Ruby was found: (1) as xenocrysts up to 2 cm across in the alkali basalt (Figure 25E); (2) in clinopyroxenite xenoliths composed of plagioclase (labradorite and bytownite), scapolite (meionite), garnet (pyrope), clinopyroxene (Al-diopside), orthopyroxene (85 mol % En), and pargasite; and (3) in banded metagabbro (Figure 25D,F) composed of alternating dark and greyish ruby-bearing bands. The dark bands contain pyrope (Figure 26A,C), plagioclase (labradorite-bytownite), Mg (±Cr)-spinel, scapolite (meionite), Al-diopside, and orthopyroxene (85 mol % En). The greyish band is made up of labradorite-bytownite, diopside, magnesian spinel, sapphirine (Figure 26D), some calcite, and pyrope. Diopside formed as a clean and clearly visible corona around both corundum and Mg (±Cr)-spinel (Figure 26D). Two generations of clinopyroxene are discriminated (Figure 26A,B): (1) early Al-rich crystals (13–17.5 wt % Al_2O_3), 1–2 mm across, associated with plagioclase, garnet, and scapolite (Cpx1), and (2) late and less Al-rich crystals (5–8.8 wt % Al_2O_3, Cpx2), 10–40 μm across, formed around Cpx1 (Figure 26B) or as inclusions in ruby.

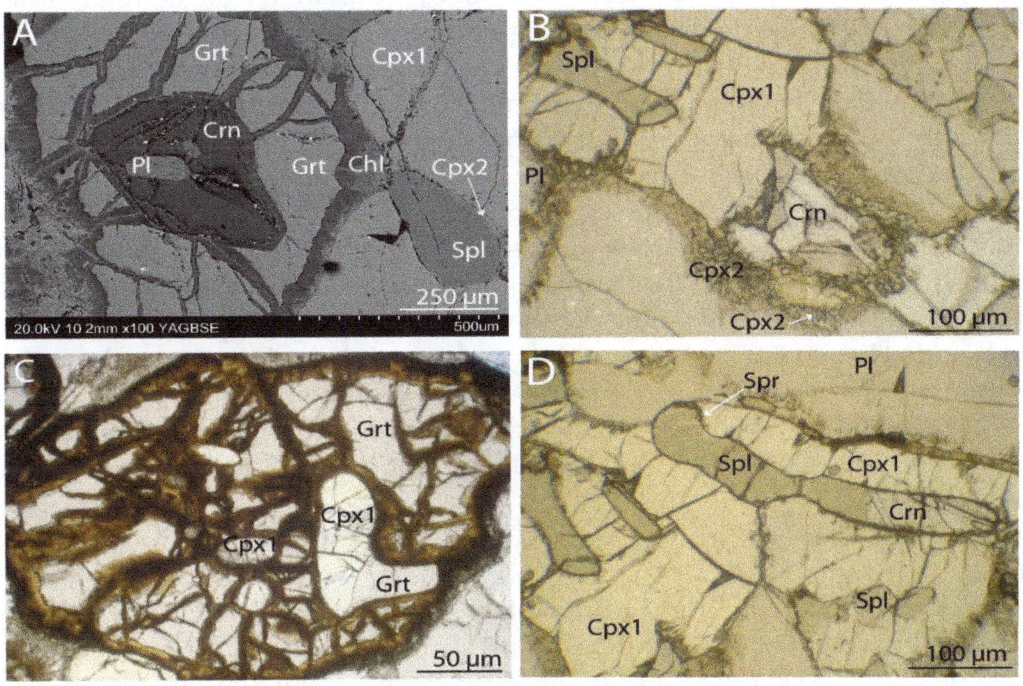

Figure 26. Petrographic aspects of the metagabbroic xenoliths in the Soamiakatra ruby deposit. (**A**) Plagioclase (Pl)-bearing ruby (Crn) within garnet (Grt). Garnet (pyrope) associated with an Al-rich clinopyroxene (Cpx1) and showing a keliphitic texture. The keliphitic halo is composed of small grains of chlorite (Chl). The spinel (Spl) and Cpx1 are bordered by a second generation of clinopyroxene (Cpx2); (**B**) association ruby-Cpx1-plagioclase (Pl)-spinel. The ruby formed upon Cpx1 while small crystals of Cpx2 formed later; (**C**) inclusions of Cpx1 in pyrope (Grt); and (**D**) association spinel-ruby with Cpx1 with plagioclase. Spinel is bordered by sapphirine (Spr). Photos: Saholy Rakotosamizanany.

Mineral inclusions in the ruby consist of phlogopite, rutile with some lamellar exolutions of ilmenite, zircon, albite, Al-pyroxene, garnet (pyrope), and Mg-spinel. The trace element chemistry of the ruby is characterized by low Ga_2O_3 (between 70 and 110 ppm) and high Cr_2O_3/Ga_2O_3 ratios (see Figure 16). Vanadium varies from less than 22 to 860 ppm. Titanium ranges between 60 and 940 ppm, and Cr between 350 and 3830 ppm. The Soamiakatra rubies fall into the metamorphic domain defined by [68] when plotted on the Fe_2O_3/TiO_2 versus Cr_2O_3/Ga_2O_3 diagram (see Figure 16).

Petrographic examinations showed two main stages for the mineral assemblages: (1) the first stage is typified by the association Cpx + garnet + scapolite + ruby, and spinel + ruby + garnet. The spinel is always surrounded by a ring of ruby, clinopyroxene, and sapphirine, and the mineral association is in turn haloed by coronitic garnet. Thermodynamic calculations constrained the P–T estimates of the assemblages to around 1100 °C and 2 GPa; (2) the second stage results in the destabilization of scapolite to anorthite + calcite with the destabilization of some Al-diopside in pargasite, and also the reaction spinel + ruby to sapphirine (T 1100 °C and P 15 GPa).

The P–T conditions of around 1100 °C and 2 GPa clearly show that the xenoliths originated from the upper mantle and that the metagabbros experienced the eclogite facies. During decompression the mineral assemblage was marked by the destabilization of Cpx1 and the formation of Cpx2 (in the granulite facies). Rakotosamizanany et al. [77] hypothesized that the ruby formed in M-UMR at the base of the lower crust-upper mantle and was later transported to the surface by the alkali basalt. The oxygen isotopic composition of the rubies (1.25‰ < $\delta^{18}O$ < 5‰, n = 5) is within the range of values defined for ruby in basaltic environments (1.25‰ < $\delta^{18}O$ < 5.9‰, n = 4; mean $\delta^{18}O$ = 3.1 ± 1.1‰, n = 80) and in metamorphosed M-UMR (0.25‰ < $\delta^{18}O$ < 6.8‰, n = 19; [163]. Rubies from Chantaburi-Trat volcanics have similar O-isotopic ranges (1.3‰ < $\delta^{18}O$ < 4.2‰; [152]) and the P–T conditions are identical to those proposed for the Soamiakatra rubies.

(5) Occurrences of ruby in the Neoproterozoic Vohibory formation from Southern Madagascar (see Figure 1E) are found in M-UMR complexes of meta-peridotite, -gabbro, and -troctolite transformed into amphibolites and serpentinites with anorthositic dykes [116]. The ruby occurrences include Anavoha, Gogogogo, Maniri, Vohitany, Ianapera, and Marolinta [164]. The ruby frequently occurs in sapphirine or gedrite and locally garnet (pyrope-almandine)-bearing amphibolites and locally anorthosite [122]. These rocks are associated with garnet-bearing granulites (salite + almandine + hornblende + andesine). At Anavoha, ruby occurs with (1) sapphirine or gedrite + garnet, and (2) anorthosite with ruby + garnet + anorthite. These ruby-bearing rocks are associated with a coronitic metatroctolite.

The transition is sharp between the metatroctolite and the sapphirine or gedrite amphibolites and gradual between the garnet (50% almandine-35% pyrope)-ruby-bearing anorthosite and amphibolites. The ruby is surrounded by a corona of green spinel and anorthite and the mineralogical association is gedrite + anorthite + garnet (60% pyrope) + ruby ± spinel. Sometimes the ruby is totally replaced by spinel. The P–T conditions of ruby formation during the Neoproterozoic metamorphism was estimated at P = 0.9–1.15 GPa and 750–800 °C [116], similar to those reported by [75] for the ruby-bearing xenolith of Ruby Hill in Australia.

At Vohitany and Anavoha, ruby is also found in desilicated pegmatites at the contact of amphibolites.

We previously examined the origin of ruby transported to the surface by basaltic volcanism by studying examples from Thailand, Australia [81], and Madagascar. Mafic cumulate xenoliths are rare worldwide and [154] showed that mantle and deep crustal xenoliths from the Bo Rai ruby deposit were the probable source of "basaltic rubies" mined in placers from Thailand and Cambodia. The Cpx-Spl-Grt ± (Ol) ± (Opx) and Pl-Cpx-Spl ± (Grt) ± (Crn) assemblages were considered to be two different types of mafic cumulates with high Al_2O_3 contents that suffered metamorphic overprints [154]. P–T constraints based on textural observations gave estimated depths of formation between 30 and 80 km. These xenoliths are similar to those reported for ruby-related alkali basalts from Madagascar [77]. At Soamiakatra, the estimated temperature 1100 °C and pressure 2 GPa of ruby formation are comparable to those proposed for mafic cumulates from Bo Rai. Al-rich clinopyroxene-pyrope-anorthite is the main assemblage of these ruby-bearing metagabbros and clinopyroxenites. Spinel and ruby are associated but ruby also formed at the expense of clinopyroxene.

The probable original source of the ruby in the mafic cumulates carried by Cenozoic alkali basalt plugs in the Ankaratra massif is visible in the Neoproterozoic metamorphic rocks of the Vohibory formation in Southern Madagascar. The ruby-bearing amphibolites and anorthosites are the fruit of the metamorphic transformation of a magmatic cumulate complex.

The presence of melt inclusions as described by [105,148,149] opened the debate about the possibility of "magmatic" ruby. The genetic model proposed by [105], i.e., an anorthosite converted at higher pressure to a garnet-clinopyroxenite, is debatable but not applicable for ruby in Madagascar. The melt inclusions can result either from an incongruent melting of plagioclase during metamorphism [165] or directly through a peritectic melting reaction involving plagioclase or other al-rich minerals. In consequence, most rubies in alkali basalts can be considered actually as metamorphic in origin but the presence of primary melt inclusions in ruby [105] indicates that a free magma phase existed during ruby growth. Additional studies characterizing carefully the inclusions in more basalt-related rubies would be helpful to shed more light on the two possible origins, i.e., metamorphic versus magmatic. The ruby-bearing mineral assemblages described in M-UMR, metamorphosed and recrystallized at depth within the mantle, at Beni Bousera, Morocco [166,167] and Ronda, Spain [97], and considered of metamorphic origin, can be described as a potential restitic component of a partially melted mafic M-UMR protolith?

The $\delta^{18}O$ values of ruby from the Soamiakatra, Bo Rai, Pailin, Barrington, Anavoha, and Beni Bousera occurrences fit within the worldwide range defined for ruby in basaltic environments (1.25‰ $< \delta^{18}O <$ 5.9‰, mean $\delta^{18}O$ = 3.5 ± 1.3‰, n = 57) and ruby associated with M-UMR (1.25‰ $< \delta^{18}O <$ 7.0‰, mean $\delta^{18}O$ = 5.0 ± 1.2‰, n = 32; [84], Figure 27). Both ranges of $\delta^{18}O$ values overlap and are closed to the $\delta^{18}O$ mantle value of 5.5 ± 0.4‰ [168]. Nevertheless, the genetic source of ruby, i.e., strictly metamorphic or magmatic cannot be specified as the oxygen isotopic composition of ruby is largely buffered by that of its protolith, whether metamorphic or magmatic [92].

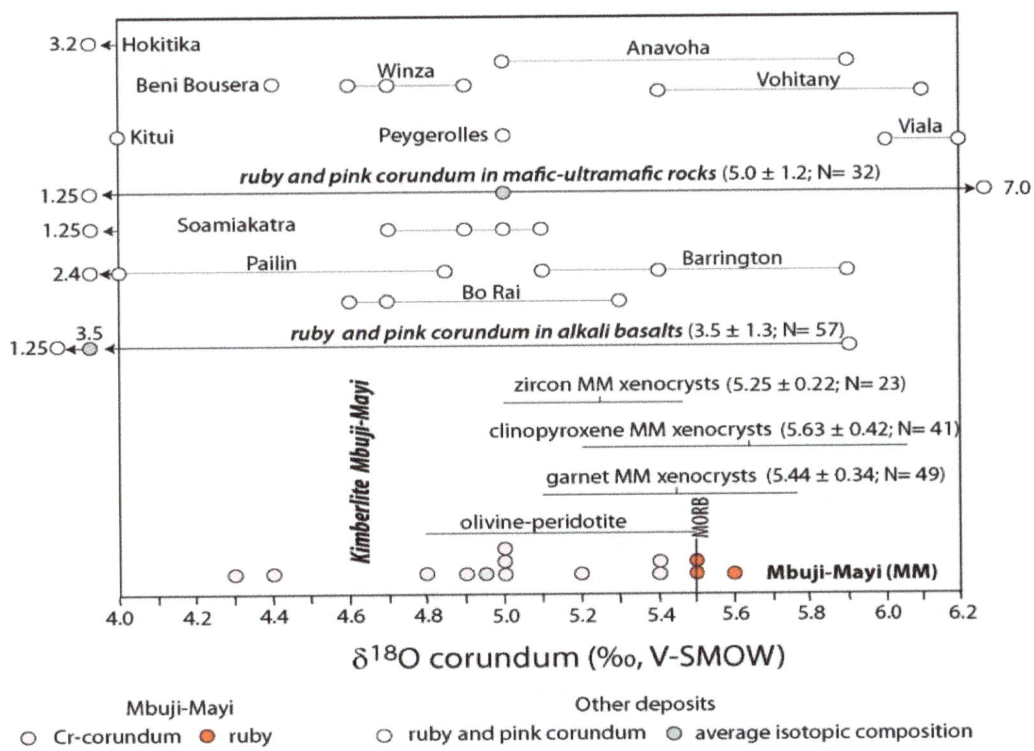

Figure 27. Oxygen isotope ranges defined for different ruby and pink corundum deposits and occurrences related to metamorphosed mafic-ultramafics and alkali basalts worldwide as well as the kimberlite at Mbuji-Mayi, modified from [84]. The data are reported in the conventional delta notation relative to V-SMOW (Vienna Standard Mean Ocean Water). The $\delta^{18}O$ values for zircon, clinopyroxene, and garnet megacrysts from the Mbuji-Mayi kimberlites are from [163]. The deposits that occur in amphibolites are Anavoha and Vohitany (Madagascar), Viala and Peygerolles (France), Kitui (Kenya), and Winza (Tanzania). The Hokitika (New Zealand) and Beni Bousera (Morocco) occurrences are linked to metamorphosed M-UMR.

6.1.2. Sub-Type IB: Metamorphic Rubies as Xenocrysts in the Mbuji-Mayi Kimberlite (Democratic Republic of Congo; See Figure 2, Table 3)

Ruby has been described as micrometric crystal inclusions in diamond at Juina in Brazil [43,169–171]. Crystals of ruby and Cr-bearing corundum (up to 2 cm) xenocrysts have been found in the Mbuji-Mayi kimberlite in the Democratic Republic of Congo [84]. The kimberlite contains megacrysts of garnet and diopside [172], zircon and baddeleyite, and also Mg-rich ilmenite, Nb- or Cr-rich rutile, corundum, kyanite, Mg-rich chlorite, and rutile-silicate intergrowths [173]. The 0.5–2 cm-sized corundum megacrysts are:

(1) Gem rubies from red to fuchsia and flashy-pink deep pink rubies (n = 3) that contain Cr-bearing spinel (with Cr_2O_3 up to 1.5 wt %), chromite, and rutile. The rubies have very high contents of Cr_2O_3 (0.71–4.59 wt %) and FeO (0.22–0.66 wt %). Vanadium is between 986 and 1867 ppm, and Ti less than 68 ppm. The contents of MgO (31–100 ppm) and Ga_2O_3 (50–193 ppm) range to moderate levels with a low Ga/Mg ratio (0.6–2.6, see Figure 19).

In the diagram FeO-Cr_2O_3-MgO-V_2O_3 versus FeO + TiO_2 + Ga_2O_3 [84], the three rubies plot in the domain R3 of M-UMR (Figure 28):

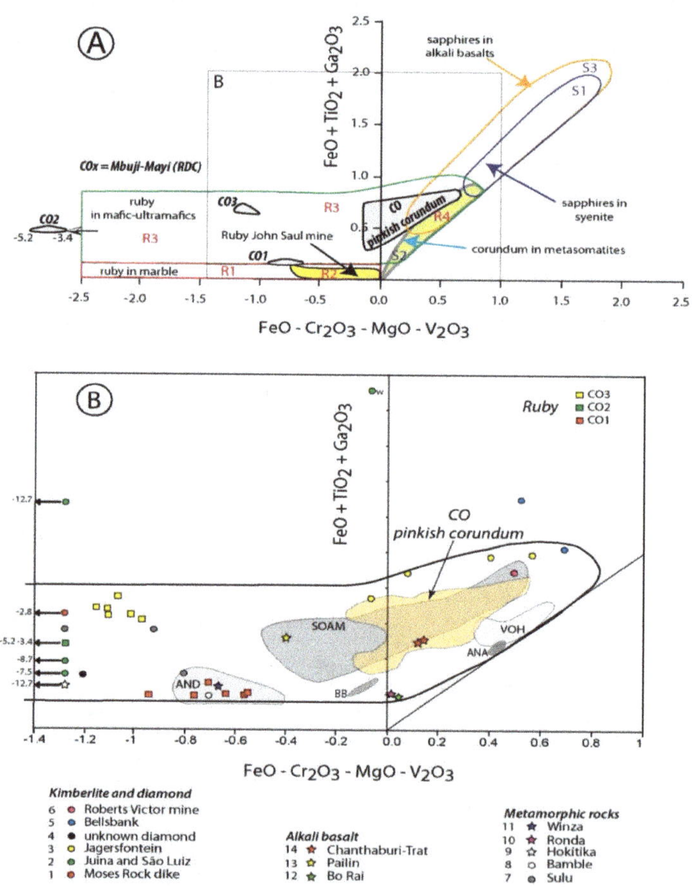

Figure 28. Chemical variation of ruby and pinkish corundum from the Mbuji-Mayi kimberlite in the Democratic Republic of Congo. (**A**) The FeO-Cr_2O_3-MgO-V_2O_3 versus FeO + TiO_2 + Ga_2O_3 (analysis in wt %) classification diagram, modified from [84]. The different chemical fields of the studied corundum are reported with respectively those for ruby (samples CO1, CO2, and CO3) and pinkish corundum (CO, yellowish field); (**B**) Field of chemical compositions of the ruby and pink sapphires from Mbuji-Mayi. Different chemical fields are reported for ruby from Beni Bousera in Morocco (BB), Soamiakatra (SOAM), Anavoha (ANA), Vohitany (VOH), and Andriba (AND) from Madagascar. Other representative compositions of ruby from occurrences in kimberlite and diamond, alkali basalt, and metamorphic rocks are also shown; modified from [84,164].

(a) The field of sample CO1 overlaps the plots of rubies from Andriba in Madagascar hosted in ultramafics [174], the Winza deposit in Tanzania located in amphibolite and metagabbro [120], high-grade Proterozoic orthoamphibolites of the Bamble area in Norway [43], and ultrahigh pressure ruby-rich garnetite layers in garnet lherzolite in the Sulu terrane of China [175].
(b) The field of sample CO2 is defined by the very high Cr_2O_3 content of ruby up to 4.6 wt %. Chromium-rich ruby is rare and found in the Brazilian São Luiz alluvial deposit and in the Juina kimberlite [43,170], the Moses Rock Dyke kimberlite in Utah [176], the metagabbros at Karelia in Russia [41], and in the serpentinite at Hokitika in New Zealand [42].
(c) The field of sample CO3 is comparable to the chemical composition of ruby-bearing garnetite xenolith in garnet peridotite from the Sulu terrane in China [175].

The rubies have a restricted range of $\delta^{18}O$ values from 5.5‰ to 5.6‰ ($n = 3$; Figure 27). They fit within the worldwide O-isotopic range of rubies associated with metamorphosed M-UMR.

(2) Cloudy pinkish to greyish corundum ($n = 10$), which are host to inclusions of Mn-rich ilmenite, margarite, and rutile. They are richer in FeO (0.38–0.75 wt %) than rubies and they always contain some Cr_2O_3 (0.08–0.58 wt %) and V (318–1128 ppm). The Ga/Mg ratio (1.95–3.95) is in the range of, or above, the Ga/Mg ratio of ruby. They have $\delta^{18}O$ values in the range 4.3–5.4‰ with a mean $\delta^{18}O$ value of 4.9 ± 0.4‰ ($n = 10$; Figure 27), within the accepted range of mantle values [168]. These oxygen isotope values are within the range of values defined for ruby xenocrysts or megacrysts carried to the surface by alkali basalts [84]. They are also similar to those found for ruby in eclogites derived from oceanic crust [166] and hosted by clinopyroxene xenocrysts from Bo Rai alkali basalt alluvials as well as those from the Soamiakatra metagabbros.

Kimberlites, like alkali basalts, are conveyors of gems (respectively diamond and corundum) from the asthenosphere and/or lithosphere to the upper crust [86,177]. Ruby and sapphire inclusions in diamond are rare and corundum xenocrysts or megacrysts in kimberlites are exceptional. On the contrary, diamond has never been found as inclusions in corundum or within xenocrysts in alkali basalts. The chemistry and oxygen isotope composition of ruby from Mbuji Mayi allow for different possible origins:

(1) An upper mantle-derived material (garnet lherzolite, garnetite and pyroxenite) that could have been infiltrated and metasomatized by the kimberlite melt, or a high-pressure orogenic peridotite; or (2) eclogite-type xenoliths (omphacitic clinopyroxene-bearing metagabbro and/or clinopyroxenite) affected by a retrograde metamorphic stage in the granulite facies at the base of the continental crust (Figure 29).

The pinkish to greyish corundum could also have different probable origins: either as subducted oceanic crust transformed into eclogite, or as clinopyroxenite layers/horizons in peridotite.

6.2. Metamorphic-Related (Type II)

6.2.1. Sub-Type IIA: Metamorphic Deposits Sensu Stricto with Two Different Types Formed in Amphibolite to Granulite Facies (See Table 3)

Sub-Type IIA_1 includes ruby in metamorphosed mafic and ultramafic rocks (M-UMR) as found at Montepuez in Mozambique (see Table 3).

Ruby and pink corundum in metamorphosed M-UMR called "ruby in amphibolite" by some gemmologists originate from the metamorphism of gabbroic and dunitic rocks from ophiolites or volcano-sedimentary sequences. These deposits are the main sources of pink and red corundum but generally not of gem quality material.

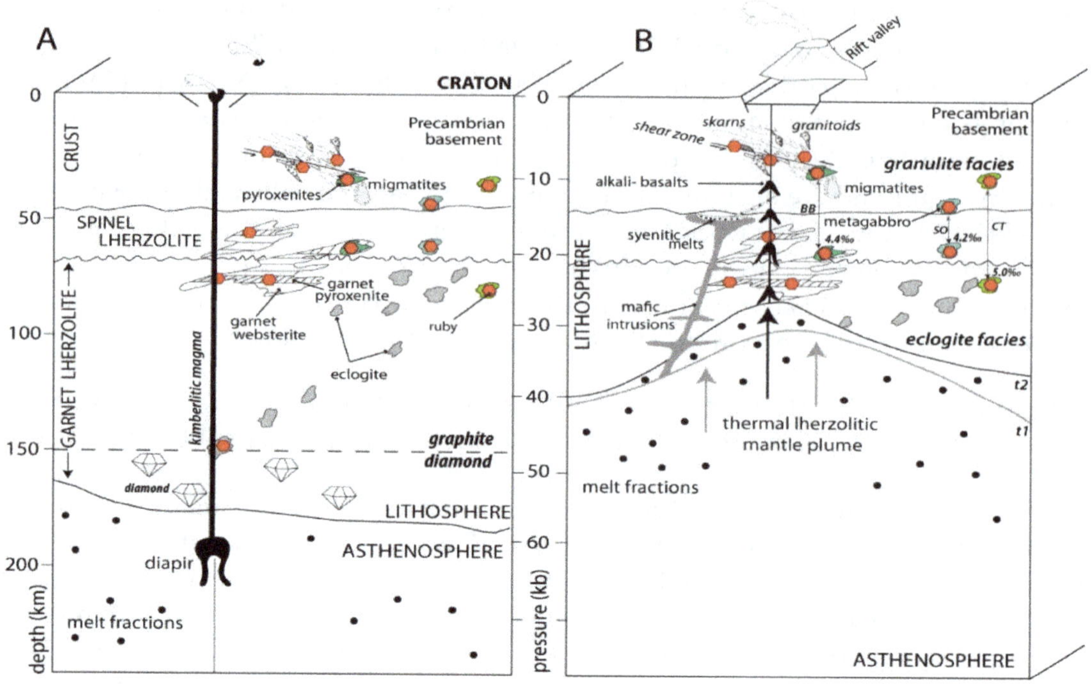

Figure 29. Schematic genetic models for gem corundum related to kimberlite (**A**) and alkali basalt (**B**) environments, modified from [178–180]. The drawings show the different types of corundum occurrences hosted in Precambrian basements in the crust, and lherzolites in the lithosphere. Selected ruby occurrences are indicated (with their corresponding $\delta^{18}O$ composition, in ‰) including the Beni Bousera (BB) garnet clinopyroxenite, the Soamiakatra (SO) garnet metagabbro and -clinopyroxenite, and the Chanthaburi-Trat (CT) pyroxene inclusion in ruby, with their retrograde metamorphic location in the granulite facies. Panel (**B**) shows ruby deposits associated with alkali basalt in continental extension domains characterized by thermal anomalies linked to mantle plumes and rift valley fracturing and volcanism in the upper part of the crust. t1 and t2 are two successive ascent stages of the alkali basalt magmatism.

The crystals are used for cabochon cut and ornamental artifacts such as the famous 'anyolite' from Longido in Tanzania [181–183] (see Figure 4B). Such ruby deposits in amphibolite and gabbro are known (see Figure 2) from Kittilä in Finland [184], Sittampundi in India [185], Chantel in France [186], Paranesti in Greece [95,187], Lossogonoi in Tanzania [181], Kitui in Kenya [188], Chimwadzulu in Mozambique [189,190], the Vohibory formations in Madagascar [119,122,164] (see Figure 1E), the Harts Range in Australia [191,192], Dir in Pakistan [193], Hokkaido in Japan [194], and some of the deposits in North Carolina in the USA [5,121,195]. In Africa, the ruby deposits discovered in the last decade are highly economic for their color, clarity, and quantity; these are Winza in Tanzania [120,196,197] (see Figure 4C), and at Montepuez, Ruambeze, and M'sawize in Mozambique [21–23,25,26,96,198] (see Figure 1C).

The common mineral assemblage of the ruby-bearing amphibolite includes corundum, anorthite, amphibole (gedrite, pargasite), and margarite [21,22,121,122]. Others minerals including sapphirine, pyrope-almandine garnet, spinel, kornerupine, plagioclase, phlogopite, and zoisite may be present. The initial basic composition of the protolith includes gabbro, dunite, and troctolite [121,122,194,199].

(1) The ruby deposits at Longido and Lossogonoi in Northeastern Tanzania are hosted in amphibolite dykes that crosscut metamorphosed M-UMR (Figure 30). At Longido, the ruby is located in amphibolite dykes up to 1 m thick that are oriented parallel to the metamorphic foliation [118]. The dykes crosscut a serpentinite with relics of olivine and orthopyroxene. They are retromorphosed to a rock composed of zoisite and ruby called "anyolite" (green stone in Massai language). The contact of the dyke with the serpentinite is outlined by the presence of phlogopite and Mg-amphibole schist.

The mineral assemblage in the amphibolite is anorthite + corundum + amphibole + green spinel while the anyolite contains zoisite + corundum ± tschermakite. The age of formation of the ruby obtained on zircon by the U/Pb method is 610 Ma [124].

(2) At the Winza deposit in central Tanzania, high-quality ruby and sapphires have been recovered from an eluvial deposit. Mining revealed that the gem corundum crystals are embedded in dark amphibolite [120,196,197]. The corundum is locally associated with areas of brown to orangey garnet ± feldspar, with accessory Cr-spinel, mica, kyanite, and allanite. The ruby and sapphire crystals are closely associated with dykes of a garnet- and pargasite-bearing rock crosscutting amphibolite (Figure 31A). The central part of the dyke is composed of garnet (pyrope-almandine) + pargasite ± plagioclase ± corundum ± spinel ± apatite. Metamorphic conditions estimated from the garnet-amphibole-corundum equilibrium assemblage are T 800 ± 50 °C and P 8–10 kbar [120]. The rubies and pink corundum are zoned with rhombohedral twin planes. The rubies exhibit a suite of inclusions with amphibole, garnet, apatite, and opaque minerals (chalcocite). Medium and vivid red rubies have orange-brown hair-like inclusions of unknown composition [120]. The chemical compositions of the Winza ruby and pink sapphires (Figure 31B) include moderate contents of Cr (0.1–0.8 wt % Cr_2O_3) and Fe (0.2–0.8 wt % Fe_2O_3), very low to low amounts of Ti (55–192 ppm TiO_2) and V (up to 164 ppm V_2O_3), and low to moderate Ga (64–146 ppm Ga_2O_3).

Figure 30. The ruby-bearing amphibolite from Longido, Northern Tanzania. (**A**) The amphibolite dyke, which is up to 1 m thick and parallel to the metamorphic foliation, crosscuts a serpentinite. The contact zone between the two rocks is marked by the presence of a phlogopite and amphibole-bearing schist. Photo: Elisabeth Le Goff; (**B**) anyolite with patches of ruby (Crn) and retrograde zoisite (Zo). Remnants of the amphibolite (Amph) highlight the metamorphic foliation. Photo: Elisabeth Le Goff; (**C**) representative samples of anyolite with ruby and black amphibole (tschermakite). Photo: Vincent Pardieu; (**D**) ruby crystals extracted from new veins that are very different to the classical massive ruby of Longido. Photo: Vincent Pardieu.

(3) The majority of gem-quality rubies from Africa are from deposits located in metamorphosed M-UMR from the Neoproterozoic Metamorphic Mozambique Belt (NMMB). The Rockland mine was the main producer of cabochon-grade ruby until the discovery of the Winza deposit in 2007. Then,

the confirmed presence of ruby in the Niassa National Reserve in Mozambique in 2008 opened the economic reign of Mozambique over the production of very high quality rubies worldwide as in the Montepuez mining district (Figure 32) [21,22,198,200,201]. A review and summary of the history, geology, and gemmological properties of ruby from Mozambique was recently published by [26].

The NMMB is subdivided into several complexes grouped in four geological entities, which are from west to east [202–204]: (1) the Ponta Messula complex; (2) the Unango and Maruppa complexes; (3) the Nampula complex; and (4) the east–west Pan-African overthrusts (e.g., Nairoto, Meluco, Muaquia, M'sawize, Xixano, Lalamo, and Montepuez), which form the Cabo Delgado complex overlying the Marrupa complex (see Figures 2–42 from [40]). The ruby deposits are within the Cabo Delgado complex:

(a) The Ruambeze (or Luambeze) deposit, in Niassa province, is located between Marrupa and Meculo along the Luambeze River. It was apparently discovered in 1992 [21] and it has produced dark red to brown cabochon grade material.

(b) The M'sawize deposit, in Niassa province, is located in the Niassa National Reserve, about 43 km southeast of M'sawize village [21]. The M'sawize complex is composed of orthogneiss and paragneiss. The M'sawize ruby deposit is hosted by metagabbro and gabbroic gneiss [22]. The ruby is associated with actinolite, anorthite, scapolite, epidote, and diopside, as well as with eluvial ruby-rich soil [22]. Ruby has Fe and Cr oxide contents between 0.28 and 0.33 and 0.12 and 0.48 wt %, respectively, which fall in the range of the Winza ruby from Tanzania [23].

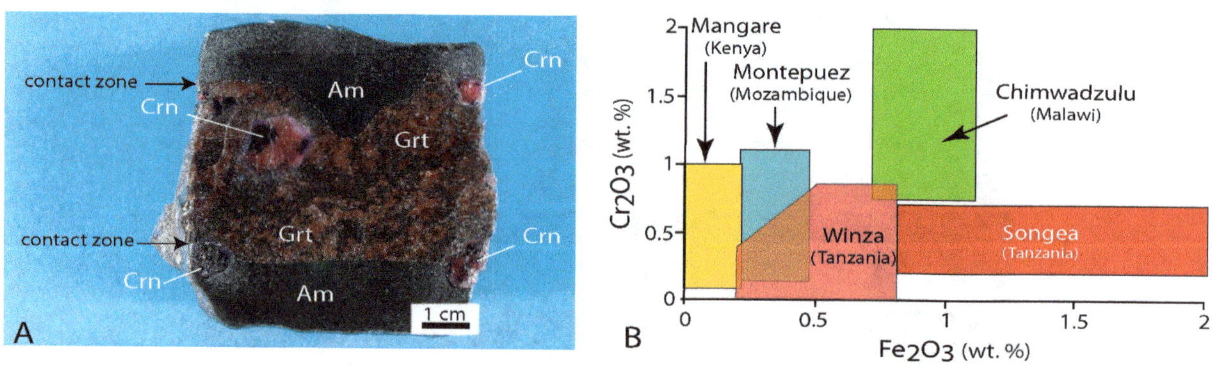

Figure 31. The Winza ruby deposit in Tanzania. (**A**) The primary ruby and sapphire (Crn) mineralization is hosted by an orangey brown garnet-rich rock (Grt) intercalated in a fine-grained amphibolite (Am). Photo: Patrick Lagrange; (**B**) the Cr_2O_3 versus Fe_2O_3 diagram showing the chemical composition of the Winza rubies compared with those from other ruby deposits hosted in metamorphosed M-UMR in the Neoproterozoic Metamorphic Mozambique Belt: Chimwadzulu (Malawi), Mangare (Kenya), Montepuez (Mozambique), and the placers at Songea (Tanzania).

(c) The Namahumbire and Namahaca deposits in Delgado province are together called the Montepuez deposits. The Cabo Delgado nappe complex represents the remnant of Neoproterozoic volcanic arcs and a microcontinent after 596 ± 11 Ma [205]. The lithological succession is very complex due to intense deformation but structural trends can be drawn from satellite images (Figure 32; [25]). The ruby area includes high-grade metamorphic rocks such as gneisses, migmatites, marbles, and amphibolites [96]. At the end of 2018, mining was being done by companies such as Fura Gems, Beleza Ruby, Gemfields, Mustang, and Redstone [25], and by numerous small-scale miners working either in community areas such as at Nacaca or illegally inside other company concessions.

Two types of deposits are known: (1) extensive secondary deposits, i.e., colluvials, which are the main source of faceted-grade material [25] (also see Chapter 6.3.2. Sub-Type IIIB) and (2) primary deposits in amphibolite [96,198,200].

The primary deposit is made up of ruby- and pink corundum-bearing lenses of amphibolites in migmatitic gneiss. They contain amphibole (pargasite), anorthite, and mica [22,96,200] (Figure 33).

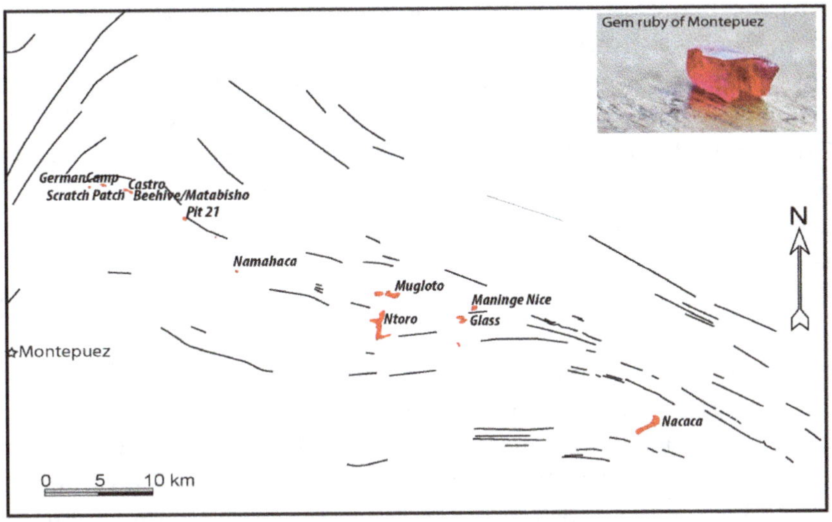

Figure 32. The Montepuez mining district with the different ruby mines of primary and mostly secondary deposits (drawn as red spots). The structural trends observed from both satellite imagery and field works are shown. The ruby mineralization defines a N110°E trending zone that extends for 40 km; modified from [25]. Crystal of ruby (size 1 cm long).

Figure 33. The Mozambican ruby deposits. (**A**) Ruby-bearing amphibolite (Am) at the M'sawize deposit, Niassa Province, 2009. The weathered amphibolite (wAm) is crosscut by whitish plagioclase veins (Plv); (**B**) the Maninge Nice deposit, Concession Gemfields, Montepuez, 2012. The ruby (Crn) is surrounded by a plagioclase corona (Pl, anorthite) in a weathered amphibolite (wAm); (**C**) euhedral pargasite (Prg) in ruby. Field of view 1.5 mm; (**D**) inclusions of actinolite (Act) in ruby. Field of view 2.5 mm; (**E**) crystal of pargasite associated with mica (Mca) in ruby. Field of view 1.1 mm; (**F**) a melted mica (Mca) surrounded by a halo of fluid inclusions (Flinc) arranged in a "butterfly wing" indicative of decrepitation and recrystallization phenomena. The pargasite is euhedral and devoid of deformation. Field of view 1.7 mm. Photos A and B: Vincent Pardieu.

The texture is granoblastic with 120° triple junctions between amphibole and anorthite. The mineral assemblage formed during prograde metamorphism. The ruby often shows a corona of spinel + anorthite formed during the retrograde stage [96]. The ruby porphyroblasts have sizes between 1 mm and 1 cm. The P–T conditions of ruby formation (P = 10–11.5 kbar and T = 450–600 °C), calculated from P–T pseudosections, show the lowest temperature of formation compared to similar ruby-bearing amphibolites (see Figure 21) such as at Mangare [119], Winza [120], North Carolina [121], and Vohibory [122]. These calculated P–T conditions obtained by pseudo-sections [97] for the formation of ruby of Montepuez is somewhat questionable because clinochlore (and quartz) is a retrograde mineral in M-UMR and not in equilibrium with ruby. Geochemical analyses of amphibolites indicated that the protolith was basaltic and associated with subduction-related arc magmatism [96], following the model proposed by [206] for the arc assembly and continental collision in the Neoproterozoic East African Orogen.

Sub-Type IIA_2 includes rubies in marble such as those of the Mogok Stone Track (Myanmar), and others from central and eastern Asia, Canada, and Tanzania, (see Table 3).

Ruby and pink corundum deposits hosted in marbles are known from Asia [5,30,65,103,112,115,134,207–210], North America [113,211], Europe [95,110], and Africa [48,110,114,182] (see Figure 2). These ruby deposits are hosted by meta-pure or impure limestone of carbonate platforms and are of two types: (1) in marble where the gem corundum crystallized as a result of prograde and retrograde isochemical metamorphic reactions, mainly in a closed system, and where the ruby occurs as disseminated crystals in marble, in Central and Southeast Asia and other places worldwide [65]; and (2) in impure marble containing gneiss and silicate layers where the ruby crystallized at the peak of prograde metamorphism, such as at Revelstoke in Canada [113,212], Snezhnoe in Tadjikistan [115,134], and Morogoro in Tanzania [114,163].

(1) Ruby mines in marble in Central and South-East Asia.

These are one of the main worldwide sources of high-quality ruby with intense color and high transparency [5]. There is extensive literature on the geology and gemmology of these deposits; see [5,29,40,65,80,87,88,110,133,143,156]. Below we will present a short overview of the geology of these ruby deposits and focus on the significance of the mineral associations and geochemistry of the paleofluids in order to discuss the genesis of the corundum.

These deposits occur from Afghanistan to Southern China (see Figure 2) in marbles that are intercalated with garnet biotite-sillimanite- or biotite-kyanite-bearing gneisses, amphibolites, and quartzites (Figure 34A), which are sometimes intruded by granitoids, as in the Mogok district [87,88,213]. The marble units consist of discontinuous horizons up to 300 m in thickness, oriented parallel to the main regional foliations, thrusts, or shear zones related to the Cenozoic Himalayan orogenesis between 45 and 5 Ma [88,133]. The ruby mineralization is restricted to impure marble horizons. The protolith of the ruby-bearing metamorphic rocks comprises of carbonates enriched in detrital clays and organic matter, and intercalated evaporitic layers (Figure 34B). Ruby crystals occur: (1) disseminated within marbles and associated with phlogopite, muscovite, scapolite, margarite, spinel, titanite, pyrite, and graphite, as at Jegdalek, Afghanistan; Chumar and Ruyil, Nepal; Hunza and Nangimali, Pakistan; Mogok and Mong Hsu, Myanmar; Luc Yen, Vietnam; and Yuan Jiang in China; (2) in veinlets or gash veins, as in some occurrences in Northern Vietnam, and associated with phlogopite, margarite, titanite, graphite, and pyrite, and sometimes related to micro-shear zones, as at Nangimali in Pakistan (Figure 34C); and (3) in pockets associated with K-feldspar, phlogopite, margarite, graphite, and pyrite in some occurrences in Northern Vietnam.

The thermobarometric conditions of mineral equilibrium have shown that ruby could have grown during both prograde and retrograde metamorphic events. The most common reaction is the formation of ruby by the destabilization of spinel (Figure 35A):

$$\text{spinel} + \text{calcite} + CO_2 \leftrightarrow \text{corundum} + \text{dolomite}$$

Ruby formed during the retrograde metamorphism stage at T 620–670 °C and P 2.6–3.3 kbar [65]. The aluminum and the chromophorous elements in the ruby originated from marbles (Al up to 1000 ppm, V and Cr between 5 and 30 ppm for the Nangimali deposit).

The variations of local chemistry of the marbles along several decimeters are due to lateral facies variations of the protoliths which resulted in a succession of different F-Na-SO_4^{2-} -rich mineral associations: (1) F-aspidolite (Na_2O-rich phlogopite; Figure 34D) associated with F-phlogopite and F-paragonite in the different mineralized zones, as seen in the Nangimali deposit in Pakistan [214]; (2) anhydrite either associated with F-tremolite, F-edenite, F-pargasite, and carbonates in samples from the Hunza Valley, Pakistan and Luc Yen, Vietnam, or included in ruby with relics of spinel (Figure 35A,B); (3) anhydrite and salt crystals ($CaCl_2$, NaCl, and KCl) as solids in the Nangimali, Luc Yen, and Mogok deposits; and (4) F-bearing dravite-uvite tourmaline and apatite, indicating a low activity of water in the fluid.

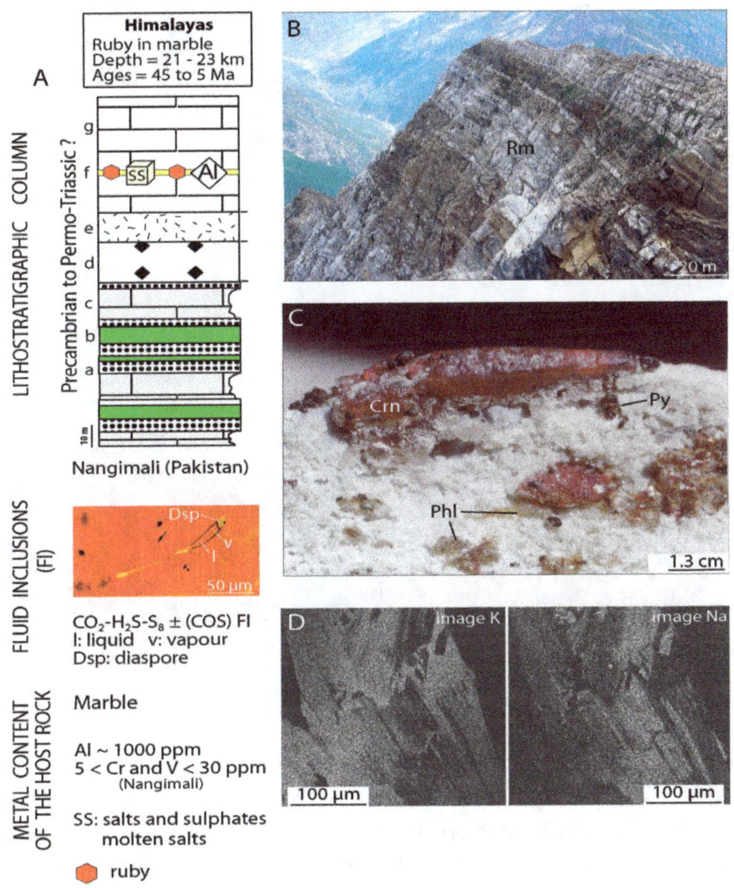

Figure 34. Ruby in metamorphosed carbonate platforms from Central and South-East Asia. (**A**) The Nangimali formation, Pakistan, showing the location of the ruby deposit in the metamorphic pile. Abbreviations: a: quartzite; b: amphibolite; c: mica-bearing marble; d: pyrite-bearing marble; e: calc-schist and garnet-biotite micaschist; f: ruby-anhydrite-bearing yellow Mg-marble; g: white dolomite and/or calcite marble; (**B**) the succession of marble (white) and garnet-sillimanite-bearing calc-schist and micaschist (brown); the unique ruby-bearing marble is reported as Rm; (**C**) gem ruby-bearing gash vein: (Crn), phlogopite (Phl) and pyrite (Py) in marble; and (**D**) back-scattered electron images of coexisting phlogopite and aspidolite with their X-ray images for potassium (K) and sodium (Na) [214]. Photos: Gaston Giuliani.

Paleo-fluids trapped as primary fluid inclusions in the minerals during their growth allow access to the composition of the mineralizing fluids. The combination of these data with those obtained for the stable isotope ratios of elements such as oxygen, hydrogen, sulphur, and boron on minerals coeval with the gems allows for characterization of the origin and source of the different chemical elements. Microthermometry studies combined with Raman spectroscopy of primary fluid inclusions in ruby revealed the coeval trapping of two types of carbonic fluid inclusions [62,63]: mono- to two-phase fluid inclusions (Figure 35D) in the system CO_2-H_2S ± COS ± S_8 ± (1 < H_2O < 10 mol %), and polycrystalline fluid inclusions (Figure 35C) in the system Na-K-Ca-CO_3-SO_4-NO_3-Cl-F ± CO_2-H_2S ± H_2O. The different solids in the polycrystalline fluid inclusions are mixtures of carbonates with Ca-Na-Al cations, such as shortite and dawsonite, sulphates (mainly anhydrite and barite), phosphates (F-apatite), nitrates, fluorides (fluorite), and chlorides (halite, Ca and K chlorides). These solids are daughter minerals of ionic liquids formed during the metamorphism of evaporites and limestones [65]. In ruby, these polycrystalline fluid inclusions are rare because if the salts are not immediately trapped by the crystals, and they dissolve because of their high solubility in water-rich fluids.

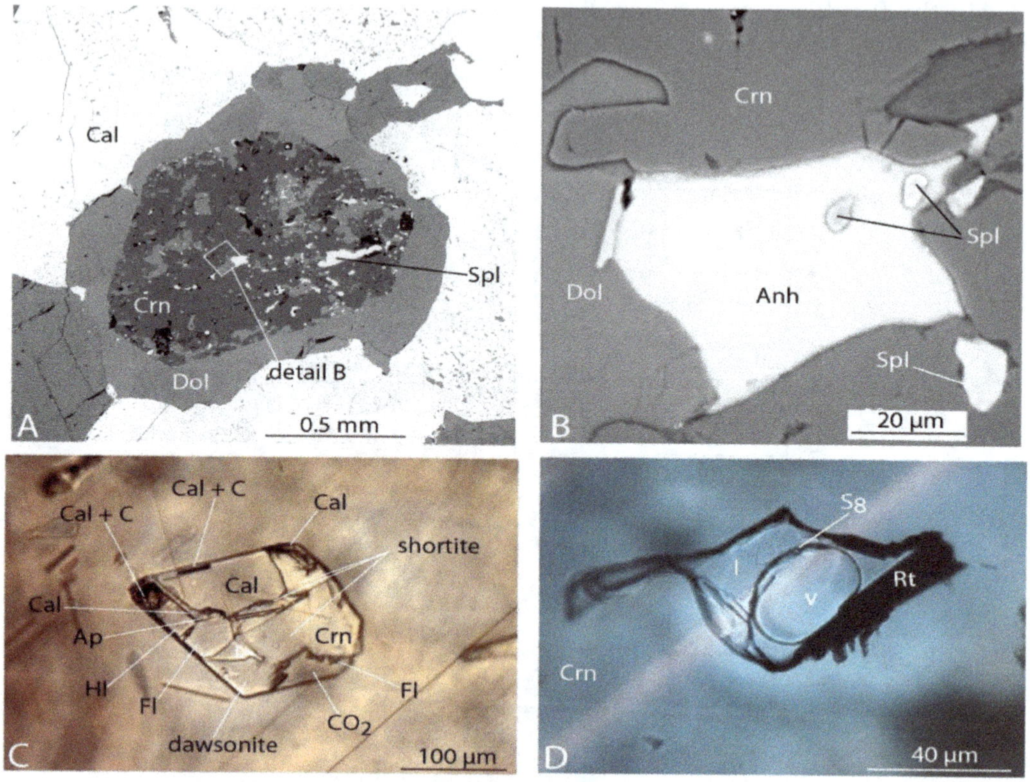

Figure 35. Mineral assemblage and fluid inclusions in ruby-bearing marble deposits from Central and South-East Asia. (**A**) Association ruby (Crn), dolomite (Dol), calcite (Cal), and spinel (Spl) illustrating the formation of ruby by the chemical reaction: spinel + calcite + CO_2 ↔ corundum + dolomite, Hunza valley (Pakistan); (**B**) inset of Figure 35A (detail B) showing the presence of anhydrite (Anh) with spinel inclusions (Spl). Spinel is a pre-ruby phase that has high Cr (up to 19 wt % Cr_2O_3) and Zn (up to 10 wt % ZnO) contents; (**C**) primary polycrystalline fluid inclusion cavity in ruby containing different solids (molten salts), which are mixtures of carbonates, with Ca-Na-Al cations such as calcite (Cal), dawsonite (Dw), shortite, and apatite (Ap), fluorite (Fl), halite (Hl), graphite (C), and a CO_2-H_2S-bearing fluid phase (CO_2), Mogok, Myanmar; and (**D**) three phase fluid inclusions with CO_2 and H_2S liquid (l) + vapor (V) phases and native sulphur (S_8) trapped by ruby in a rutile (Rt)-bearing cavity, Mong Hsu deposit Myanmar. Photos A and B: Virginie Garnier, [65]; Photos C and D: Gaston Giuliani, [62,63].

Crushing and leaching of rubies [62] showed that chloride is the dominant anion (25 to 53 mol %) with sulphate (2 to 36 mol %), nitrate (2 to 17 mol %), and fluoride (0 to 25 mol %). Sodium is the dominant cation (16 to 42 mol %). The presence of nitrate is a strong argument for a continental input to the original sediment as, generally, nitrate salts precipitate in closed basin playas or salars. In addition, the isotopic variation of sulphur in anhydrite included in ruby and marble defines two sets of $\delta^{34}S$ (V-CDT) values: the first, between 27‰ and 23‰, for a marine anhydrite; the second, between 4.8‰ and 1.6‰, for a continental source [65].

The presence of sulphate-carbonate-fluorine mixtures decreased the melting temperature of halite and other salts, and allowed the formation of chlorine- and fluorine-bearing ionic liquids. Fluorine of continental origin probably played an important role in the extraction of the Al present in impurities (clays) from the impure limestone. The existence of an ionic liquid trapped in the form of polycrystalline solids in ruby explains the color and clarity of the ruby by (1) the mobilization of Al, Cr, and V contained in the meta-limestone, and (2) their incorporation in an isotropic environment allowing crystalline growth with a minimum of defects.

The presence of evaporites of either continental or marine origin in the Ca-Mg protoliths is the key feature in the metamorphic model for ruby in marble. Ruby formed during the metamorphism, in the amphibolite facies, of carbonates interbedded with mudstones and containing intercalations of sulphates-chlorides-nitrates-borates of impure evaporites. The lithological control of the mineralization is a function of the paleogeography of the carbonate platform environment. The original sedimentary landscape converges to an epeiric carbonate platform succession with a combination of salt and evaporite mudflats of gypsum and anhydrite [62].

(2) Ruby and pink corundum in marble with intercalated silicate and gneiss layers.

(i) The Revelstoke occurrence is in the Monashee Complex of the Omineca belt of the Canadian Cordillera, British Columbia [113,212]. Ruby ($Cr_2O_3 \leq 0.21$ wt %) occurs in folded and stretched layers associated with green Cr-V-Ti-Ba-rich muscovite + Ba-bearing K-feldspar + anorthite ± phlogopite ± Na-poor scapolite. Other silicate layers within the marble are: (1) diopside + tremolite ± quartz and (2) almandine-rich garnet + Na-rich scapolite + diopside + tremolite + (Na,K)-amphiboles. Corundum formed at the peak of metamorphism (T 650–700 °C and at P 8.5–9 kbar) and the sources of the Cr are the silicate layers and gneiss.

(ii) The Snezhnoe deposit in Tadjikistan [115,134] is localized in the Muzkol-Rangkul anticlinorium within the Cimmerian zone of the Central Pamir. The Muzkol metamorphic series consists of calcitic marbles present as layers intercalated with amphibole-pyroxene and scapolite calciphyres, gneisses, Ti-Ba-rich muscovite schists, and quartzites. These calcitic marbles host the Snezhnoe deposit, formed by micaceous ruby-bearing lenses arranged as discrete en echelon lenses along the marble foliation. Ruby is found in four mineral assemblages as (i) scapolite + phlogopite + muscovite + margarite; (ii) plagioclase + muscovite + margarite; (iii) muscovite + phlogopite + margarite; and (iv) calcite. The main economic paragenesis is associated with ruby + calcite. The temperature of the metamorphic processes was estimated at 760 ± 30 °C using Zr-in-rutile geothermometry [134]. Ancient Al-enriched sediments such as metapelite or meta-bauxite intercalated in the marbles are suggested to be the possible protolith for the ruby-parent rocks.

(iii) At the Ngong'Oro mine, Morogoro district in Tanzania [182], ruby is located in metasomatic zones concordant to the metamorphic foliation, and formed at the contact between marble and biotite gneiss layers. The ruby is associated with spinel and sapphirine.

6.2.2. Sub-Type IIB: Metamorphic-Metasomatic Deposits Formed by Fluid–Rock Interaction and Metasomatism (See Table 3)

1. **Sub-Type IIB$_1$** corresponds to desilicated pegmatites (plumasite) and associated metasomatites in M-UMR such as at the Rockland mine (Kenya) or Polar Urals and Karelia (Russia), and others (see Table 3). This sub-type also concerns deposits associated with M-UMR intercalated with Si-Al rocks (gneiss, leuco-gabbro, or anorthosite), which have suffered fluid percolation and consequent metasomatism such as at the Aappaluttoq ruby deposit in Greenland.

These deposits formed at medium to high temperatures during hydrothermal–metamorphic fluid circulation and bi-metasomatism that developed at the contact of two contrasting lithologies, i.e., either granitoid or pegmatite or gneiss adjoining different rocks, such as metamorphosed M-UMR or marble [83,110]. The metasomatic reactions are related to the infiltration of post- or magmatic or metamorphic solutions, which originated from the same granite or from other magmatic or metamorphic events. The geochemical mechanism of dissolution of quartz (desilication) by a loss of a SiO_2-bearing rock, involves diffusion of Si from a pegmatite vein to a M-UMR for example (which plays the role of SiO_2 sink) at a rate more rapid than the diffusion of Al. The Al:Si ratio increases in the pegmatite vein, and a metasomatic plagioclase-corundum association in which the mass of alumina per unit volume is much greater than in the initial rock may develop if the volume of rock decreases at the same time and is underlain by a zone of quartz dissolution. These desilicated pegmatite or granitoidic veins are called plumasite following the definition of [215] that where formerly described as an oligoclase-corundum rock [216] forming a dyke in a serpentinized peridotite at Plumas, California. Gordon [215] noted that they were described from several countries (United-States, South Africa, and Russia) as "an abnormal group of granitic pegmatites, composed either of plagioclase, usually albite, alone (albitite), of plagioclase and corundum or largely corundum, all of which occcuring exclusively in peridotites or their altered equivalents". Gordon [215] indicated that the desilication occurred by fluid–rock interaction along the contact between the pegmatite and the intruded M-UMR. The mineralogical composition of the desilicated pegmatite and contact rocks depends of the degree of desilication. Some desilicated pegmatites present metasomatic zoning characterized by different mineral assemblages from the centre to the periphery of the dyke as described by [195] at the ruby Corundum Hill mine, in Macon County (United-States).

World-class gem corundum samples, including tabular ruby in desilicated pegmatites, have been described from the Umba and Kalalani deposits, which are located in serpentinite bodies [217] in Tanzania. World-class rubies were mined in the Mangare area in southern Kenya, including from the well-known (and formerly) John Saul mine [118,119]; and the Polar Urals in Russia [218] and Karelia [41]. Another unusual case of bi-metasomatism that is not a desilicated pegmatite, is related to the intrusion of ultramafic dykes in marble as observed at the Kitwalo ruby occurrence in the Mahenge district (Tanzania). At Kitwalo the ruby is hosted by phlogopite schists and formed via fluid–rock interaction at the contact between marble and an ultramafic dyke (Figure 36) [182].

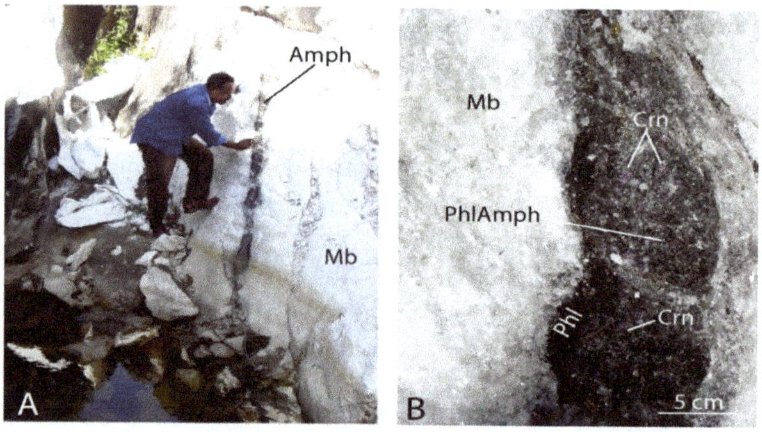

Figure 36. The ruby occurrence at Kitwalo, Mahenge district, Tanzania. (**A**) Amphibolite vein (Amph) crosscutting a dolomitic marble (Mb) and (**B**) detail of the contact between the two rocks showing a phlogopitite (Phl), a result of fluid–rock interaction. The amphibolite exhibits a high degree of phlogopitization (PhlAmph) associated with the ruby deposition (Crn). Photos: Elisabeth Le Goff.

We proposed to examine different deposit cases in order to characterize the infiltrational fluid metasomatism of M-UMR during medium to high grade metamorphism:

(1) In the United-States, these deposits are associated with M-UMR intercalated with Si-Al rocks metamorphosed in the amphibolite to granulite facies. The contact between meta-dunite and amphibolite with gneisses at different scales is underlined by metamorphic-metasomatic zones. The Corundum Hill mine at Macon County in North Carolina presents different types of contacts and distributions of ruby [195]:

(a) A gradual transition between ruby accumulations and meta-dunite with some remnants of M-UMR in the ruby zone (Figure 37A).

(b) The contact zone is characterized by a metasomatic column, of 6–7.5 m width, symmetrical from the periphery to the centre (Figure 37B) with dunite, talc schist, fibrous enstatite, green chloritite, chloritite-bearing ruby, and spinel (1.8–2.4 m wide). At Hunters mine, the amphibolite presents a plagioclase vein with symmetrical metasomatic zoning, 0.9–1.2 m wide, from the outer to inner zones the following mineral assemblage: amphibolite, actinolitite, ruby-bearing margarite-vermiculite, and ruby-bearing feldspar vein altered into kaolin (Figure 37C). These different types of contact indicate the circulation of metamorphic-metasomatic fluids between two rocks of different competency and chemistry. The considerable width of some metasomatic columns, up to 7.5 m, indicates a clear infiltrational metasomatic process as described for the Polar Urals [99].

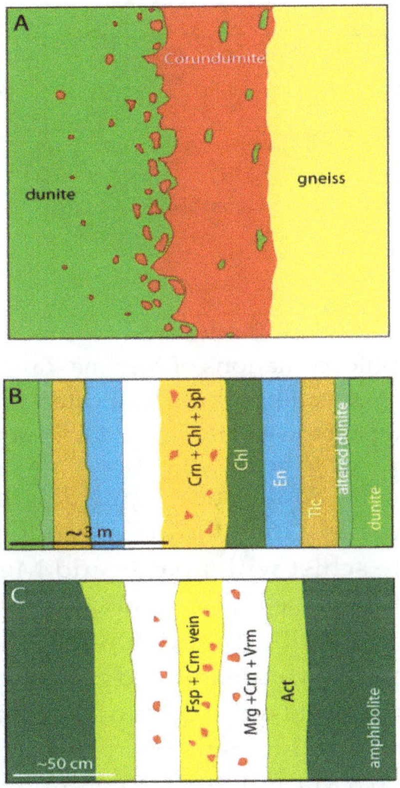

Figure 37. Metasomatic zoning in ruby-bearing metamorphosed dunites and amphibolites in the deposits of North Carolina, USA, modified from [195]. (**A**) The contact between dunite and gneiss is outlined by corundumite, Corundum Hill mine, Macon County. The scale is not qualitative; (**B**) ruby-bearing metasomatic zones in dunite at the Corundum Hill mine. The central part contains green chlorite (Chl) + ruby (Crn) + spinel (Spl) and the other zones are symmetrically disposed from this centre part. En = enstatite; Tlc = talcose rock; (**C**) ruby-bearing feldspar vein in amphibolite at Hunters, Iredell County. A metasomatic-metamorphic symmetrical zoning is developed around the vein. From the centre to the periphery: (1) a zone composed of vermiculite (Vrm)-bearing ruby with some margarite (Mrg) and (2) an actinolite (Act) zone at the contact with the amphibolite.

(2) In Kenya, rubies were discovered in the Mangare area, Voi mining district, in 1973 by John Saul [183]. Mining in the John Saul mine (recently renamed the Rockland mine) provided a regular supply of either facet grade or cabochon rubies from 1993 to 2012. The John Saul, Penny Lane, and Hard Rock deposits occur in lenses of serpentinized ultramafic rock (UMR) bound by shear zones in a metasedimentary series composed of graphitic gneiss and felsic gneiss (Figure 38) [118]. Fe-poor ruby (FeO < 0.06 wt %) and pink corundum is located in (i) micaceous lenses and pockets along the contact of a shear zone with the metasomatized UMR. The ruby is associated with ± spinel ± sapphirine + mica or in some cases with phlogopite + Mg-chlorite; (ii) in plumasitic veins that crosscut the serpentinites

where the ruby is associated with zoisite and anorthite ± muscovite [119]. The P–T conditions for ruby formation are between 700 and 750 °C and 8 and 10 kbar in the metamorphic granulite facies.

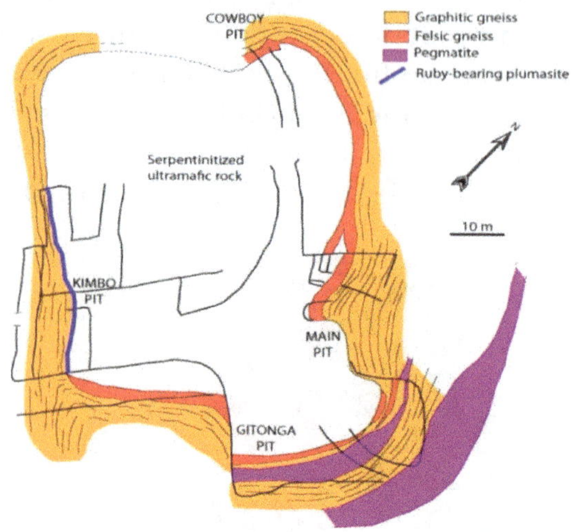

Figure 38. Geological map of the John Saul (now Rockland) mine and its serpentinized ultramafic rock body located in the metamorphic formations of Mgama-Taita, modified after [118]. The Kimbo Pit extension follows the structural direction of the desilicated pegmatite vein, i.e., a ruby-bearing plumasite, which formed at the contact between the graphitic gneisses and the ultramafic body.

As seen in the Kimbo Pit at the John Saul mine (Figure 38), the ruby-bearing-plagioclasite vein is always present at a depth of 40 m and has a longitudinal extent of 100 m. The contact of the vein with the UMR is outlined by a phlogopite schist with apatite and Mg-chlorite [118]. The plagioclasite is coarse-grained and is composed of oligoclase-andesine (Figure 39A). Locally, a huge concentration of ruby (around 70% in volume), green tourmaline (uvite-dravite), and blue apatite with some phlogopite is observed (Figure 39B). In other places, the ruby exhibited zoned scalenohedron crystals (Figure 39C) or "mushroom" (so-called by John Saul) habits up to 3 cm across (Figure 39D). The production of ruby from July 1995 to April 1997 was estimated to have been 36 kg/m^3 [219]. The U/Pb age obtained on zircons from the plumasite was 612 ± 0.6 Ma and the P–T conditions for ruby formation were estimated to have been 500 °C and 5 kbar [118].

(3) The ruby occurrences in Northwestern Karelia in Russia, located between the Kola and Karelia Archean blocks, is another case of desilication of pegmatite and gneisses due to the fluid–rock interaction [143,220]. The deposits, which are located in a 3.0–2.5 Ga polymetamorphic Archean complex, are in ruby-bearing paragneisses, amphibolites, and amphibole schists. Several collisional episodes occurred but at 1.9–1.8 Ga metamorphism and regional intrusions with pegmatites were responsible for the ruby formation through metasomatic processes. The ruby-bearing plumasite veins crosscut biotite-hornblende gneiss and amphibolite. The veins range from several cm to 10 m thick and up to 60 m long, and are parallel to the metamorphic foliation. The plumasite is composed of plagioclase (oligoclase-andesine), phlogopite, tourmaline, zoisite, epidote, and chlorite. The corundum crystals have prismatic and bladed shapes, 5 cm long and up to 3 cm in diameter. The ruby contains up to 9.4 wt % Cr_2O_3. Ruby and pink corundum from the Proterozoic gneisses have the lowest-known $\delta^{18}O$ compositions (−1.5‰ to −27‰) for terrestrial corundum [221], while the corundum in the amphibolite has $\delta^{18}O$ contents between −2.3‰ and 7‰ [1] (Figure 40). The plagioclase and amphibole have similar O-isotope values as the ruby. The highly depleted $\delta^{18}O$ values are due to the circulation of meteoric glacial subwaters in a rift zone at around 2.4 Ga, which affected the initial isotopic composition of the protolith prior to ruby formation [41,222].

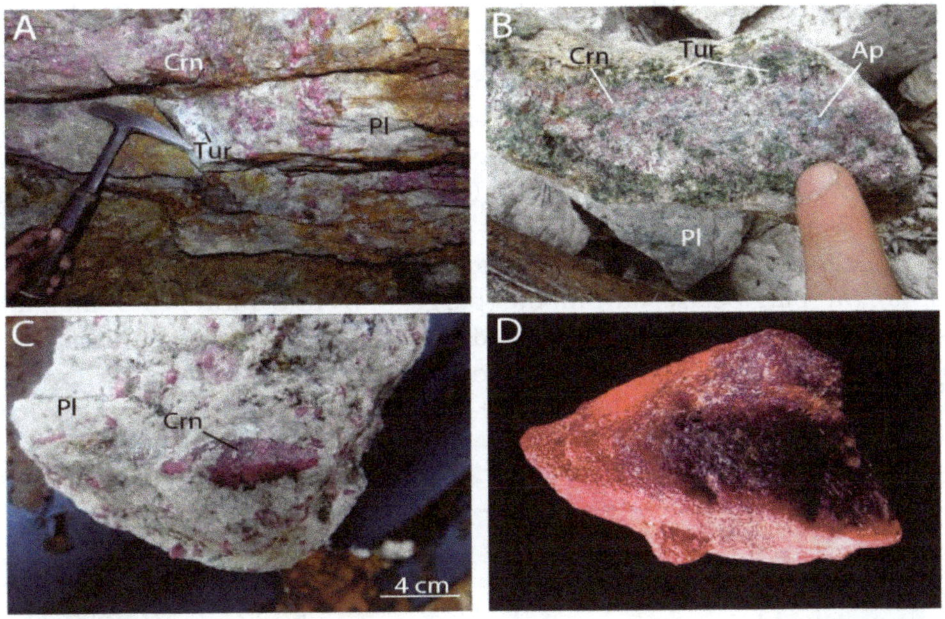

Figure 39. The ruby-bearing rocks of the John Saul mine, Voi area, Kenya [118,163]. (**A**) The Kimbo Pit and its ruby-bearing plumasite showing the mineral assemblage ruby (Crn), plagioclase (Pl), and some tourmaline (Tur); (**B**) concentration of ruby, green tourmaline, and blue apatite (Ap) in the plagioclase vein of the Kimbo Pit; (**C**) ruby showing the scalenohedron habit in the plagioclase matrix of the plumasite, Kimbo Pit; and (**D**) mushroom habit of a ruby, 3 cm across, from the John Saul mine. Photos A–C: Gaston Giuliani; Photo D: Courtesy John Saul.

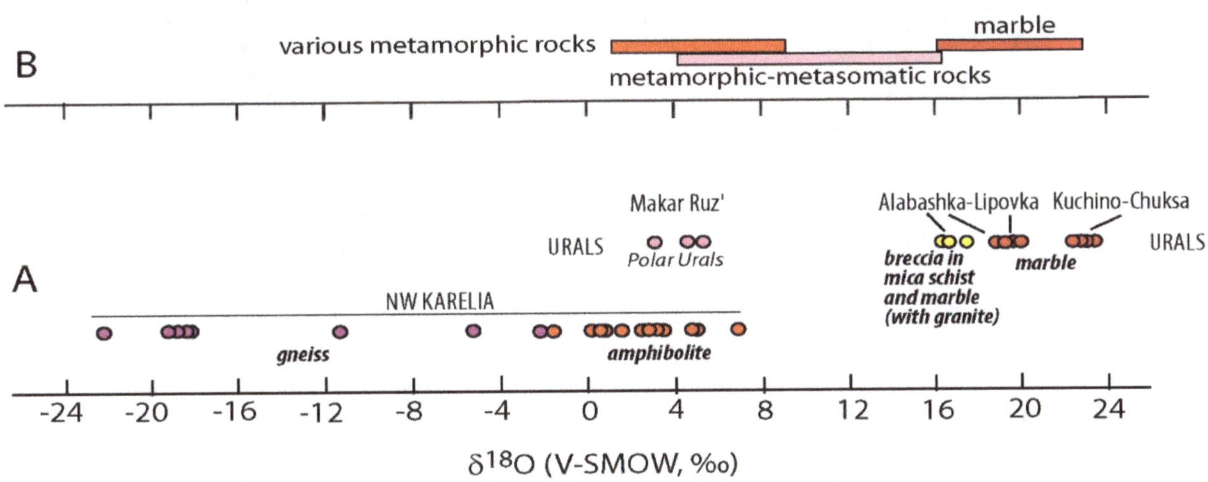

Figure 40. Oxygen isotopic composition of ruby from various Russian occurrences. (**A**) Oxygen isotopic composition of ruby in marble, breccia in mica schist and marble, and plagioclase-dominated metasomatic rocks from the Urals, and gneisses and amphibolites from northwest Karelia, Russia, modified from [153] and (**B**) the oxygen isotopic range of values for ruby worldwide associated with primary deposits [40,41,93,163].

(4) Ruby-bearing plagioclasite vein at Makar Ruz' in the Ray-Iz ophiolitic belt of Polar Urals, Russia (see Figure 2) [143,223]. The Paleozoic ophiolitic massif consists mainly of dunite and harzburgite that underwent polyphase metamorphism with a high degree of serpentinization. Chromitite in dunite yielded an Ordovician Re-Os age of 470 Ma [224] while zircon from the ruby-bearing plagioclasite gave a Silurian U/Pb between 380 [218] and 398 Ma [225]. The vein was completely mined out and is no longer visible. Scherbakova (1975, in [99]) recognized two types of plagioclasite: (i) plagioclase-rich and phlogopite-rich. The plagioclasite vein, 10–12 m thick, presented a clear symmetric metasomatic

zoning as also observed in emerald-bearing plagioclasite hosted in metamorphosed M-UMR [226]. From the centre to the outer metasomatic zones (Figure 41) this is: (i) plagioclasite (andesine to oligoclase), 1.5–2.5 m wide, with amphibole (actinolite-tremolite, pargasite) and without ruby; (ii) a ruby ($0.3 < Cr_2O_3 < 7.5$ wt %)-bearing phlogopite-rich zone (with Na-rich phlogopite, i.e., aspidolite in Cr-spinel; [99]), 0.3–1 m thick, with plagioclase, Cr-spinel (0.46 < Cr# < 0.86). The phlogopite is Ba-rich (0.8 < BaO < 2.7 wt %, [218]). Paragonite (0.8 < Sr < 2.3 wt %), margarite, and zoisite always occur between the ruby and plagioclase, and are probably retromorphic phases. The mean modal composition is plagioclase (32–67 vol %), paragonite (7–31 vol %), phlogopite (10–22 vol %), and Cr-spinel (5–7 vol %); (iii) plagioclase, amphibole (pargasite), clinochlore, Cr-spinel, talc, and augite in a 7–8 m thick zone; (iv) finally the dunite, which exhibited three sub-zones, from the inner to the outer, respectively: (a) chlorite + talc + Cr-spinel; (b) chlorite + talc + carbonate + serpentine + olivine; and (c) amphibole (actinolite-tremolite) + chlorite + Cr-spinel + talc + serpentine + olivine.

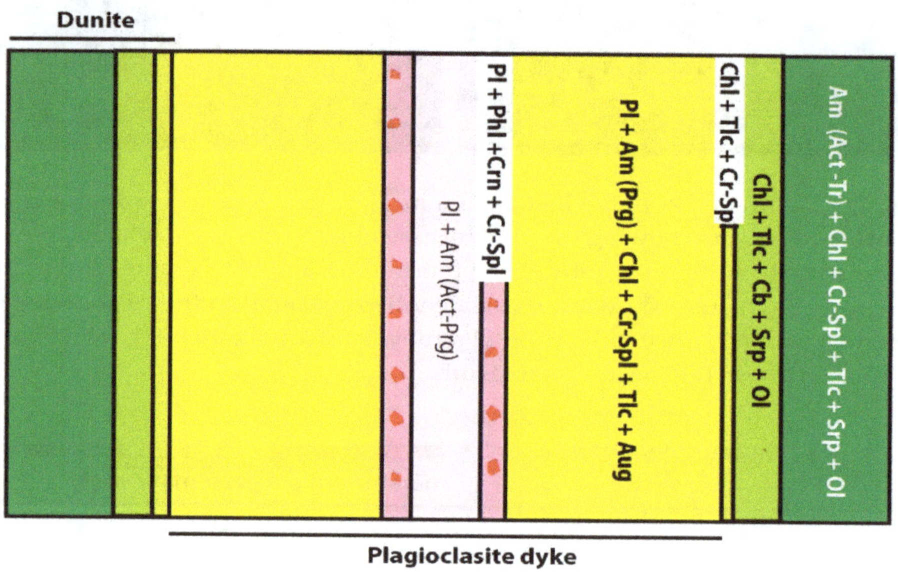

Figure 41. The ruby-bearing plagioclasite dyke in a serpentinized dunite from the Makar Ruz' deposit in the Ray-Iz ophiolitic belt of Polar Urals, Russia, modified from [99]. The plagioclasite shows a symmetrical metasomatic zoning with sharp contact zones characteristic of metasomatic zoning and percolation of fluids. Ruby (Crn) is hosted in the plagioclase (anorthite)-phlogopite (Phl)-Cr-spinel (Cr-Spl) inner zone of the metasomatic column. Abbreviations: Am = amphibole, Act = actinolite, Tr = tremolite, Prg = pargasite, Tlc = talc, Srp = serpentinite, Ol = olivine, Pl = plagioclase, Chl = chlorite, Aug = augite, and Cb = carbonate.

Aqueous fluid inclusions in the plagioclase are rich in HCO_3^-, SO_4^{2-}, and Cl^-. Zircon εHf values are between −11 and +13; the calculated εNd (380 Ma) value is +3.3 for the whole-rock plagioclasite [227] and the εNd (400 Ma) values are +6.4 and +6.6, respectively for dunite and harzburgite (Ronkin, 2000 in [99]).

Based on these considerations, different authors agree that (i) the vein minerals resulted from the fluid–rock interaction with the dunite and not from the fractional crystallization of ultramafic magma as proposed by [228]; (ii) the source of Cr is the chromitite in the dunite; and (iii) the metasomatic columns are due to fluid circulation, which is considered subduction-zone-derived and from metasomatism in the mantle wedge. The association plagioclase-corundum gave P–T estimates at 1.0–1.1 GPa and between 600 and 700 °C [99]. The formation of ruby and pink corundum is attributed to desilication of the dunite by the Na-Al-Si fluid circulation [218] rather than derivation from precursor pegmatite [100]. Such metasomatic zones that display clear zonation with very sharp metasomatic fronts, i.e., metasomatic columns (see Figure 37C), formed by infiltrational processes following the rules of [229], are also seen in the Haut-Allier, France [40,186], at Corundum Hill mine in North Carolina [195], and in the Vohibory unit in Southern Madagascar [164].

(5) In Southern Madagascar, desilicated rocks called "sakenites" by [230] occur at Sakeny, northwest of Ihosy. Sakenite is a whitish to greenish rock composed of anorthite ± corundum (mostly sapphire and sometimes ruby) ± spinel ± sapphirine ± phlogopite ± amphibole (edenite) ± clinopyroxene ± zircon, and is found in boudinaged bands or benches in the Precambrian high-grade granulitic domain of Southern Madagascar. The "sakenites" are intercalated with Al-rich paragneisses, amphibolites, and clinopyroxenites. The benches can attain 10 m width and display petrographic variations within a single bench, with local predominance of one mineral species. [230] distinguished sapphirine, spinel, spinel + sapphirine-bearing "sakenites", and sometimes "anorthitic corundumite" (anorthite + corundum or ± spinel) and "corundumite" occurring as pebbles and rounded blocks in rivers [231] (Figure 42A). Raith et al. [232] revisited the occurrence and described sakenites associated with phlogopitites, calc-silicate rocks, and marbles. The P–T conditions of formation of the intercalated Al-rich paragneisses were estimated at 800 °C and 6–7 kbar [232].

Figure 42. The sakenites from Sakeny and other sites in Southern Madagascar. (**A**) Corundumite composed of ruby (Crn) and spinel (Spl). Collection Lacroix, MNHN, Muséum National d'Histoire Naturelle de Paris; (**B**) sakenite, i.e., plumasite from the Anavoha occurrence. The ruby crystals are disseminated in a matrix of anorthite (An) and pyroxene (Px). Collection Lacroix, MNHN Paris; (**C**) the corundumite from Sakeny. A corona of sapphirine (Spr) and spinel is present around the whitish sapphire (Crn). Collection Lacroix, MNHN Paris; (**D**) sakenite vein with anorthite, corundum + spinel at Bekinana (Tranomaro area). The corundum generally shows a corona of black spinel. Photos: Gaston Giuliani.

At lower P–T conditions (750–700 °C and 6 Kbar) the sakenite was affected by the circulation of potassic fluids. The corundum + anorthite assemblage was transformed into phlogopite and Mg-Al minerals producing corundum-spinel-sapphirine-bearing phlogopitites. Subsequently the corundum was replaced by spinel, spinel-sapphirine, or sapphirine at lower temperatures and pressures [232].

Other occurrences in southern Madagascar include Vohidava, Anavoha (Figure 42B), and Bekinana [164]. These rocks are anorthitites showing corona textures due to the partial or complete replacement of corundum porphyroblasts by spinel and sapphirine (Figure 42C,D), or spinel + hibonite, spinel + sapphirine, K-feldspar + spinel, and anorthite + sapphirine. Lacroix considered "sakenites" to be a product of high-grade metamorphism of clay-rich marls. [233] favored metamorphism/metasomatism of clay-rich limestones. Raith et al. [232] suggested kaolinite-rich sediments or calcareous bauxites as a protolith. The chemical composition of the "anorthitic corundumite" [230], i.e., 12.3 wt % SiO_2 and 73.6 wt % Al_2O_3, implies a loss of silica with drastic enrichment of alumina, with mafic rocks (amphibolites and clinopyroxenites) or marble and calc-silicate rocks playing the role of SiO_2 sink at a rate more rapid than the diffusion of Al.

(6) In Southwest Greenland, numerous ruby occurrences have been discovered and mapped in the Fiskenæsset Anorthosite Complex [33,34,83,234,235]. Fagan [34] stated that there are at least 534 individual corundum occurrences in the area and an open-pit ruby mine recently opened at the Aappaluttoq locality (Figure 43A). The deposit has defined minable reserves of 59.2 million grams of pink corundum and ruby [35].

Figure 43. (**A**) The ruby mine at Aappaluttoq. Photo: Vincent Pardieu; (**B**) gem corundum in outcrop. Photo: Vincent Pardieu; (**C**) ruby rough. Photo: Vincent Pardieu; (**D**) a selection of cut gems (0.5–0.75 ct) from the Greenland deposits [34], courtesy of True North Gem Inc.

The 2.97 Ma Fiskenæsset Complex is one of the best preserved layered Archean intrusions in the world ([236] and references therein). It consists of a ca. 550-m-thick association of anorthosite, leucogabbro, gabbro, and metamorphosed UMR (dunite, peridotite, pyroxenite, and hornblendite), which are found in the mining area at Aappaluttoq (Figure 44). Despite amphibolite to granulite facies metamorphism (that peaked at 2.80 Ga ([235] and references therein)) and polyphase deformation, primary cumulate textures and igneous layering are well-preserved [236].

The rubies occur in a reaction metasomatic zone (Figure 43B) formed at the contact between an anorthite-rich rock, such as anorthosite or leucogabbro, and an UMR, such as a metamorphosed dunite or peridotite. The metasomatic zones resulting from fluid percolation are typically 0.5–5 m wide and contain ruby, phlogopite, cordierite, spinel, sapphirine, calcic amphibole, sillimanite, and kornerupine [83]. The ruby reaction zone consists of three main groups: (i) phlogopitite with 80% mica (eastonite-phlogopite) and up to 20% corundum; (ii) sapphirine-gedrite rocks, and (iii) anthophyllite-cummingtonite (-pargasite) rocks. The rubies formed at approximately 640 °C and 7 kbar, when pegmatites intruded, at 2.71 Ga, the anorthite-rich rock and juxtaposed UMR [83,235]. This caused the anorthite to alter to calcic amphibole, releasing Al, and Cr from the UMR to be mobilized by the pegmatitic fluids. The ruby-forming reaction occurred at upper amphibolite facies conditions and under high fluid flux interaction between two rocks of contrasting chemistry.

Keulen et al. [83] proposed to name this kind of deposit as ruby-bearing ultramafic complexes. The description of the deposit confirms that it formed through an intense fluid–rock interaction that resulted in large amounts of K-metasomatism (ruby-phlogopitite zone). In consequence, the Aappaluttoq deposit might be classified in the Sub-type IIB_1 proposed by the enhanced classification for ruby deposits (see Table 3). The nature of the protoliths can vary worldwide from mafic to ultramafic rocks for low Si and Cr-bearing rocks, and from gneiss, granitoid to anorthosite rocks for Si-Al-K-rich rocks.

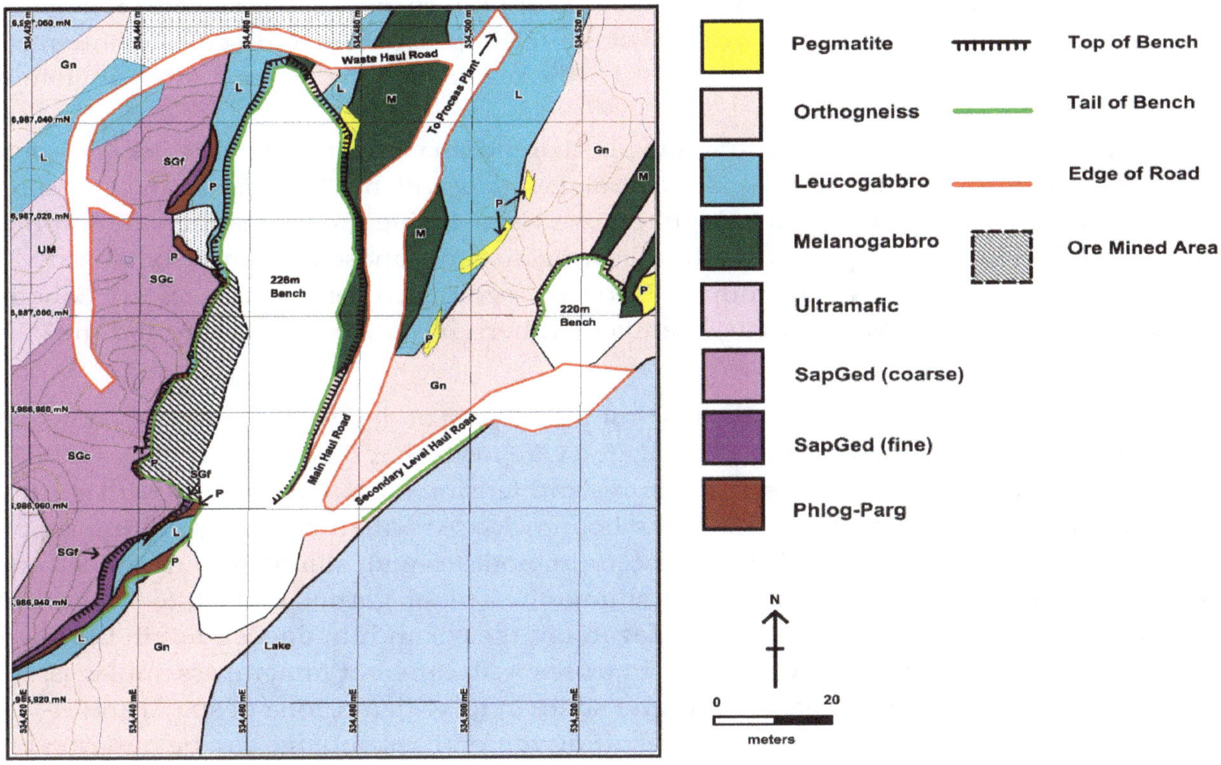

Figure 44. Geological map of the Aappaluttoq ruby mine [34]. SapGed: sapphirine-gedrite; Phlog-Parg: phlogopite-pargasite; Gn: orthogneiss; SGc and SGf: coarse (c) and fine (f) sapphirine-gedrite; L: leucogabbro; M: melanogabbro; UM: ultramafic rock; and P: pegmatite (yellow) or phlogopitite (marron).

According to [83] rubies from the Fiskenæsset Complex (Figure 43C) are distinguished by their high Cr, intermediate Fe, low V, Ga, and Ti, low O isotope values (1.6–4.2‰), and rare mineral inclusions (e.g., anthophyllite + biotite) and growth features. Numerous faceted pink corundum to ruby more than 1 carat have been cut (Figure 43D) and the largest piece of Cr-corundum recovered to date is the carved opaque 440 ct Kitaa Ruby [35].

There are also a number of occurrences in Greenland north of the Fiskenæsset Complex. Keulen et al. [83] listed (approximately from north to south): Ujarassuit Nunaat, Maniitsoq NW, Kangerluarsuk, several occurrences on the island of Storø, and Kapisillit.

At Kangerluarsuk, ruby occurs in schist with plagioclase, both surrounded by biotite, with minor orthopyroxene, amphibole, and olivine ([237] and references therein). The ruby formed by a metamorphic reaction between a metaperidotite and a kyanite-quartz-garnet schist (with minor plagioclase, spinel, and mica) at amphibolite-facies conditions.

At Ujarassuit Nunaat, an ultramafic enclave experienced amphibolite-grade metamorphism resulting in gabbro anorthosite with amphibole-calcite plagioclase and a small amount of red corundum overgrowing the latter [238].

On Storø Island, similar to Kangerluarsuk, the ruby occurs in a mica schist layer up to 3 m thick and formed by a metamorphic reaction between a metaperidotite and a sillimanite-biotite-garnet schist (with minor plagioclase) at amphibolite-facies conditions [237].

At both Kangerluarsuk and Storø Island, the ruby is thought to have resulted from the breakdown of kyanite and sillmanite at elevated temperatures due to the removal of SiO_2. The reaction was facilitated by a chemical potential gradient between the relatively Si- and Al-rich metasedimentary rocks and the low-Si ultramafic rocks. According to [237], the additional required Al_2O_3 was supplied via metasomatism, with the Al mobilized by complexation with hydroxide at alkaline conditions or transported as K-Al-Si-O polymers at deep crustal levels.

Poulsen et al. [239] described several occurrences of pink corundum near Nattivit in Southeast Greenland. The corundum occurs where late-stage felsic pegmatites crosscut and interact with metamorphosed ultramafic rocks. The source of Al is the plagioclase in the pegmatite with removal of Si by interaction with the UMR [239].

2. **Sub-Type IIB$_2$** is characterized by ruby and pink corundum in shear zone-related or fold-controlled deposits in different substrata (M-UMR, gneiss, schist, and marble) with fluid circulation and metasomatism.

These deposits formed at medium to high temperatures during metasomatic-metamorphic fluid circulation. They include ruby or pink corundum-bearing metamorphosed and metasomatized M-UMR and marble (see Table 3) such as those from the Neoproterozoic occurrences of Zazafotsy, Sahambano, and Ambatomena (Southern Madagascar), Kerala (southern India), and Mahenge and the Uluguru Mountains (Tanzania).

(1) For Madagascar, the geology and genesis of these types of deposits was described by [40,164]. These studies focused on the Sahambano and Zazafotsy deposits that produced sapphires (rather than rubies) from feldspar-rich mafic gneisses. Fractures and tension gashes associated with shear zones favored fluid circulation and biotitization of the host rocks. At Sahambano, sapphires and pink to reddish corundum ($1000 < Cr_2O_3 < 2000$ ppm) occur in biotitites with sillimanite and spinel, and also in gneisses with sapphirine, K-feldspar, biotite, sillimanite, spinel, garnet, and albite. The corundum formed at T 650 °C and P 5 kbar [240,241]. At Zazafotsy, the corundum deposit lies in the Zazafotsy shear zone system [242]. The corundum occurs in lenses with different mineral assemblages:

(i) Almandine garnet, biotite, plagioclase, and sillimanite, with a corona of spinel and K-feldspar formed around corundum (Figure 45A). The resulting mineral assemblage can be described by the following reaction:

$$\text{biotite} + \text{corundum} \rightarrow \text{spinel} + \text{K-feldspar} + H_2O$$

The association of corundum, garnet, spinel, and sillimanite in some samples results from the reaction:

$$\text{garnet} + \text{corundum} \rightarrow \text{spinel} + \text{sillimanite}$$

(ii) In biotite schist with biotite, corundum, spinel, and minor grandidierite (Figure 45B).

Figure 45. The Zazafotsy ruby and sapphire deposit, northeast of Ihosy, Southern Madagascar. (**A**) Purple sapphires (Crn) in a garnet (Grt)-bearing biotite schist (Bt). The crystals are embedded in a corona of K-feldspar (Kfs) and spinel (Spl) and (**B**) pink corundum hosted by a biotitite (Bt). Photo: Gaston Giuliani.

The Ambatomena ruby deposit was exploited by a private mining company between 2000 and 2001. The production was of very good quality with euhedral crystals up to 2 cm across and 3 cm long. The ruby deposit is located in the Androyan granulitic metamorphic series, which is composed of para-and orthogneisses, marble, granite, clinopyroxenite, and quartzite. The four ruby deposits (called D1 to D4) formed lenses of clinopyroxenite (D1 and D2), sapphirinite (D3), and plagioclasite (D4) intercalated with garnet-bearing granofels [243] (Figure 46). The formation of ruby is linked to the percolation of fluids at the contact between the lenses and the granofels. The ruby (2180 < Cr_2O_3 < 6670 ppm) is found in phlogopitized clinopyroxenites (G2), sapphirinites (G3 and G4; Figure 47A), and plagioclasites (G4; Figure 47B). The mineral assemblage of the ruby is spinel, K-feldspar, plagioclase, clinopyroxene, sapphirine, phlogopite, sillimanite, cordierite, titanite, rutile, and hibonite (Figure 47C,D). The retrograde metamorphic stage is characterized by the destablization of ruby into spinel in the plagioclasites and sapphirine in the clinopyroxenites, each mineral forming a corona around ruby crystals.

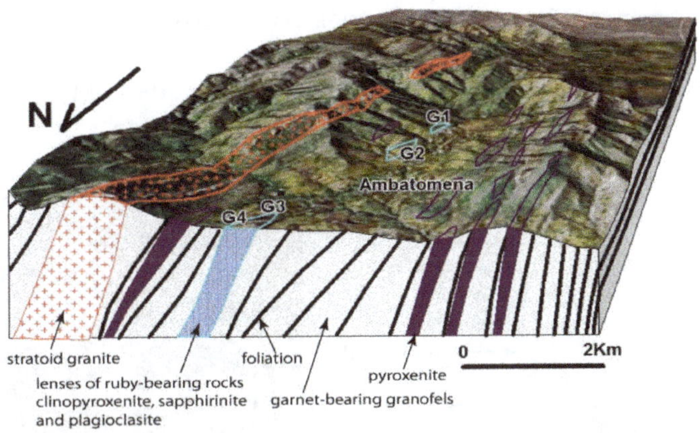

Figure 46. Panoramic view and geological cross-section of the Ambatomena ruby deposit, Betroka region, Southern Madagascar, modified from [243]. The correlation between the ruby-bearing occurrences (G1–G4) and the garnet-bearing hornfels and pyroxenites can be seen in the cross-section.

Figure 47. The Ambatomena ruby deposit, south of Betroka, Southern Madagascar. (**A**) Ruby (Crn)-bearing sapphirinite (Spr); (**B**) plagioclasite (Pl) with whitish corundum (Crn) showing a corona of spinel (Spl); (**C**) microscopic image of ruby-bearing sapphirinite with a granoblastic texture of K-feldspar (Kfs), sapphirine (Spr), sillimanite (Sil), and rutile (Rt). The ruby sometimes has a corona of sapphirine and K-feldspar; (**D**) microscopic image of the corundum-bearing plagioclasite. The texture is granoblastic decussate with plagioclase (Pl) and corundum. The plagioclase is associated with some sapphirine and clinopyroxene (Cpx). The corundum is embedded in a corona of green spinel with some hibonite (Hib). Photos: Alfred Andriamamonjy.

(2) For Tanzania, the ruby deposits in the Uluguru and Mahenge Mountains are located, respectively, in dolomite- and calcite-dominated marbles but deformation controlled the mineralization, which is located in saddle reef structures within fold hinges [114] (Figure 48A). The ruby is metamorphic and formed between 620 and 580 Ma. At Mahenge the deposits are Kitwalo, Ipango, Matote, and Lukande. The Morogoro mining district consists of the Kibuko, Mwalazi, Nyama Nyama, Visaki, and Msonge deposits. The marbles belong to the Neoproterozoic cover sequence that was thrust onto the gneissic basement of the Eastern Granulites. At Uluguru, the mineral assemblage is dolomite, phlogopite ± spinel, pargasite, scapolite, plagioclase, margarite, chlorite, and tourmaline.

At Mahenge, the paragenesis is calcite, plagioclase, phlogopite ± dolomite, pargasite, sapphirine, titanite, and tourmaline. Two episodes of ruby formation were defined by [114]:

(i) Ruby formed during the prograde metamorphic stage at T 750 °C and P 1.0 GPa by the breakdown of either diaspore or via the chemical reaction:

$$\text{margarite} \leftrightarrow \text{anorthite} + \text{corundum} + H_2O \quad (1)$$

The metamorphic mineralizing fluids circulated along the zones of higher permeability at granulite facies conditions.

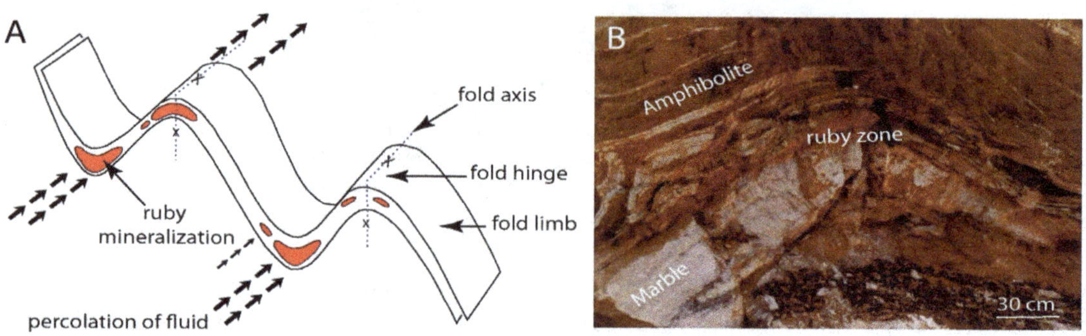

Figure 48. (**A**) The schematic structural control of ruby mineralization in the Uluguru and Mahenge mountains, Southern Tanzania, modified from [114]. Ruby formed in zones of higher permeability near fold hinges; (**B**) the structural control of ruby formation by folding and opening of fractures is also observed at the Lukande ruby deposit in the Mahenge Mountains. The ruby mineralization is located in the hinge of a fold that affected both marble and amphibolite. The contact between the two contrasting rocks hosts a metasomatic zone composed of carbonates, phlogopite, and Mg-amphibole. Photo: Gaston Giuliani.

(ii) Changes in the T-XCO_2 conditions of the metamorphic fluid triggered the ruby deposition through two main reactions:

$$\text{spinel} + \text{calcite} + CO_2 \leftrightarrow \text{corundum} + \text{dolomite} \quad (2)$$

$$\text{margarite} + \text{calcite} \leftrightarrow \text{corundum} + \text{scapolite} + H_2O \quad (3)$$

The sources of Cr (and V) were not discussed by [114] but two possibilities can be proposed: (i) a sedimentary source, i.e., limestone or Mg-limestone (muds or clays interbedded with impure carbonates) from the carbonate platform of the Neoproterozoic Mozambican ocean. Such a source was also proposed for the tsavorite mineralization in Northern Tanzania [244]. Both tsavorite at Merelani and ruby in the Mahenge and Uluguru Mountains formed in a similar geological setting and during the East African orogeny; (ii) a mafic source such as amphibolite that was infiltrated by the fluids circulating along zones of higher permeability during folding and the formation of saddle reef structures.

At the Lukande deposit, the marbles are intercalated with Cr-bearing amphibolites that were folded and fractured (Figure 48B). Both types of rocks have suffered fluid percolation and ruby-bearing metasomatic zones formed at the marble-amphibolite contact. The mineral assemblage is ruby-carbonate-phlogopite-magnesian amphibole. The style of deformation and the nature of the protoliths are similar to those proposed for the formation of tsavorite vein deposits from the Merelani area [244]. The metamorphic formations were affected by tight ductile isoclinal folding and shearing, and the tsavorite occurs at the hinges of sheared isoclinal folds within saddle-reef structures.

6.3. Sedimentary-Related (Type III): Placer Deposits

Placers are hosted in sedimentary rocks (soil, rudite, arenite, and silt) that formed via erosion, gravity effect, mechanical transport, and sedimentation along slopes or basins related to neotectonics. Climate is a major factor in the formation of secondary deposits. In tropical areas, rocks are exposed to meteoric alteration, resulting in an assemblage of clay minerals, Fe and Mn oxides, and other supergene phases. Corundum and zircon are resistant minerals found in soils, laterites, and gravelly levels overlying bedrock. In high-altitude areas such as in Nepal, Pakistan, and Azad-Kashmir the alteration is mechanical and ruby is found in boulders accumulated on the lower part of the slopes as for the Nangimali deposit [245]. Present-day placers are illustrated in Figure 49, which follows the classification proposed by [138]:

(i) Eluvial concentrations derived by in situ weathering or weathering plus gravitational movement or accumulation. Eluvial-deluvial deposits are on slopes and in karst cavities (marble type). Colluvial deposits correspond to decomposed primary deposits, which have moved vertically and laterally down-slope as the hillside eroded.

(ii) Alluvial deposits resulting from the erosion of the corundum-bearing host rock and transport by streams and rivers. Concentration occurs where water velocity drops at a slope change in the hydrographical profile of the river, e.g., at the base of a waterfall, broad gullies, debris cones, meanders, and inflowing streams. Terminal placers [138] formed by deltaic, aeolian, and marine accumulations are unknown for ruby deposits.

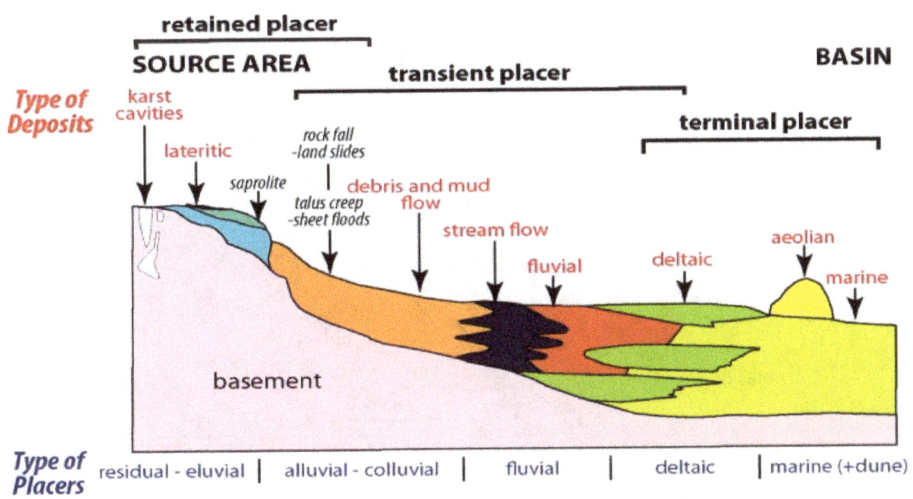

Figure 49. The different types of placer deposits formed on the continental surface and deltaic or marine basins, modified from [138]. The majority of these types of placer, except for aeolian and marine placers, occur for gem corundum deposits worldwide.

The placer deposits are divided in three main sub-types (Table 3).

6.3.1. Sub-Type IIIA: Gem Placers in an Alkali Basalt Environment (Eastern Australia, Pailin-Chanthaburi-Trat, Eastern and Central Madagascar, and Others)

Ruby-bearing placers linked to volcanic rocks are found in continental alkali basalt extrusions [68] (Figure 2). These deposits are sporadic and economically quite rare when compared to sapphire deposits of the same type, except for the ruby deposits at Pailin and Chanthaburi-Trat [145]. In Australia, Ray et al. [246]

showed that within the New South Wales deposits, in 420 listed gem corundum occurrences, 10% had ruby. However, the production realized on the short-term depended on worldwide demand.

Ruby placers in alkali basalt environments include the deposits in eastern Australia [81,106,247–250]; Asia, with deposits at Pailin, Cambodia [5,74,82,248] and the provinces of Chanthaburi-Trat and Kanchanaburi [145,146,148,149,154,251]; China with the Muling deposit in Shandong province [252]; Eastern Africa, in Kenya with the Baringo deposit, which is linked to the volcanism of the East African rift [118]; and in Madagascar at Antsirabe in Antananarivo province [95,161,231] and Vatomandry in Toamasina province [77,78].

(1) In Eastern Australia, ruby placers linked to volcanic rocks are found in continental alkali basalt extrusions [68] (Figure 2). They occur as scattered deposits that are subordinate to sapphires [81,85] (Figure 50). Rubies are found in alluvial placers shed from Cenozic lava fields within an eruptive zone extending from Northeast Queensland to Southeast Tasmania. The main deposits are located in New South Wales, i.e., Barrington Tops, Macquarie-Cudgegong Rivers, and New England (Figure 50A). Occurrences are also found in South Australia at Adelaide, Victoria around Melbourne, and Tasmania.

In New South Wales the deposits are found [81,85,86]:

(i) At Barrington Tops, erosion and transport by drainages of the eroded Barrington Tops volcano gave rise to the formation of mineable placers in the river. Ruby was found in 1851 and alluvial prospecting was done over a long period between the 1960s and 1990s. The geology of the Barrington Tops Plateau was described in the early 2000s [253] and mining at Upper Manning River took place from early 2005 until 2008 (Figure 50B). The ruby is associated with other pink, purple, and pastel blue corundum used for jewelry (Figure 50C).

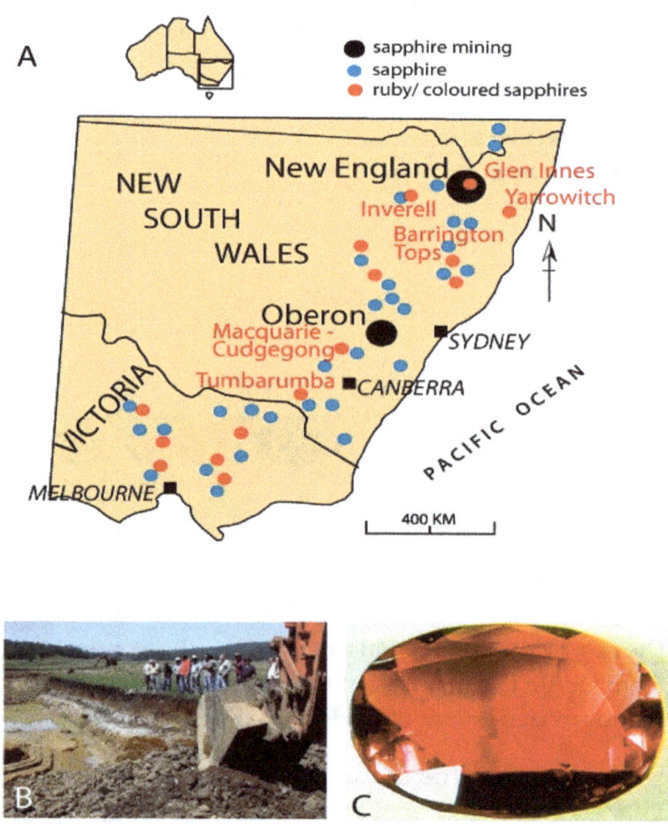

Figure 50. (**A**) Location of ruby deposits in New South Wales, modified after [81,86]; (**B**) opening of the Barrington Tops mine, March 2005, upper Manning flats. Photo: Cluff Resources Pacific M/L, Courtesy Lin Sutherland; (**C**) faceted 1 carat ruby, 6 mm × 4 mm across, Gloucester Tops, SE Barington Tops plateau. Photo. Gayle Webb.

(ii) At Yarrowitch, the alluvial gem field is more restricted than at Barrington Tops and the ruby is associated with sapphires, red zircon, and spinel. The alluvial occurrences have not been mined.

(iii) At Macquarie-Cudgegong Rivers, extensive alluvial exploitation took place in the rivers from 1851 to 1941. The placers contain ruby, sapphires, zircon, gold, and diamond; their original source is unknown, despite the presence of numerous basaltic vents in the area.

(iv) At Tumbarumba, ruby is minor and the placers contain blue to green sapphires, which are corroded and worn.

(v) At New England, the Inverell-Glen Innes area is known for its sapphire placers. Ruby is scarce but is recovered from all the rivers. The crystals grade from red into pink and purple corundum [86]. Sometimes, the crystals are zoned with red to pink cores, a white peripheral zone, and a violet rim. Several large gem crystals have been cut, such as the 48 carat "Aurora" zoned stone found near Glen Innes in 1988, which was cut into a 17 carats red, a 9 carats blue, and a 2 carats pink stones [249]. Other stones have been faceted, such as a 2.13 carats purple red oval cut, a 0.31 carats light red baguette cut, and a 0.22 carats red marquise cut [81].

In Queensland, rubies are very rare and the deposits are mainly composed of sapphires. In Victoria and Tasmania, pink to red corundum are rare and the high content of Fe turns the color to brown [254].

(2) The presence of corundum in alluvial placers from Madagascar has been known since the beginning of the 20th century [231]. The deposits formed by the erosion of corundum-bearing Cenozoic lava fields within eruptive zones extending from the centre of the island, in the Ankaratra massif, Antananarivo province, to the north in the Montagne d'Ambre, Antsiranana province (Figure 51). The production of ruby in Madagascar became significant after the discovery in 2000 of two ruby deposits in the province of Toamasina named Andilamena [164] and Vatomandry [255].

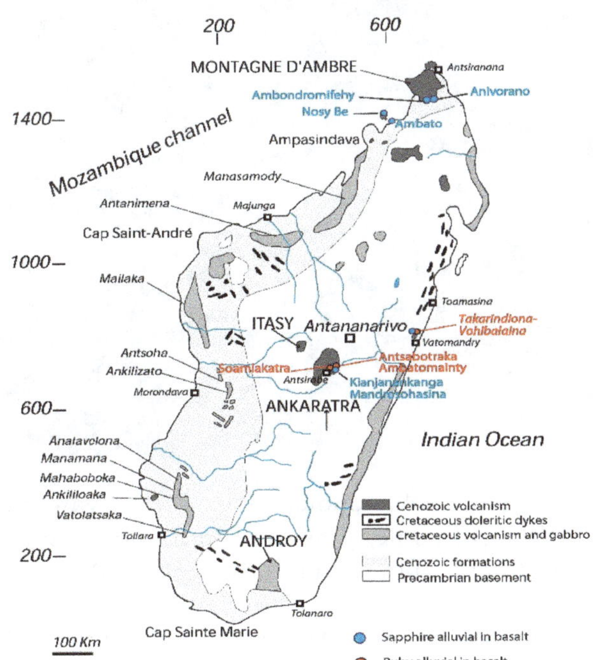

Figure 51. The volcanic provinces of Madagascar, modified from [256] with the localities of the basaltic corundum deposits related to the Cenozoic basalts in the provinces of Antsiranana, Antananarivo, and Toamasina [161,162]. The deposits of gem corundum: north in Antsiranana Province, the mining districts of Ambondromifehy and Anivorano in the Montagne d'Ambre, Ambato peninsula and Befotaka in Nosy Be island; Central Madagascar in Antananarivo Province, the mines of Soamiakatra, Mandrosohasina–Kianjanakanga (Antsirabe area) in the volcanic Ankaratra massif; and east in Toamasina Province, with the deposits of Vatomandry in the Takarindiona and Vohibalaina basaltic areas. The geographic coordinates are related to the Laborde projection.

In 2005 Madagascar was the third largest ruby-exporting country after Kenya and Tanzania (see Figure 3). The secondary deposits, i.e., alluvial placers and paleoplacers are located (i) to the north and east of Antsirabe city in Antananarivo province (Ambatomainty, Antsabotraka, and Soamiakatra deposits) and (ii) to the north of Vatomandry in Toamasina province. In this last province the alluvial and paleoplacer deposits border the Takarindiona and Vohibalaina basaltic massifs (Figure 51). The mining sites are Amfao, Sahanonoka, and Antsidikana near the commune of Amboditavolo. They are located in a number of watersheds and major river systems downstream of the Vohibalaina volcanic massif, and those of Antanambao, Mahatsara, and Ankazombanga on the southern and eastern borders of the Takarindiona massif [94].

At Vatomandry, the ruby is associated with crystals of sapphire and red zircon in some paleoplacers but mostly in alluvials. Generally, the gem gravels consist of 0.5–2 m thick sediments overlying bedrock in the riverbeds. The ruby crystals are 5 mm to 3 cm across, rounded and with different colors (red, red purple, red brown, and pink), but a red-purple color is dominant [77].

The most common inclusions in ruby from the Sahanonoka deposit are rutile and zircon (Figure 52A), which are cogenetic and arranged in clusters [161,162,255]. Ilmenite, rutile, and Al-silicates also occur as inclusions (Figure 52B). The other solids are talc (Figure 52C), phlogopite (Figure 52D), pentlandite, plagioclase, titanite, ilmenorutile, hematite, and Al-silicates. Apatite was observed by [78]. Rutile crystals were identified as a daughter phase in the cavities of fluid inclusions trapped by the ruby. The fractures in the rubies are filled by limonite and kaolinite [78].

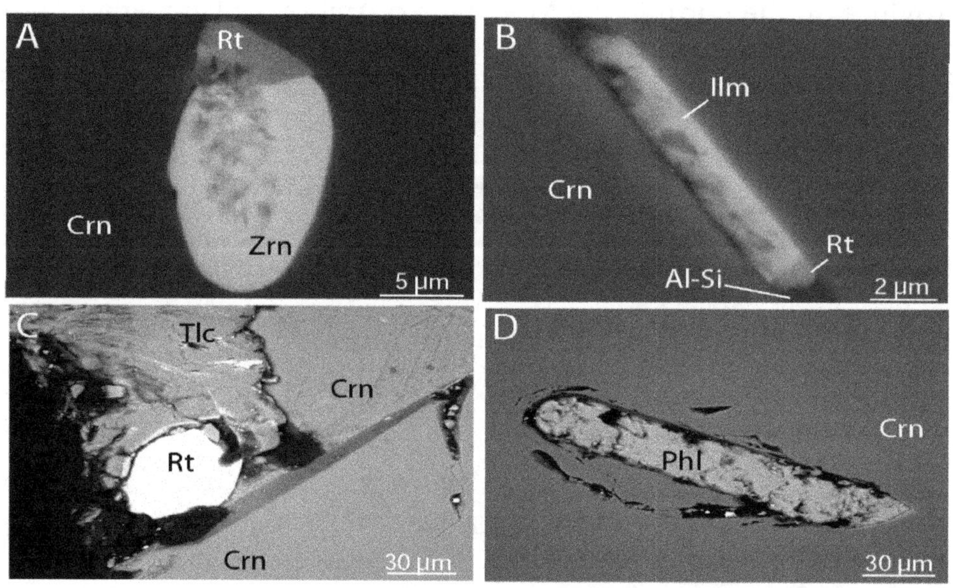

Figure 52. Inclusions in ruby from the Sahanonoka deposit in the Vatomandry area [94]. (**A**) Zircon (Zrn) associated with rutile (Rt); (**B**) rutile, ilmenite (Ilm), and Al-silicate (Al-Si); in ruby (Crn); (**C**) association of rutile and talc (Tlc) in ruby; and (**D**) phlogopite (Phl) crystal included in a ruby.

The original source of these rubies is unknown neither xenocrysts nor corundum xenoliths have ever been observed in the basalts. The placers that are proximal and downstream of the different basaltic volcanics, ruby crystals associated with fragments of basalt, and red to orangey zircon and sapphires leave a high probability that the rubies were connected to the different basaltic flows of the Takarindiona and Vohibalaina basaltic massifs.

The Fe contents of the rubies (1500–8460 ppm) are always higher than Cr and V. The Cr_2O_3 contents differ with color: (i) red brown (4760–6670 ppm), (ii) red (3500–3800 ppm), (iii) pink (1390–11,670 ppm), and (iv) red to purple (720–2000 ppm). The chemical diagram for classification of corundum deposits (Figure 53) shows a distinctive bimodal chemical distribution for the rubies from Vatomandry, which plot in the metamorphosed M-UMR (R3) and metasomatic (R4) domains. The first field in the R3 domain (M-UMR signature) contains red and red brownish rubies; the second field in the R4 domain (metasomatic such as plumasite or metasomatite in M-UMR) contains red to purple and pink stones. The bimodal distribution is also present for the rubies from Soamiakatra in Ankaratra and Andimalena, while the rubies from Antsabotraka and Ambatomainty plot in the metasomatic domain.

The $\delta^{18}O$ values for rubies from the Vatomandry mining district span the range 2.7–6.7‰ ($n = 8$) with a mean value of $\delta^{18}O = 5.1 \pm 1.4$‰. They fit within the worldwide range defined for ruby in basaltic environments (1.3‰ < $\delta^{18}O$ < 5.9‰, mean $\delta^{18}O = 3.1 \pm 1.1$‰, $n = 57$; [80]) and ruby associated with M-UMR (0.25‰ < $\delta^{18}O$ < 6.8‰, $n = 19$). Solid inclusions, chemical composition, and oxygen isotopes are compatible with a metamorphic origin for the Vatomandry rubies.

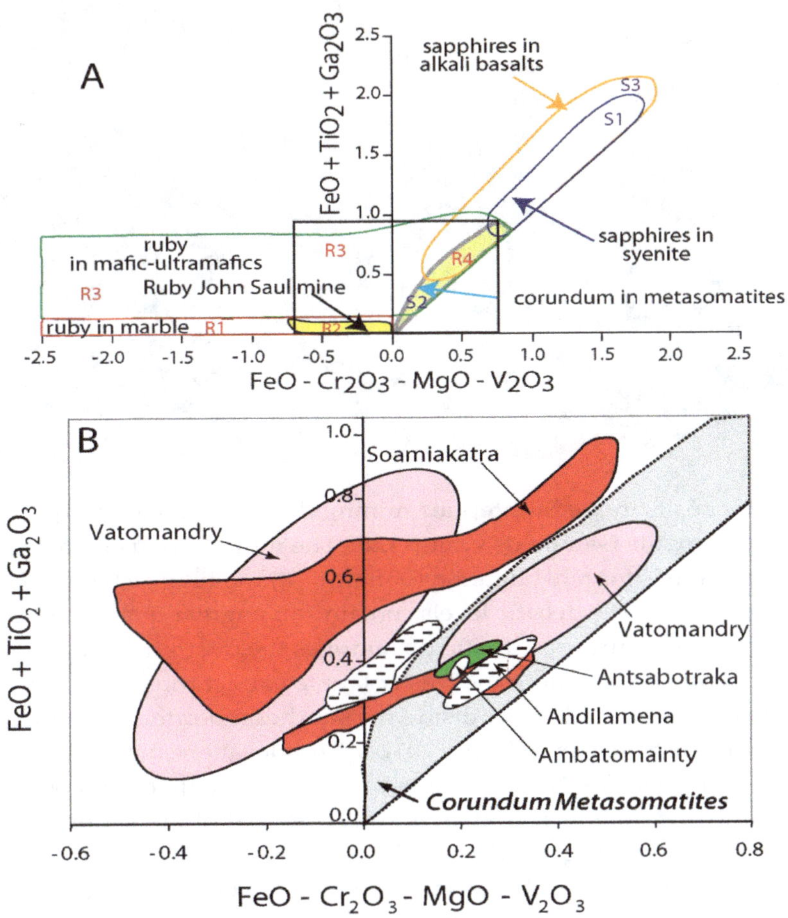

Figure 53. (**A**) Chemical diagram FeO-Cr_2O_3-MgO-V_2O_3 versus $FeO + TiO_2 + Ga_2O_3$ showing the main chemical fields defined for different types of corundum deposits worldwide [72,77]; (**B**) field of chemical compositions of ruby from the Vatomandry deposits (Toamasina Province) compared with the primary deposit at Soamiakatra, and the placer deposits of Antsabotraka and Ambatomainty (Antananarivo Province). The Vatomandry and Soamiakatra deposits show a distinct bimodal chemical distribution in the M-UMR (R3) and metasomatite (R4) domains, respectively. The deposit at Andilamena in Toamasina Province is apparently not related to alkali basalts but is likely derived from plumasite in M-UMR because field work has shown the presence of in situ highly altered mafic rocks, biotite schists, and kaolinite-bearing rocks.

6.3.2. Sub-Type IIIB: Gem Placers in Metamorphic Environment in M-UMR (Montepuez, Mozambique) and Marble (Mogok Stone Track, Myanmar, and Others)

(1) At Montepuez in Mozambique, the main source of facet-grade material in the Gemfields Group Ltd. properties is secondary deposits in M-UMR environments [25,26] (see Figures 32 and 54). The placer deposits were previously interpreted as alluvial and related to a flood event in the river

systems [27,200,257,258]. However, Simonet [25] concluded that the deposits were not alluvial since there was no evidence of systematic rounded pebbles, variations in the grain sizes and graded bedding in the sediments.

Figure 54. The placer deposits in the Montepuez mining district in Mozambique. (**A**) Aerial view of the Mugloto Pit operated by the Gemfields Group Ltd. The miners are using a back-filling technique (top) and using tree plantations for rehabilitation (bottom); (**B**) the ruby-rich gravel layer at Mugloto in the Montepuez mining district. The detritic levels contain sub-angular quartz and clay pebbles located between 4 and 10 m under the surface. Gemfields removes the overburden and uses an excavator to collect this gravel layer, that is then sent to the wash plant to extract rubies; (**C**) small scale miners near Nacaca village, an area reserved for locals outside the southeast border of the Gemfields concession, where a few thousand people have worked since 2012. At this location small rubies are found in gravels under to 2–5 m of overburden. The Nacaca stones are usually smaller and contain more inclusions than those from Mugloto and their iron content is similar to that of the material from Maninge Nice; (**D**) parcel of rough rubies from Mugloto Pit near Montepuez (about 0.5 g per stone) at a Gemfields rough ruby auction in Singapore. The rubies coming from a gravel layer range from bright to dark red (depending on their iron contents), and present some abrasion features. The area was discovered in early 2014 and probably more than 90% of Gemfields income since then has been extracted from this deposit.

The secondary deposits are mainly colluvial with limited horizontal transport (more sub-angular fragments) and a few alluvial pebbles (smooth and round appearance). Most of the mineralized horizons are stone lines, i.e., weathering-resistant rock fragments formed over millions of years of erosion [25]. The stone lines, which are composed of angular fragments of quartz and pegmatite, are excellent traps for smaller grains of high-density minerals such as ruby. The colluvial deposits (Figure 54) are either associated with the proximal primary deposits (as at the Maninge Nice/glass mines) or disconnected (as at the Mugloto mine). Simonet [25] proposed an exploration scheme for the colluvial deposits in the Montepuez area based on detailed mapping of the primary deposits (stressing the importance of the strike and dip of the mineralized structures) and erosion through time (Figure 55). The distance between the primary deposit in outcrop and the edge of the colluvial accumulation increases during the erosion process. Mapping of the stone lines and outcrops of ruby-bearing rocks could be an important tool for future ruby placer exploration in the Montepuez mining area.

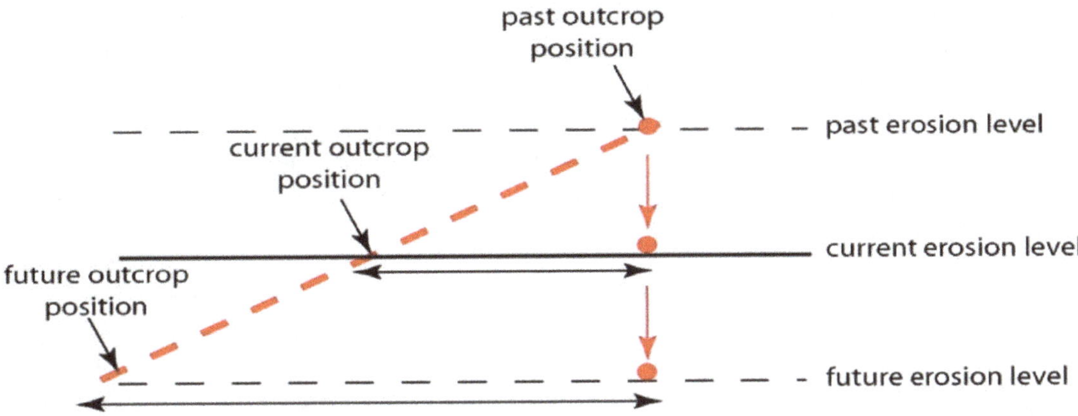

Figure 55. Schematic time-relationship of the formation and position of a colluvial placer through erosion of an angle-dipping primary deposit [25].

(2) Economic ruby placer deposits in marble environments of South-Eastern Asia (Figure 2).

In the Mogok Stone Track in Myanmar (Figure 56A), gem-bearing levels enriched in pebbles, sand, silt, clay, and iron oxides are called 'byon' [259,260] (Figure 56B,C).

Figure 56. Different aspects of the gem placers in the Mogok Stone Track, Myanmar (in January 2015). (**A**) Panoramic view of the primary deposits mined in marble on the flank of the hills (foreground view) and secondary deposits, i.e., placers, mined in the valley (background); (**B**) image of the gem-bearing material called 'byon' enriched in gravels and sand. The concentration of heavy minerals is visible and is composed of iron oxides, corundum, quartz, and pieces of marble; (**C**) concentration of gem minerals (ruby, spinel, and sapphire) in the pan of a miner; and (**D**) sedimentary breccia formed in a cave placer karst. The breccia formed from an argillite containing crystals of spinel, plates of graphite, and fragments of marble and quartz. Size of the breccia sample around 15 cm long. Photo: Gaston Giuliani.

Similar deposits are found at Namya in the northern part of the country where the monsoon is an effective erosion mechanism in low-altitude environments [5,108,261]. The gem concentration is closely related to the formation of karst and chemical weathering of marble and associated rocks, i.e., schists, gneisses, and granitoids (Figure 56D). The cave placers in karst contain sedimentary breccias formed by argillites with ruby, sapphire, spinel, graphite, and fragments of marble and quartz. The ruby and other gems (sapphire and spinel) are trapped in eluvial, colluvial, alluvial, fracture-filling, and cave deposits (Figure 57) [29].

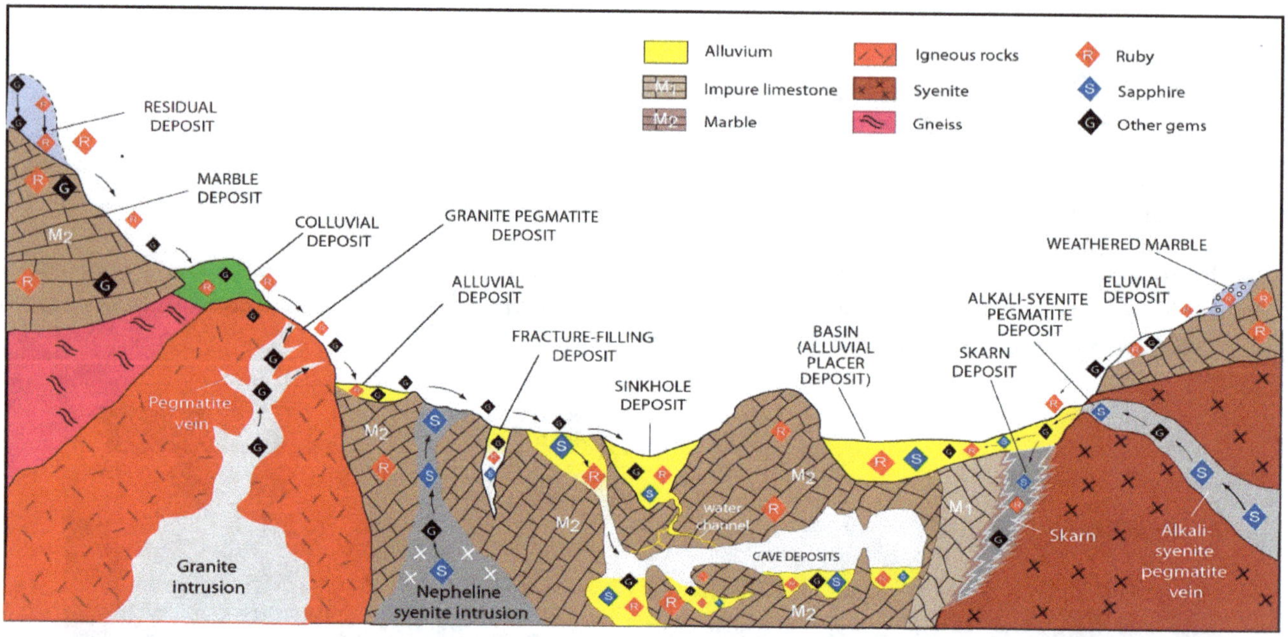

Figure 57. The main types of ruby-bearing placers in the Mogok Stone Track, Myanmar; modified from [29]. The primary ruby deposits are located in marble and the secondary deposits are residual, colluvial, alluvial, fracture-filling, and cave placers in karst morphologies. The figure represents the ruby deposit of Dattaw in marble (on the left) and the sapphire deposit of Thurein-taung (on the right). At Mogok, the different types of primary deposits of ruby and sapphires resulted in the mingling of gems from different sources in single secondary accumulations.

In Vietnam, the placers are proximal to the main primary deposits of ruby and colored sapphires in marble. They are located in the Luc Yen and Yen Bai provinces in the Red River shear zone area and in the Quy Chau ruby deposit (Nghe An province) in the Bu Kang metamorphic zone in Central Vietnam [98,210]. The placers are exploited at Luc Yen, An Phu, Khoang Thong, and Nuoc Ngap (Luc Yen district); at Truc Lau, Yen Thai, Tan Huong, Tan Dong, Hoa Cuong, and Tran Yen (Yen Bai district); and at Quy Hop and Quy Chau (Nghe An district). The paleoplacer at Truc Lau consists of 10 m of sediments overlying bedrock. The rubies and blue sapphires are contained in 5 m thick gravel layer overlain by a 3.5 m thick bed of Quaternary sediments and 1–1.5 m of soil. In 2002, up to two boulders (1–2 kg) of pink sapphire and star ruby were recovered every month in the paleoplacer. The placer deposits consist of gravel concentrations in karst pockets (Figure 58A) and alluvial fans (Figure 58B) found all over the Luc Yen region [261,262]. The gem-bearing valleys are often narrow and correspond to small depressions, which typically range in area from 2 to 3 km^2 (Figure 58D). These placers furnish a variety of gem-quality materials for a market, which has been open daily in Luc Yen city since 1987 (Figure 58C). The rubies in the placers have a proximal origin as confirmed by the oxygen isotopic values of the crystals: $\delta^{18}O$ ranges between 15.5‰ and 22.3‰, which overlaps the worldwide oxygen isotopic range (16–23‰) defined for ruby hosted in marble [40,93].

Mining is conducted on a small scale, from Bai Lai in the northeast of Luc Yen to the Na Ha area in the south, near the banks of Thac Ba Lake. The most active sites are along the streams of Cong Troi, in Nuoc Ngap (An Phu area). Material is also removed from old mines in areas such as Minh Tien, Khoan Thong, and Bai Chuoi, where big companies mined the most accessible and profitable material, often missing areas between marble pinnacles (Figure 58E) that excavators could not easily reach. As a result, some gem-rich ground was left behind; this is now collected by locals with shovels and buckets.

Figure 58. Ruby-bearing placers hosted in marble from northern Vietnam (Luc Yen area; 2009). (**A**) Exploitation by locals of ruby-bearing gravel concentration in karst pockets; (**B**) artisanal exploitation of ruby-bearing alluvial fans concentrated in narrow valleys bordered by rice fields; (**C**) the famous gem market at Luc Yen city; authorization for the photograph of the Luc Yen market by filed gemmologist M. Shuan, and (**D**) the gem-bearing valleys are often narrow and correspond to small depressions, which typically range in area from 2 to 3 km^2. These valleys are defined by shear zones and faults that crosscut the marble units. The karst geomorphology is characterized by an alignment of marble domes forming cones and lines of karst hills; (**E**) excavation of ruby-bearing areas continued up to the marble pinnacles, but the missed gem areas between the pinnacles is exploited by locals with shovels, picks and buckets.

6.3.3. Sub-Type IIIC: Gem Placers with Ruby Originating from Multiple and Unknown Sources

Gem placers of economic interest are also found in Madagascar (Ilakaka, Andilamena, Bemainty, Didy, and Zahamena) and Tanzania (Tunduru, Songea, and Umba). The historical, geological, and mineralogical features of these deposits are described in different publications; see [5,19,20,108,161,164,263–266]. These placers contain ruby, sapphires, and other minerals (garnets, alexandrite, topaz, zircon, spinel, and tourmaline) that originate from multiple unknown sources. We can note that at Songea and Andilamena, corundum is found both in placers and in situ in highly weathered rocks.

In the Ilakaka area (see Figure 4), gem corundum is found at three gravel levels in two main alluvial terraces deposited on the Triassic Isalo sandstone [108]. The terraces are weakly consolidated, but correspond to paleoplacers. The terraces are the result of the erosion and stripping of the sandstones. The deposit is not well consolidated and is composed of quartz-rich sands containing pebbles of ferruginous laterite, rounded blocks of Isalo sandstone and quartz, and quartzite and schist pebbles.

The Ilakaka deposits produce very fine blue, blue-violet, violet, purple, orange, yellow, and translucent sapphire crystals along with pink corundum and rare rubies, zircon, alexandrite, topaz, garnet, spinel, andalusite, and tourmaline. In 2002, test mining by Gem Mining Resources Company over 38 days resulted in approximately 43 kilos of gems, comprising of 4% semi-precious stones, volcanic glasses, and ruby, and 96% sapphire comprising 58% pink gem corundum, 30% sapphire, and 8% other colors of sapphire (e.g., yellow, orange, padparadscha, and green). Microscopic examination of respectively 30 ruby and 58 blue sapphire crystals showed that all of the rubies have the same gemmological features, i.e., growth, twinning, color bands, and solid inclusions, but that the blue sapphires define two different mineralogical groups [40].

The ruby crystals have $\delta^{18}O$ values between 2.6‰ and 3.8‰ ($\delta^{18}O_{mean} = 3.5 \pm 0.5$‰, $n = 4$) that are within the isotopic range defined for primary ruby deposits in metamorphosed M-UMR from Southern Madagascar (2.5‰ < $\delta^{18}O$ < 6.8‰; [128]). The Phanerozoic sedimentary sequences and the Isalo Group cover the northern part of the Vohibory unit, where intercalations or complexes of M-UMR hosting ruby and pink corundum are common [164]. Erosion of the Vohibory unit during the Late Carboniferous to Middle Jurassic contributed to the concentration of ruby in the basin. A proximal source for ruby in Ilakaka from the Vohibory region is likely. Quaternary erosion remobilized paleoplacers.

In the eastern part of Madagascar, several ruby-bearing placers have been discovered over the last 20 years as the Vatomandry deposit [255] (see Figure 4). The ensuing rush of locals and immigrants opened up several mining zones in the areas of Andilamena (2000–2004), Andrebabe (2002), Moramanga (2010), Didy (2012), Zahamena (2015), and Bemainty (2016). The deposits are generally exploited by either unlicensed miners when working outside of protected areas or illegal miners when working inside the areas (Figure 59). The miners operate in quagmires with shovels and buckets, washing gravels in pans and removing the stones from the sieves by hand.

Figure 59. Placer deposits in the eastern part of Madagascar. (**A**) General view of the ruby and sapphire deposit in the jungle of Didy. Photo: Nirina Rakotosaona; (**B**) crystals of ruby and sapphires from the Didy deposit. The crystals are around 8 mm in size; photo: Vincent Pardieu and (**C**) sapphire and pink corundum placer deposits in the jungle near Bemainty. Photo: Vincent Pardieu.

Ruby is generally associated with sapphires at Vatomandry [77,78], Didy [265] (Figure 59), and Bemainty [265] but is exploited as a main gem mineral from the Andilamena and Zahamena deposits [266]. Up until the present, geological information is lacking, but visits to the mining district of Andilamena by V. Pardieu allowed him to observe the corundum host rocks, which were highly weathered. They comprised fuchsite-bearing altered mafic rocks, biotite schists, and kaolinite-bearing rocks. The rubies are found in mafic rocks and/or biotite schists. Sapphires are restricted to veins of quartz-free kaolinite-bearing rocks that crosscut the mafic rocks.

Debate on the geologic and geographic origins of rubies from Andilamena and Vatomandry involved:

(i) The chemistry of these rubies in the diagram FeO-Cr_2O_3-MgO-V_2O_3 versus $FeO + TiO_2 + Ga_2O_3$ is similar and characterized by an overlap of two chemical domains (bimodal distribution; see Figure 53). On the contrary, the diagram V versus Fe (Figure 60) shows a clear difference in the V contents of ruby for both deposits that allows the allocation of the geographic origin of the Vatomandry rubies relative to Andilamena and also the main ruby deposits worldwide;

(ii) The $\delta^{18}O$ values for rubies from the Vatomandry mining district span the range 2.7–6.7‰ (n = 8) with a mean value of $\delta^{18}O = 5.1 \pm 1.4‰$ (n = 5). Those for rubies from the Andilamena deposit are between 0.5‰ and 4‰ with a mean value of $\delta^{18}O = 2.9 \pm 1.5‰$ [128]. Both ranges of $\delta^{18}O$ values overlap and fit within the worldwide range defined for ruby in basaltic environments ($1.3‰ < \delta^{18}O < 5.9‰$, mean $\delta^{18}O = 3.1 \pm 1.1‰$, n = 57); [80]) and ruby associated with metamorphosed M-UMR ($0.25‰ < \delta^{18}O < 6.8‰$, n = 19). Both rubies are compatible with a metamorphic origin.

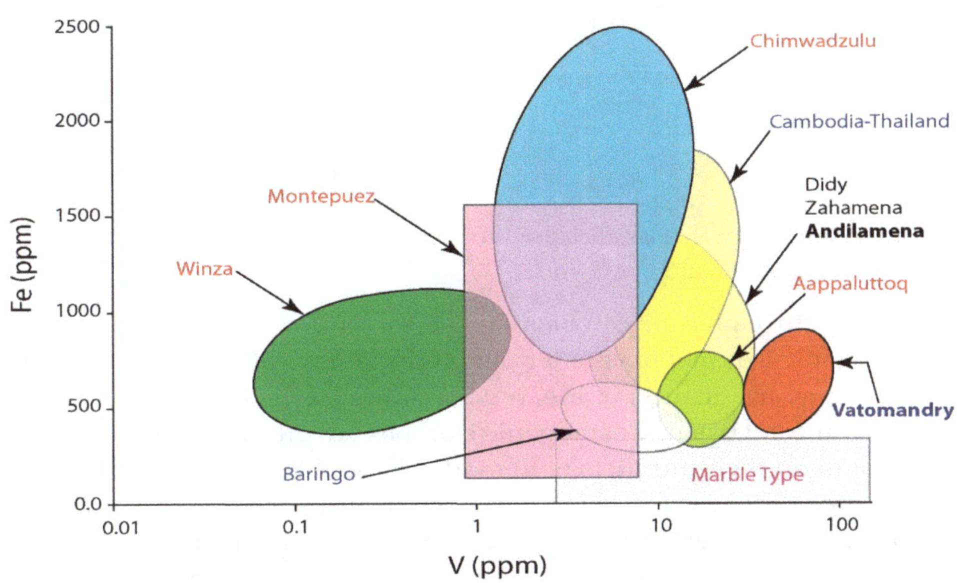

Figure 60. V versus Fe (in ppm) for the main ruby deposits of Madagascar (Vatomandry, Andilamena, Didy, Zahamena) and other deposits worldwide (Kenya, Baringo; Greenland, Aappaluttoq; Tanzania, Winza; Mozambique, Montepuez; Malawi, Chimwadzulu; Cambodia and Thailand; and finally deposits of marble type from Central and Southeastern Asia). Chemical data from [200,267] and Pardieu unpublished data. The chemical domain of ruby from Vatomandry is clearly separated from all the other ones and this parameter permits to precise its geographic origin. The name of ruby deposits related respectively, to M-UMR is written in red (Types IIA_1, IIB_1), to marble deposits in violet (Type IIB_2), to alkali-basalt in blue (Type IIIA), and to metamorphic environment in black (Type IIIB).

7. Conclusions and Perspectives

In the present paper, the ruby occurrences and deposits were classified into three main types: (Type I) magmatic-related, (Type II) metamorphic-related, and (Type III) sedimentary-related (Figure 61).

The economic ruby deposits, in terms of volume and quality, are the placers that in the new ruby deposits classification are called "Sedimentary-related (Type III)". The hardness and density of ruby are the main physical parameters that enable it to survive the erosion of primary deposits, mechanical transport, and accumulation along slopes or sedimentation in basins. Today Africa is the top gem ruby producer in the world, via secondary deposits in metamorphic environments (Sub-Type IIIB) resulting from the erosion of primary deposits in metamorphosed mafic and ultramafic rocks (M-UMR; Sub-Types IIA_1 and IIB_1).

Over the last decade the main discoveries of ruby have been in Mozambique and Madagascar. It is not the end but the beginning of a new era for understanding the formation of such primary deposits and prospecting their associated placers. At the Montepuez mining district, the colluvial deposits are highly economic and characterized by limited transport. Most of the mineralized horizons are stone lines but the placers are proximal to the primary deposits. The exploration scheme proposed by [25] for the colluvial deposits based on detailed mapping of the primary deposits and erosion through time is of high importance. It includes the study of the paleogeography of eroded areas (plateaux and basins) in Eastern Africa where ruby occurrences are known, and the geology of mineralized structures in metamorphic belts.

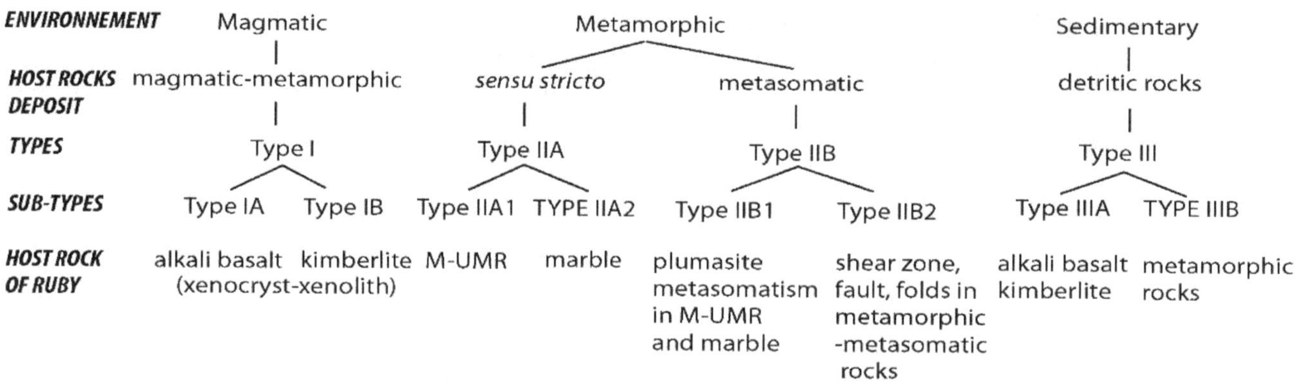

Figure 61. New typological classification for ruby occurrences and deposits worldwide. M-UMR = metamorphosed mafic-ultramafic rocks.

The extent of erosion through time can be assessed by dating the colluvial or alluvial material by measuring ^{10}Be (cosmogenic isotope) in recent quartz-rich sediments or in pebbles of quartz (alluvial or not), which allows the quantification of mean denudation rates, in alluvial terraces and abrasion, across a watershed or an ancient basin. Cosmogenic isotopes are produced on the Earth's surface by the interaction of cosmic rays and their products with matter (^{10}Be, T1/2 1.5 Ma). Such isotopes accumulate progressively in rocks near the surface, depending on altitude and erosion rates. The cosmogenic isotopic concentration of the sediment provides access to the average erosion rate of the watershed at its source. Fission track studies on zircon (and apatite) may be used in sediment provenance determination and in terms of the time of cooling below 300 °C in the zircon-bearing protolith before erosion.

The new and major discoveries of ruby and sapphire placers in Eastern Madagascar at Didy, Zahamena, and Bemainty, located in tropical forests, are economic. They were found by local people in areas, which are crosscut by numerous NE-SW faults separating small alluvial and diluvial basins. These zones are difficult to access and are protected forest areas; therefore prospecting has not been done but the combination of photosatellite images with geological maps, looking for metamorphic M-UMR formations, will help in the discovery of new placers.

Other ruby-bearing Neoproterozoic metamorphic M-UMR in East Africa have been productive and some of them still produce a few rubies as in Tanzania (Winza, Longido), Kenya (Mangare area, Kitui), Malawi (Chimwadzulu), and Southern Madagascar (Vohibory region). In 2005 these countries assumed most of the worldwide production. These primary deposits in M-UMR can be found in all metamorphosed volcano-sedimentary sequences from the Archean to the Cenozoic. The discovery and production of ruby at the Aappaluttoq deposit (south of Nuuk; Fiskenæsset district) and the presence of new occurrences north of Nuuk (Kangerdluarssuk) is an excellent example that must encourage a better understanding of their formation and their prospecting as a function of the latitude of the different countries (tropical versus glacial and temperate climates).

The second important type of ruby deposit includes those located in marble (Sub-Type IIA_2) in Central and Southeastern Asia. These Cenozoic deposits are always one of the main worldwide sources of high quality ruby with intense color and high transparency, such as the famous Mogok pigeon blood rubies. The mining district of Mogok is rich in different types of gems including red spinel and ruby in marble. Genetic models are proposed for their genesis. They formed during the metamorphism, in the amphibolite facies, of carbonates interbedded with mudstone and evaporitic (salts and sulfates) intercalations. Lithological control of the mineralization is a function of the paleogeography of the carbonate platform environment before metamorphism. The role of evaporites is proposed as key for the genesis of ruby in marble but Al-enriched sediments such as metapelite or meta-bauxite intercalated in the marbles are in some cases also possible protolith for the ruby-parent rocks (Revelstoke and Snezhnoe deposits). The Mogok mining district remains a suitable site for developing further geological research considering the complexity of the relationships between the metamorphic rocks and granitoid intrusions. The formation of ruby-bearing skarns is possible, given the recrystallization of some carbonates, with crystals up to 10 cm across, present in the Dattaw mine. Multi-stage mineralization and telescoped episodes of thermal anomalies related to regional or contact metamorphism are highly probable. Future efforts will move towards developing more effective exploration guidelines that consider new geological factors responsible for the formation of ruby at Mogok and the timing of formation of these multi-stage gem deposits.

The timing between deformation, metamorphism, and ruby formation in these marble deposits is of importance, as shown in some deposits from Tanzania that are associated with folding, saddle-reef structure, and fluid circulation in zones of higher permeability. This scheme is applicable for all deposits, whatever their ages, but must be prospected in the Neoproterozoic Metamorphic Belt.

The classification proposed for ruby occurrences and deposits is a necessary step to build a new scheme for understanding ruby-mineralization from the field to the laboratory. The determination of the different sub-types based on the host rock of ruby and the knowledge on the genesis of most of the deposits will help us to understand the mineral associations and the chemistry of rubies. The combination of all these data linked to a specific type of deposit will serve as a basis for the determination of the geographical origin of rubies.

The diversification of known ruby sources in the world during these last two decades has increased considerably the challenge of gem ruby geographic origin determination [11–15]. The gemmological laboratories amassed rubies with samples obtained by professional field gemmologists [9,21,37,90,200]. The precise gem localization combined with accurate chemical analysis obtained by LA-ICP-MS calibrated by natural standards [92], enriched the collections of standard stones for different types of ruby deposits worldwide. The combination of inclusions and trace element chemistry makes the geographic origin determination more reliable currently, although it can remain inconclusive [11]. The case of ruby and sapphire in placers from basaltic environments is illustrative and the geographic origin is always matter of debate; the geologic origin is often confronted to different possible possibilities, i.e., metamorphic sensu stricto, metamorphic–metasomatic, and magmatic for gem material [11,268]. The V versus Fe diagram of ruby from several economic deposit (see Figure 60) confirms clearly the separation of the two main types of primary ruby deposits worldwide, i.e., those hosted in M-UMR (Types IIA_1 and IIB_1) and in marble (IIB_2), which are typically metamorphic. The origin of ruby from secondary deposits, i.e., placers in the alkali basalt environment (Type IIIA) is debatable because melt inclusions are present in some rubies from Thailand, Cambodia, and Montana. A magmatic origin is probable but the V versus Fe diagram shows that their chemical domain overlaps that of rubies from metamorphic origin. The discussion presented here provides a basis for further investigations into the origin of these rubies keeping in mind that they were incorporated by ascending basaltic magmas as xenocrysts from possible different geological sources.

Rubies are classified by most of the gemmological laboratories into two groups based on their trace element chemistry [11]: (i) low (below 200 pm) and high iron (above 400 ppm) rubies. The marble-hosted metamorphic rubies (Type IIA_2) have a very low Fe content when compared to all the other types of rubies except those from the former John Saul mine (see Figure 19, e.g., plumasite of Type IIB_1). Nevertheless rubies in marble deposits have similar inclusion characteristics and their chemical domain overlap as well as their oxygen isotopic values. This remark opens the debate about the reality of determining the geographic origin of a gem when disconnected from its geological environment. Some stones and sometimes of top-quality are mingled with other ones and finally whenever the ruby is Fe-poor, the geographic determination is inconclusive. Myanmar stones are always highly prized on the market for their color and transparency. The Mogok and Mong Hsu rubies are identified based on their internal features and trace element profiles [11]. High-iron rubies (Thailand, Cambodia, Madagascar, Mozambique, Kenya, and Greenland) are usually basically identified by their inclusions and trace element profiles (V, Mg, and Fe but not Cr, see Figure 20; [11,82,83]). The main question of geographic attribution remains for rubies between these two ranges of Fe contents.

The new geological classification of ruby based on the geological environment and the gem host-rock proposed three main types and eight sub-types of deposits. The classification system reflects the interplay between tectonic with magmatic and metamorphic events during the history of the Earth, and through continental plate tectonics within Wilson cycles, over circa 150 millions of years. The final point concerns the basic erosion of ruby-bearing rocks at the surface, and the transport and sedimentation of the detrital material in basins. The different sub-types for primary deposits are based on the gem host-rock of ruby, which is different from place to place in the crust, the final petrographic product depending on the nature of the fluids, the timing of their circulation and the intensity of fluid–rock interaction. Desilication and/or metasomatism phenomena are effective processes able to extract not only the chromophores Cr, V, and Fe but also Al from M-UMR for the formation of ruby. The different sub-types of deposits will help either the geologist to understand the genesis of new deposits and to define zones of interest for exploration or the gemmologist for geographic origin determination.

Fingerprinting of ruby used by gemmologists is a tool derived from geological and mineralogical investigations: (i) Trace elements (V, Fe, Ga, and Mg) are used as previously reported. Ti is also an interesting element that can be used for determining the geological history of the gem-bearing sample as shown by [269] for corundum. The combination of different ratio such as Mg/Mg+Fe versus Cr/Cr+V will permit to trace the source of these elements, i.e., ultramafic versus mafic rocks as done for emerald by [270]; (ii) the combination of Fe, Ti, Ga, Cr, V, and Mg with oxygen isotope ratios of ruby provide a powerful tool for the determination of geologic origin [40,82,87,93,144,151,152]. The oxygen isotopes separate clearly three types of ruby deposits: (a) marble, (b) the John Saul mine type, and (c) metamorphosed M-UMR. The analysis of oxygen isotopes by SIMS is semi-destructive as is LA-ICP-MS for trace elements analysis but can be applied on cut and polished rubies; and (iii) inclusions are helpful in origin determination but similar geologic environments can produce gems of similar inclusion suites [269]. Determining the geological origin of corundum mingled in huge metamorphic placers remains a very difficult task, as for the Ilakaka deposit in Madagascar.

The objective geological criteria synthesized in the present review through scientific works and experience open a new framework for mineral exploration guidelines. The final conclusion is that the main economic ruby deposits are the result of continental plate tectonics. Ruby deposits are found where continental plate collision formed mountain ranges, as was the case in the Neoproterozoic with the Metamorphic Mozambican Belt, or with the Himalayas. Continental collision increases the thickness of the continental lithosphere, which is accompanied by an increase of pressure and temperature producing a zone of higher grade metamorphism (amphibolite to granulite facies), and sometimes melting rocks to form felsic magmas. Extensive deformation at high temperature with kilometer-scaled migration of material and fluids along shear zones (e.g., in the case either of the Red River shear zone in Southern China and Northern Vietnam or the Mogok Metamorphic Belt, where ruby is found in placers defining the trace of shear zones in Northern Vietnam) and faults results in the remobilization and precipitation of elements, and tectonic melange zones. These processes are heterogeneous, and proceed by dissolution and precipitation, as well as by fusion and crystallization, where elements (such as chromium) are unusually concentrated in the crust. The return to isostatic equilibrium (30 km thick for the continental crust) and erosion processes through time brings to the surface the metamorphic roots (belts) of these ancient mountain chains. This is currently the case with the Mozambican metamorphic belt where high-temperature gems such as ruby, spinel, tsavorite, and other garnets are found.

The main production of gem ruby is provided by this Pan-African metamorphic belt. All ruby deposits are of metamorphic origin in the broad sense. These mineral deposits are associated either with regional metamorphism sensu stricto or through fluid–rock interactions accompanied by metasomatism, contemporary with, or slightly later than, metamorphism. Their dismantling by current erosion creates important reserves and the exploration of continental placers in tropical landscapes where regoliths have been present for several thousand years should be studied with new approaches.

Author Contributions: All the authors contributed with their field and laboratory work and experience, and the works of worldwide researchers for over one century with the first detailed description, at the beginning of the 20th century, of the Malagasy deposits by A. Lacroix (MNHN Paris), to the realization of this description of the state-of-the-art of the geology and genesis of ruby deposits worldwide. G.G., L.A.G. and V.P. collected the studied samples in the field as well as realized photographs on specific geological and gemmological cases. G.G. conducted geological mapping in several areas worldwide and G.G., L.A.G., I.P. and V.P. obtained and evaluated part of the mineralogical and geochemical data. I.P. conducted cristallography on corundum crytals as well as the study on trapiche rubies from differents deposits. A.E.F. and G.G. obtained and evaluated the oxygen isotope data. V.P. obtained and evaluated part of the gemmological data. Investigations in the different specific domains were realized by the authors in their own laboratory. G.G. realized some software for corundum chemical analysis as well as the original draft preparation and the supervision of the paper project with approval of the co-authors. The writing was realized by all the autors in their area of competence with the supervision by G.G and language corrections by A.E.F. Review and editing was done by G.G. All authors have read and agreed to the published version of the manuscript.

Acknowledgments: The authors want to thank the Editor of *Minerals* and Guest Editors, Frederick Lin Sutherland and Khin Zaw, for invitation to contribute in a special issue on "Mineralogy and Geochemistry of Ruby", and for their valuable suggestions. We are also thankful to four anonymous reviewers for improving the text by pertinent and constructive comments on the article. The authors want also to thank D. Schwarz, E. Le Goff, C. Simonet for their discussion and collaborations on the scientific topic. G.G. wants to thank the Institut de Recherche pour le Développement (IRD, France) and the French CNRS and the Unversities of Lorraine and Paul Sabatier (CRPG/CNRS-Université de Lorraine and GET/IRD-Université Paul Sabatier) for academic and financial support for research on the geology of gems over the past 30 years.

References

1. Eichholz, D.E. *Theophrastus, De Lapidibus*; Oxford Clarendon Press: Oxford, UK, 1965.
2. Littré, E. Tome Second. In *Histoire Naturelle de Pline*; Chez Firmin-Didot et Cie: Paris, France, 1877.
3. Ernault, L. *Marbode, Évêque de Rennes, Sa Vie et Ses Oeuvres*; Hte Cailleres: Rennes, France, 1890; pp. 1035–1123.
4. *Albertus Magnus Liber Mineralium*; Jacob Köbel: Oppenheim, Germany, 1518.
5. Hughes, R.W. *Ruby and Sapphire*; RWH Publishing: Boulder, CO, USA, 1997.
6. Pardieu, V. Thailand, the undisputed ruby kingdom: A brief history. *InColor* **2019**, *42*, 14–22.
7. Themelis, T. *The Heat Treatment of Ruby and Sapphire*; Gemlab Inc.: New York, NY, USA, 1992.
8. Shigley, J.E.; Dirlam, D.M.; Laurs, B.M.; Boehm, E.W.; Bosshart, G.; Larson, W.F. Gem localities of the 1990s. *Gems Gemol.* **2000**, *36*, 292–335. [CrossRef]
9. Dussart, S. The version of field gemologist Vincent Pardieu. *IGR It alian Gemol. Rev.* **2019**, *7*, 58–64.
10. Raden, A. *Stoned: Jewelry, Obsession, and How Desire Shapes the World*; Ecco: Manhattan, NY, USA, 2015.
11. Palke, A.C.; Saeseaw, S.; Renfro, N.D.; Sun, Z.; McClure, S.F. Geographic origin determination of ruby. *Gems Gemol.* **2019**, *55*, 580–612. [CrossRef]
12. Gübelin Gem Laboratory: The roots of origin determination. *Jewel. News Asia* **2006**, 52–62.
13. Gübelin Gem Laboratory: The limits of orgin determination. *Jewel. News Asia* **2006**, 66–71.
14. Gübelin Gem Laboratory: A hoolistic method to determining gem origin. *Jewel. News Asia* **2006**, 118–126.
15. McClure, S.F.; Moses, T.M.; Shigley, J.E. The geographic origin dilemma. *Gems Gemol.* **2019**, *55*, 457–462.

16. Gübelin, E.J.; Koivula, J.I. *Photoatlas of Inclusions in Gemstones*; ABC Edition: Zurich, Switzerland, 1986.
17. Gübelin, E.J.; Koivula, J.I. *Photoatlas of Inclusions in Gemstones*; Opinio Publishers: Basel, Switzerland, 1985; Volume 3.
18. Sotheby's Magnificient Jewels and Noble Jewels, Auction 502. 2005. Available online: http://www.sothebys.com (accessed on 20 June 2020).
19. Pardieu, V. Ruby and Sapphire Rush near Didy, Madagascar. April–June 2012. Available online: https://www.gia.edu/doc/Ruby-and-Sapphire-Rush-Near-Didy-Madagascar.pdf (accessed on 20 June 2020).
20. Peretti, A.; Hahn, L. Record-breaking discovery of ruby and sapphire at the Didy mine in Madagascar: Investigating the source. *InColor* **2013**, *21*, 22–35.
21. Pardieu, V.; Jacquat, S.; Senoble, J.B.; Bryl, L.P.; Hughes, R.W.; Smith, M. Expedition Report to the Ruby Mining Sites in Northern Mozambique (Niassa and Cabo Delgado Provinces), Bangkok, GIA Laboratory. 2009. Available online: https://www.gia.edu/doc/Expedition-report-Ruby-mining-sites-Northern-Mozambique (accessed on 20 June 2020).
22. Pardieu, V.; Jacquat, S.; Bryl, L.P.; Senoble, J.B. Rubies from northern Mozambique. *InColor* **2009**, *12*, 32–36.
23. Pardieu, V.; Thanachakaphad, J.; Jacquat, S.; Senoble, J.B.; Bryl, L.P. Rubies from the Niassa and Cabo Delgado Regions of Northern Mozambique. A Preliminary Examination with an Updated Field Report Annex. Bangkok, GIA Laboratory. Available online: https://www.gia.edu/doc/Niassa_Mozambique_Ruby_September13_2009.pdf (accessed on 20 June 2020).
24. Pardieu, V. Gemfields' November 2017 ruby auction yields record results. *InColor* **2018**, *37*, 62–63.
25. Simonet, C. The Montepuez ruby deposits, what's next? *InColor* **2018**, *37*, 32–40.
26. Vertriest, W.; Saeseaw, S. A decade of ruby from Mozambique: A review. *Gems Gemol.* **2019**, *55*, 162–183. [CrossRef]
27. *A Competent Persons Report on the Montepuez Ruby Project, Mozambique*; SRK Consulting (UK) Limited: Cardiff, UK, 2015; 195p.
28. Yager, Y.G.; Menzie, W.D.; Olson, D.W. *Weight of Production of Emeralds, Rubies, Sapphires, and Tanzanite from 1995 through 2005. Open file Report 2008–1013*; USGS: Reston, VA, USA, 2008.
29. Themelis, T. *Gems and Mines of Mogok*; Ted Themelis: Bangkok, Thailand, 2008; ISBN 0940965-30-5.
30. Peretti, A.; Schmetzer, K.; Bernhardt, H.J.; Mouawad, F. Rubies from Mong Hsu. *Gems Gemol.* **1995**, *31*, 2–26. [CrossRef]
31. Streeter, E.W. *Precious Stones and Gems, Their History Sources and Characteristics*; George Bell & Son: London, UK, 1892.
32. Rohtert, W.; Ritchie, M. Three parageneses of ruby and pink sapphire discovered at Fiskenæsset, Greenland. *Gems Gemol.* **2006**, *42*, 149–150.
33. Fagan, A.J. True North Gems Greenland mining—The final lap. *InColor* **2015**, *29*, 36–49.
34. Fagan, A.J. The Ruby and Pink Sapphire Deposits of SW Greenland: Geological Setting, Genesis, and Exploration Techniques. Ph.D. Thesis, University of British Columbia, Vancouver, BC, Canada, 2018.
35. Smith, C.P.; Fagan, A.J.; Bryan, C. Ruby and pink sapphire from Aappaluttoq, Greenland. *J. Gemmol.* **2016**, *35*, 294–306. [CrossRef]
36. Lucas, A.; Palke, A.; Vertriest, W. Mining rubies in Greenland. The Aappaluttoq ruby mine. *InColor* **2018**, *37*, 42–52.
37. Vertriest, W.; Palke, A.C.; Renfro, N.D. Field gemology: Building a research collection and understanding the development of gem deposits. *Gems Gemol.* **2019**, *55*, 490–511. [CrossRef]
38. Cesbron, F.; Lebrun, P.; Le Cléac'h, J.M.; Notari, F.; Grobon, C.; Deville, J. Corindons et spinelles. In *Minéraux et Fossiles 15*; Editions CEDIM: Paris, France, 2002.
39. Bowles, J.F.W.; Howie, R.A.; Vaughan, D.J.; Zussman, J. *Rock-Forming Minerals. Non-Silicates: Oxides, Hydroxides and Sulphides*; The Geological Society: London, UK, 2011; Volume 5A.
40. Giuliani, G.; Ohnenstetter, D.; Fallick, A.E.; Groat, L.; Fagan, J. The geology and genesis of gem corundum deposits. In *Geology of Gem Deposits*; Groat, L.A., Ed.; Mineralogical Association of Canada: Tucson, AZ, USA, 2014; Short Course Series; Volume 44, pp. 29–112.
41. Bindeman, I.N.; Serebryakov, N.S. Geology, petrology and O and H isotope geochemistry of remarkably ^{18}O depleted Paleoproterozoic rocks of the Belmorian belt, Karelia, Russia, attributed to global glaciation 2.4 Ga. *Earth Planet. Sci. Lett.* **2011**, *306*, 163–174. [CrossRef]

42. Grapes, R.; Palmer, K. (Ruby-sapphire)-chromian mica-tourmaline rocks from Westland, New Zealand. *J. Petrol.* **1996**, *37*, 293–315. [CrossRef]
43. Hutchinson, M.T.; Nixon, P.H.; Harley, S.L. Corundum inclusions in diamonds-discriminatory criteria and a corundum composition dataset. *Lithos* **2004**, *77*, 273–286. [CrossRef]
44. Perretti, A. The Making of an International Color Standard by GRS: "Pigeon's Blood" and "Royal Blue". Definitions, Retroperspective and Future. 2015. Available online: http://www.pigeonsblood.com (accessed on 23 November 2019).
45. Smith, C.P.; Surdez, N. The Mong Hsu ruby: A new type of Burmese ruby. *Jewelsiam* **1994**, *5*, 82–98.
46. Sunagawa, I. *Crystals: Growth, Morphology and Perfection*; Cambridge University Press: Cambridge, UK, 2005.
47. Sunagawa, I.; Bernhardt, H.-J.; Schmetzer, K. Texture formation and element partitioning in trapiche ruby. *J. Cryst. Growth* **1999**, *206*, 322–330. [CrossRef]
48. Hänni, H.A.; Schmetzer, K. New rubies from the Morogoro Area, Tanzania. *Gems Gemol.* **1991**, *27*, 156–167. [CrossRef]
49. Sorokina, E.S.; Hofmeister, W.; Häger, T.; Mertz-Kraus, R.; Buhre, S.; Saul, J. Morphological and chemical evolution of corundum (ruby and sapphire): Crystal ontogeny reconstructed by EMPA, LA-ICPMS, and Cr^{3+} Raman mapping. *Am. Mineral.* **2016**, *101*, 2716–2722. [CrossRef]
50. McKague, H.L. Trapiche emeralds from Colombia. *Gems Gemol.* **1964**, *11*, 210–213.
51. Schmetzer, K.; Hänni, H.A.; Bernhard, H.J.; Schwarz, D. Trapiche rubies. *Gems Gemol.* **1996**, *32*, 242–250. [CrossRef]
52. Schmetzer, K.; Beili, Z.; Yan, G.; Bernhard, H.J.; Hänni, H.A. Element mapping of trapiche rubies. *J. Gemmol.* **1999**, *26*, 289–301. [CrossRef]
53. Win, K.K. Trapiche of Myanmar. *Austr. Gemmol.* **2005**, *22*, 269–270.
54. Schmetzer, K.; Bernhardt, H.J.; Hainschwang, T. Chemical and growth zoning in trapiche tourmaline from Zambia—A re-evaluation. *J. Gemmol.* **2011**, *32*, 151–173. [CrossRef]
55. Koivula, J.I.; Kammerling, R.C.; Fritsch, E. Gem News: "Trapiche" purple-pink sapphire. *Gems Gemol.* **1994**, *30*, 197.
56. Kiefert, L. Unusual trapiche sapphire. *Gems Gemol.* **2012**, *48*, 229.
57. Pignatelli, I.; Giuliani, G.; Morlot, C.; Pham, V.L. The texture and chemical composition of trapiche ruby from Khoan Thong, Luc Yen mining district, northern Vietnam. *J. Gemmol.* **2019**, *36*, 726–745. [CrossRef]
58. Garnier, V.; Ohnenstetter, D.; Giuliani, G.; Schwarz, D. Rubis trapiches de Mong Hsu, Myanmar. *Rev. Ass. Fr. Gemmol. AFG* **2002**, *144*, 5–12.
59. Garnier, V.; Ohnenstetter, D.; Giuliani, G.; Blanc, P.; Schwarz, D. Trace-element contents and cathodoluminescence of "Trapiche" rubies from Mong Hsu (Myanmar): Geological significance. *Miner. Petrol.* **2002**, *76*, 179–193. [CrossRef]
60. Peretti, A.; Mullis, J.; Mouawad, F. The role of fluoride in the formation of color zoning in rubies from Mong Hsu (Myanmar, Burma). *J. Gemmol.* **1996**, *25*, 3–19. [CrossRef]
61. Pignatelli, I.; Giuliani, G.; Ohnenstetter, D.; Agrosi, G.; Mathieu, S.; Morlot, C.; Branquet, Y. Colombian trapiche emeralds: Recent advances in understanding their formation. *Gems Gemol.* **2015**, *51*, 222–259. [CrossRef]
62. Giuliani, G.; Dubessy, J.; Pignatelli, I.; Schwarz, D. Fluid inclusions study of trapiche and non-trapiche rubies from Mong Hsu deposit, Myanmar. *Can. Mineral.* **2018**, *56*, 1–13. [CrossRef]
63. Giuliani, G.; Dubessy, J.; Banks, D.; Lhomme, T.; Ohnenstetter, D. Fluid inclusions in ruby from Asian marble deposits: Genetic implications. *Eur. J. Mineral.* **2015**, *27*, 393–404. [CrossRef]
64. Giuliani, G.; Dubessy, J.; Ohnenstetter, D.; Banks, D.; Branquet, Y.; Feneyrol, J.; Fallick, A.E.; Martelat, J.-E. The role of evaporites in the formation of gems during metamorphism of carbonate platforms: A review. *Mineral. Depos.* **2018**, *53*, 1–20. [CrossRef]
65. Garnier, V.; Giuliani, G.; Ohnenstetter, D.; Fallick, A.E.; Dubessy, J.; Banks, D.; Hoàng, Q.V.; Lhomme, T.; Maluski, H.; Pêcher, A.; et al. Marble-hosted ruby deposits from central and southeast Asia: Towards a new genetic model. *Ore Geol. Rev.* **2008**, *34*, 169–191. [CrossRef]
66. Garnier, V. Les Gisements de Rubis Associés aux Marbres de l'Asie Centrale et du Sud-est: Genèse et Caractérisation Isotopique. Ph.D. Thesis, University of Nancy, Nancy, France, 2003.
67. Muhlmeister, S.; Fritsch, E.; Shigley, J.E.; Devouard, B.; Laurs, B.M. Separating natural and synthetic rubies on the basis of trace-element chemistry. *Gems Gemol.* **1998**, *34*, 80–101. [CrossRef]

68. Sutherland, F.L.; Schwarz, D.; Jobbins, E.A.; Coenraads, R.R.; Webb, G. Distinctive gem corundum suites from discrete basalt fields: A comparative study of Barrington, Australia, and West Pailin, Cambodia, gemfields. *J. Gemmol.* **1998**, *26*, 65–85. [CrossRef]
69. Sutherland, F.L.; Coenraads, R.R.; Abduriyim, A.; Meffre, S.; Hoskin, P.W.O.; Giuliani, G.; Beattie, R.; Wuhrer, R.; Sutherland, G.B. Corundum (sapphire) and zircon relationships, Lava Plains gem fields, NE Australia: Integrated mineralogy, geochemistry, age determination, genesis and geographic typing. *Mineral. Mag.* **2015**, *79*, 545–581. [CrossRef]
70. Peucat, J.J.; Ruffault, P.; Fritsch, E.; Bouhnik-Le Coz, M.; Simonet, C.; Lasnier, B. Ga/Mg ratios as a new geochemical tool to differentiate magmatic from metamorphic blue sapphires. *Lithos* **2007**, *98*, 261–274. [CrossRef]
71. Sutherland, F.L.; Abduriyim, A. Geographic typing of gem corundum: A text case from Australia. *J. Gemmol.* **2009**, *31*, 203–210. [CrossRef]
72. Giuliani, G.; Caumon, G.; Rakotosamizanany, S.; Ohnenstetter, D.; Rakotondrazafy, A.F.M. Classification chimique des corindons par analyse factorielle discriminante: Application à la typologie des gisements de rubis et saphirs. *Rev. Ass. Fr. Gemmol. AFG* **2014**, *188*, 14–22.
73. Sutherland, F.L.; Schwarz, D. Origin of gem corundums from basaltic fields. *Austr. Gemmol.* **2001**, *21*, 30–33.
74. Sutherland, F.L.; Hoskin, P.W.O.; Fanning, C.M.; Coenraads, R.R. Models of corundum origin from alkali basaltic terrains: A reappraisal. *Contrib. Mineral. Petrol.* **1998**, *133*, 356–372. [CrossRef]
75. Sutherland, F.L.; Coenraads, R.R.; Schwarz, D.; Raynor, L.R.; Barron, B.J.; Webb, G.B. Al-rich diopside in alluvial ruby and corundum-bearing xenoliths, Australian and SE Asian basalt field. *Mineral. Mag.* **2003**, *67*, 717–732. [CrossRef]
76. Abduriyim, A.; Kitawaki, H. Applications of laser ablation-inductively coupled plasma-mass spectrometry (LA-ICP-MS) to gemology. *Gems Gemol.* **2006**, *42*, 98–118. [CrossRef]
77. Rakotosamizanany, S.; Giuliani, G.; Ohnenstetter, D.; Rakotondrazafy, A.F.M.; Fallick, A.E.; Paquette, J.-L.; Tiepolo, M. Chemical and oxygen isotopic compositions, age and origin of gem corundums in Madagascar alkali basalts. *J. S. Afr. Earth Sci.* **2014**, *94*, 156–170. [CrossRef]
78. Arlabosse, J.-M.; Delaunay, A.; Lenne, N. Les rubis de Vatomandry—Madagascar. *Rev. Ass. Fr. Gemmol. AFG* **2018**, *203*, 6–15.
79. Uher, P.; Giuliani, G.; Szakall, S.; Fallick, A.E.; Strunga, V.; Vaculovic, D.; Ozdin; Greganova, M. Sapphires related to alkali basalts from the Cerová Highlands, Western Carpathians (southern Slovakia): Composition and origin. *Geol. Carpathica* **2012**, *63*, 71–82. [CrossRef]
80. Giuliani, G.; Ohnenstetter, D.; Garnier, V.; Fallick, A.E.; Rakotondrazafy, A.F.M.; Schwarz, D. The geology and genesis of gem corundum deposits. In *Geology of Gem Deposits*; Groat, L.A., Ed.; Mineralogical Association of Canada: Yellowknife, NT, Canada, 2007; Short Course Series; Volume 37, pp. 27–78.
81. Sutherland, F.L.; Graham, I.T.; Harris, S.J.; Coldham, T.; Powell, W.; Belousova, E.A.; Martin, L. Unusual ruby-sapphire transition in alluvial megacrysts, Cenozoic basaltic gem field, New England, New South Wales, Australia. *Lithos* **2017**, *278–281*, 347–360. [CrossRef]
82. Sutherland, F.L.; Giuliani, G.; Fallick, A.E.; Garland, M.; Webb, G.B. Sapphire-ruby characteristics, West Pailin, Cambodia: Clues to their origin based on trace element and O isotope analysis. *Austr. Gemmol.* **2009**, *23*, 373–432.
83. Keulen, N.; Thomsen, T.B.; Schumacher, J.C.; Poulsen, M.D.; Kalvig, P.; Vennemann, T.; Salimi, R. Formation, origin and geographic typing of corundum (ruby and pink sapphire) from the Fiskenaæsset complex, Greenland. *Lithos* **2020**, *366–367*. [CrossRef]
84. Giuliani, G.; Pivin, M.; Fallick, A.E.; Ohnenstetter, D.; Song, Y.; Demaiffe, D. Geochemical and oxygen isotope signatures of mantle corundum megacrysts from the Mbuji-Mayi kimberlite, Democratic Republic of Congo, and the Changle alkali basalt, China. *C.R. Geosciences* **2015**, *347*, 24–34. [CrossRef]
85. Sutherland, F.L.; Khin, Z.; Meffre, S.; Giuliani, G.; Fallick, A.E.; Webb, G.B. Gem-corundum megacrysts from east Australian basalt fields: Trace elements, oxygen isotopes and origins. *Austr. J. Earth Sci.* **2009**, *56*, 1003–1022. [CrossRef]
86. Sutherland, L.; Graham, I.; Harris, S.; Khin, Z.; Meffre, S.; Coldham, T.; Coenraads, R.; Sutherland, G. Rubis australasiens. *Rev. Ass. Fr. Gemmol. AFG* **2016**, *197*, 13–20.
87. Khin, Z.; Sutherland, F.L.; Yui, T.F.; Meffre, S.; Thu, K. Vanadium-rich ruby and sapphire within Mogok gemfield, Myanmar: Implications for gem color and genesis. *Mineral. Depos.* **2015**, *50*, 25–39. [CrossRef]

88. Sutherland, F.L.; Khin, Z.; Meffre, F.; Thompson, J.; Goemann, K.; Kyaw, T.; Than, T.N.; Mhod, Z.M.; Harris, S.I. Diversity in ruby chemistry and its inclusions: Intra and inter-continental comparisonsfrom Myanmar and Eastern Australia. *Minerals* **2019**, *9*, 28. [CrossRef]
89. Pardieu, V. Concise field report Volume 1: Pailin, Cambodia, Dec 2008-Feb 2009, Bangkok, GIA Laboratory. 2009. Available online: https://www.gia.edu/doc/Field-ReportPailin,pdf (accessed on 25 June 2020).
90. Pardieu, V. Field gemology. The evolution of data collection. *InColor* **2020**, *46*, 100–106.
91. Groat, L.A.; Giuliani, G.; Stone-Sundberg, J.; Sun, Z.; Renfro, N.D.; Palke, A.C. A review of analytical methods used in geographic origin determination of gemstones. *Gems Gemol.* **2019**, *55*, 512–535. [CrossRef]
92. Stone-Sundberg, J.; Thomas, T.; Sun, Z.; Guan, Y.; Cole, Z.; Equall, R.; Emmett, J.L. Accurate reporting of key trace elements in ruby and sapphire using matrix-matched standards. *Gems Gemol.* **2017**, *53*, 438–451. [CrossRef]
93. Giuliani, G.; Fallick, A.E.; Garnier, V.; France-Lanord, C.; Ohnenstetter, D.; Schwarz, D. Oxygen isotope composition as a tracer for the origins of rubies and sapphires. *Geology* **2005**, *33*, 249–252. [CrossRef]
94. Rakotosamizanany, S. Les Corindons Gemmes Dans Les Basaltes Alcalins et Leurs Enclaves à Madagascar: Significations Pétrologique et Métallogénique. Ph.D. Thesis, Université Henri Poincaré, Nancy, France, 2009.
95. Voudouris, P.; Mavrogonatos, C.; Graham, I.; Giuliani, G.; Melfos, V.; Karampelas, S.; Karantoni, V.; Wang, K.; Tarantola, A.; Khin, Z.; et al. Gem Corundum deposits of Greece: Geology, mineralogy and genesis. *Minerals* **2019**, *9*, 49. [CrossRef]
96. Fanka, A.; Sutthirat, C. Petrochemistry, mineral chemistry, and pressure-temperature model of corundum-bearing amphibolite from Montepuez, Mozambique. *Arab. J. Sci. Eng.* **2018**, *43*, 3751–3767. [CrossRef]
97. Morishita, T.; Arai, S.; Gervilla, F. High pressure aluminous mafic rocks from the Ronda peridotite massif, southern Spain: Significance of sapphirine- and corundum-bearing assemblages. *Lithos* **2001**, *57*, 143–161. [CrossRef]
98. Pham, V.L.; Hoàng, Q.V.; Garnier, V.; Giuliani, G.; Ohnenstetter, D. Marble-hosted ruby from Vietnam. *Can. Gemmol.* **2004**, *25*, 83–95.
99. Ishimaru, S.; Arai, S.; Miura, M.; Shmelev, V.; Pushkarev, E. Ruby-bearing feldspathic dyke in peridotite from Ray-Iz ophiolite, the Polar Urals: Implications for mantle metasomatism and origin of ruby. *J. Mineral. Petrol. Sci.* **2015**, *110*, 76–81. [CrossRef]
100. Andriamamonjy, S.A. Les corindons associés aux roches métamorphiques du Sud ouest de Madagascar: Le gisement de saphir de Zazafotsy. Master Thesis, Université d'Antananarivo, Tananarive, Madagascar, 2006.
101. Kongsomart, B.; Vertriest, W.; Weeramonkhonlert, V. Preliminary observations on facet-grade ruby from Longido, Tanazania. *Gems Gemol.* **2017**, *53*, 472–473.
102. Krebs, M.Y.; Pearson, D.G.; Fagan, A.J.; Bussweiler, Y.; Sarkar, C. The application of trace elements and Sr–Pb isotopes to dating and tracing ruby formation: The Aappaluttoq deposit, SW Greenland. *Chem. Geol.* **2019**, *523*, 42–58. [CrossRef]
103. Harlow, G.E.; Bender, W. A study of ruby (corundum) compositions from the Mogok Belt, Myanmar: Searching for chemical fingerprints. *Am. Mineral.* **2013**, *98*, 1120–1132. [CrossRef]
104. Sorokina, E.S.; Litvinenko, A.K.; Hofmeister, W.; Häger, T.; Jacob, D.E.; Nasriddinov, Z.Z. Rubies and sapphires from Snezhnoe, Tajikistan. *Gems Gemol.* **2015**, *51*, 160–175. [CrossRef]
105. Palke, A.C.; Wong, J.; Verdel, C.; Avila, J.N. A common origin for Thai/Cambodian rubies and blue and violet sapphires from Yogo Gulch, Montana, USA? *Am. Mineral.* **2018**, *103*, 469–479. [CrossRef]
106. Graham, I.; Sutherland, L.; Khin, Z.; Nechaev, V.; Khanchuk, A. Advances in our understanding of the gem corundum deposits of the West Pacific continental margins intraplate basaltic fields. *Ore Geol. Rev* **2008**, *34*, 200–215. [CrossRef]
107. Stern, R.J.; Tsujimori, T.; Harlow, G.; Groat, L. Plate tectonic gemstones. *Geology* **2013**, *41*, 723–726. [CrossRef]
108. Garnier, V.; Giuliani, G.; Ohnenstetter, D.; Schwarz, D. Saphirs et rubis. Classification des gisements de corindon. *Le Règne Minéral* **2004**, *55*, 4–47.
109. Emmett, J.L.; Stone-Sundberg, J.; Guan, Y.; Sun, Z. The role of silicon in the color of gem corundum. *Gems Gemol.* **2017**, *53*, 42–47. [CrossRef]
110. Simonet, C.; Fritsch, E.; Lasnier, B. A classification of gem corundum deposits aimed towards gem exploration. *Ore Geol. Rev.* **2008**, *34*, 127–133. [CrossRef]

111. Kissin, A.J. Ruby and sapphire from the southern Ural Mountains, Russia. *Gems Gemol.* **1994**, *30*, 243–252. [CrossRef]
112. Okrusch, M.; Bunch, T.E.; Bank, H. Paragenesis and petrogenesis of a corundum-bearing marble at Hunza (Kashmir). *Mineral. Depos.* **1976**, *11*, 278–297. [CrossRef]
113. Dzikowski, T.J.; Cempirek, J.; Groat, L.A.; Dipple, G.M.; Giuliani, G. Origin of gem corundum in calcite marble: The Revelstoke occurrence in the Canadian Cordillera of British Columbia. *Lithos* **2014**, *198–199*, 281–297. [CrossRef]
114. Balmer, W.A.; Hauzenberger, C.A.; Fritz, H.; Sutthirat, C. Marble-hosted ruby deposits of the Morogoro region, Tanzania. *J. Afr. Earth. Sci.* **2017**, *134*, 626–643. [CrossRef]
115. Litvinenko, A.K. Nuristan-South Pamir province of Precambrian gems. *Geol. Ore Depos.* **2004**, *46*, 263–268.
116. Garde, A.; Marker, M. Corundum crystals with blue-red color zoning near Kangerdluarssuk, Sukkertoppen district, West Greenland. *Rep. Gronl. Geol. Unders.* **1988**, *140*, 46–49.
117. Key, R.M.; Ochieng, J.O. The growth of rubies in south-east Kenya. *J. Gemmol.* **1991**, *22*, 484–496. [CrossRef]
118. Simonet, C. Géologie des Gisements de Saphir et de Rubis. L'exemple de la John Saul Mine, Mangari, Kenya. Ph.D. Thesis, Université de Nantes, Nantes, France, 2000.
119. Mercier, A.; Debat, P.; Saul, J.M. Exotic origin of the ruby deposits of the Mangari area in SE Kenya. *Ore Geol. Rev.* **1999**, *14*, 83–104. [CrossRef]
120. Schwarz, D.; Pardieu, V.; Saul, J.M.; Schmetzer, K.; Laurs, B.M.; Giuliani, G.; Klemm, L.; Malsy, A.K.; Hauzenberger, C.; Du Toit, G.; et al. Ruby and sapphires from Winza (Central Tanzania). *Gems Gemol.* **2008**, *44*, 322–347. [CrossRef]
121. Tenthorey, E.A.; Ryan, J.G.; Snow, E.A. Petrogenesis of sapphirine-bearing metatroctolites from the Buck Creek ultramafic body, southern Appalachians. *J. Metam. Geol.* **1996**, *14*, 103–114.
122. Nicollet, C. Saphirine et staurotide riche en magnésium et chrome dans les amphibolites et anorthosites à corindon du Vohibory Sud, Madagascar. *Bull. Minéral.* **1986**, *109*, 599–612. [CrossRef]
123. Kröner, A. Late Precambrian plate tectonics and orogeny: A need to redefine the term Pan-African. In *African Geology*; Klerkx, J., Michot, J., Eds.; Tervuren, Musée Royal de l'Afrique Centrale: Tervuren, Belgium, 1984; pp. 23–28.
124. Le Goff, E.; Deschamps, Y.; Guerrot, C. Tectonic implications of new single zircon Pb-Pb evaporation data in the Lossogonoi and Longido ruby-districts, Mozambican metamorphic Belt of north-eastern Tanzania. *Comptes Rendus Geosci.* **2010**, *342*, 36–45. [CrossRef]
125. Jöns, N.; Schenk, V. Relics of the Mozambique Ocean in the central East African Orogen: Evidence from the Vohibory Block of Southern Madagascar. *J. Metam. Geol.* **2008**, *26*, 17–28. [CrossRef]
126. Bingen, B.; Jacobs, J.; Viola, G.; Henderson, I.H.C.; Skår, Ø.; Boyd, R.; Thomas, R.J.; Solli, A.; Key, R.M.; Daudi, E.X.F. Geochronology of the Precambrian crust in the Mozambique belt in NE Mozambique, and implications for Gondwana assembly. *Precambrian Res.* **2009**, *170*, 231–255. [CrossRef]
127. Paquette, J.L.; Nédelec, A.; Moine, B.; Rakotondrazafy, A.F.M. U-Pb single zircon Pb-evaporation and Sm-Nd isotopic study of a granulite domain in SE Madagascar. *J. Geol.* **1994**, *102*, 523–538. [CrossRef]
128. Giuliani, G.; Fallick, A.E.; Rakotondrazafy, A.F.M.; Ohnenstetter, D.; Andriamamonjy, A.; Rakotosamizanany, S.; Ralantoarison, T.; Razanatseheno, M.M.; Dunaigre, C.; Schwarz, D. Oxygen isotope systematics of gem corundum deposits in Madagascar: Relevance for their geological origin. *Mineral. Depos.* **2007**, *42*, 251–270. [CrossRef]
129. Sorokina, E.S.; Rošel, D.; Häger, T.; Mertz-Kraus, R.; Saul, J.M. LA-ICP-MS U–Pb dating of rutile inclusions within corundum (ruby and sapphire): New constraints on the formation of corundum deposits along the Mozambique belt. *Mineral. Depos.* **2017**, *52*, 641–649. [CrossRef]
130. Meert, J.G.; Van Der Voo, R.; Ayub, S. Paleomagnetic investigation of the Neoproterozoic Gagwe lavas and Mbozi complex, Tanzania and the assembly of Gondwana. *Precambrian Res.* **1995**, *74*, 225–244. [CrossRef]
131. Garnier, V.; Giuliani, G.; Maluski, H.; Ohnenstetter, D.; Phan, T.T.; Hoàng, Q.V.; Pham, V.L.; Vu, V.T.; Schwarz, D. Ar-Ar ages in phlogopites from marble-hosted ruby deposits in northern Vietnam: Evidence for Cenozoic ruby formation. *Chem. Geol.* **2002**, *188*, 33–49. [CrossRef]
132. Garnier, V.; Ohnenstetter, D.; Giuliani, G.; Maluski, H.; Deloule, E.; Phan, T.T.; Pham, V.L.; Hoang, Q.V. Age and significance of ruby-bearing marbles from the Red River shear zone, northern Vietnam. *Can. Mineral.* **2004**, *43*, 1315–1329. [CrossRef]

133. Garnier, V.; Maluski, H.; Giuliani, G.; Ohnenstetter, D.; Schwarz, D. Ar-Ar and U-Pb ages of marble hosted ruby deposits from central and south East Asia. *Can. J. Earth Sci.* **2006**, *43*, 1–23. [CrossRef]
134. Litvinenko, A.K.; Sorokina, E.S.; Häger, T.; Kostitsyn, Z.A.; Botcharnikov, R.E.; Somsikova, T.; Romashova, T.V.; Hofmeister, W. Petrogenesis of the Snezhnoe ruby deposit, Central Pamir. *Minerals* **2020**, *10*, 478. [CrossRef]
135. Khin, Z.; Sutherland, F.L.; Graham, I.; McGee, B. Dating zircon inclusions in gem corundums from placer deposits, as a guide to their origin. In Proceedings of the 33rd International Geological Congress (IGC), Oslo, Norway, 6–14 August 2008.
136. Sutherland, F.L.; Fanning, C.M. Gem-bearing basaltic volcanism, Barrington, New South Wales: Cenozoic evolution, based on basalt K-Ar ages and zircon fission track and U/Pb isotope dating. *Austral. J. Earth Sci.* **2001**, *48*, 221–237. [CrossRef]
137. Graham, I.T.; Sutherland, F.L.; Webb, G.B.; Fanning, C.M. Polygenetic corundums from New South Wales gemfields. In *Metallogeny of the Pacific Northwest: Tectonics, Magmatism and Metallogeny of Active Continental Margins*; Khanchuk, A.I., Gonevchuk, A.N., Mitrokhin, I.F., Simanenko, N.J., Cook, Seltmann, R., Eds.; Falnulka: Vladivostok, Russia, 2004; pp. 336–339.
138. Dill, H.G. Gems and placers—A genetic relationship par excellence. *Minerals* **2018**, *8*, 470–513.
139. Rhamdohr, R.; Milisenda, C.C. Neue Vorkommen von Saphir-Seifenlagerstätten auf Nosy-Bé, Madagaskar. *Z Dt. Gemmol. Ges.* **2004**, *53*, 143–158.
140. Giuliani, G.; Groat, L. Geology of corundum and emerald gem deposits: A review. *Gems Gemol.* **2019**, *55*, 464–489. [CrossRef]
141. Ozerov, K. Form of corundum crystals as dependent upon chemical composition of medium. *Dolk. Akad. Nauk SSSR* **1945**, *XLVII*, 49–52.
142. Schwarz, D. Aus Basalten, Marmoren und Pegmatiten. Spezielle Ursachen formten in der Erdkruste edle Rubine und Saphire. In *Rubin, Saphir, Korund: Schön, Hart, Selten, Kostbar*; Weise, C., Ed.; ExtraLapis: München, Germany, 1998; Volume 15, pp. 5–9.
143. Kievlenko, E.Y. *Geology of Gems*; Ocean Pictures Ltd.: Littleton, CO, USA, 2003.
144. Giuliani, G.; Ohnenstetter, D.; Fallick, A.E.; Groat, L.; Feneyrol, J. Geographic origin of gems linked to their geographical history. *InColor* **2012**, *19*, 16–27.
145. Sutthirat, C.; Saminpanya, S.; Droop, G.T.R.; Henderson, C.M.B.; Manning, D.A.C. Clinopyroxene-corundum assemblages from alkali-basalt and alluvium, eastern Thailand: Constraints on the origin of Thai rubies. *Mineral. Mag.* **2001**, *65*, 277–295. [CrossRef]
146. Saminpanya, S.; Sutherland, F.L. Different origins of Thai area sapphire and ruby, derived from mineral inclusions and co-existing minerals. *Eur. J. Mineral.* **2001**, *23*, 683–694. [CrossRef]
147. Baldwin, L.C.; Tomaschek, F.; Ballhaus, C.; Gerdes, A.; Fonseca, R.O.C.; Wirth, R.; Geisler, T.; Nagel, T. Petrogenesis of alkaline basalt-hosted sapphire megacrysts. Petrological and geochemical investigations of in situ sapphire occurrences from the Siebengebirge volcanic field, Germany. *Contrib. Mineral. Petrol.* **2017**, *172*. [CrossRef]
148. Promwongan, S.; Sutthirat, C. Mineral inclusions in ruby and sapphire from the Bo Welu gem deposit in Chanthaburi, Thailand. *Gems Gemol.* **2019**, *55*, 354–369.
149. Promwongan, S.; Sutthirat, C. An update on mineral inclusions and their composition in ruby from the Bo Rai gem field in Trat Province, eastern Thailand. *J. Gemmol.* **2019**, *36*, 634–645. [CrossRef]
150. Palke, A.C.; Hapeman, J.R. Rubies from Rock Creek, Montana. *Gems Gemol.* **2019**, *55*, 286–288.
151. Yui, T.F.; Khin, Z.; Limkatrun, P. Oxygen isotope composition of the Denchai sapphire, Thailand; a clue to its enigmatic origin. *Lithos* **2003**, *67*, 153–161. [CrossRef]
152. Yui, T.F.; Wu, C.-M.; Limkatrun, P.; Sricharn, W.; Boonsoong, A. Oxygen isotope studies on placer sapphire and ruby in the Chanthaburi-Trat alkali basaltic gemfield, Thailand. *Lithos* **2006**, *86*, 197–211. [CrossRef]
153. Vysotskiy, S.V.; Nechaev, V.P.; Kissin, A.Y.; Yakovenko, V.V.; Ignat'ev, A.V.; Velivetskaya, T.A.; Sutherland, F.L.; Agoshkov, A.I. Oxygen isotopic composition as an indicator of ruby and sapphire origin: A review of Russian occurrences. *Ore Geol. Rev.* **2015**, *68*, 164–170. [CrossRef]
154. Sutthirat, C.; Hauzenberger, C.; Chualaowanich, T.; Assawincharoenkij, T. Mantle and deep crustal xenoliths in basalts from the Bo Rai ruby deposit, eastern Thailand: Original source of basaltic ruby. *J. Asian Earth Sci.* **2018**, *164*, 366–379. [CrossRef]

155. Khin, Z.; Sutherland, F.L.; Dellapasqua, F.; Ryan, C.G.; Yui, T.Z.; Mernagh, T.P.; Duncan, D. Contrasts in gem corundum characteristics, eastern Australian basaltic fields: Trace elements, fluid/melt inclusions and oxygen isotopes. *Mineral. Mag.* **2006**, *70*, 669–687.
156. Vertriest, W.; Palke, A. Identification of opaque sulphide inclusions in rubies from Mogok, Myanmar and Montepuez, Mozambique. *Minerals* **2020**, *10*, 492. [CrossRef]
157. Limkatrun, P.; Khin, Z.; Ryan, C.G.; Mernagh, T.P. Formation of the Denchai gem sapphires, northern Thailand: Evidence from mineral chemistry and fluid/melt inclusion characteristics. *Mineral. Mag.* **2001**, *65*, 725–735.
158. Palke, A.C.; Renfro, N.D.; Berg, R.B. Melt inclusions in alluvial sapphires from Montana, USA: Formation of sapphires as restitic component of lower crustal melting? *Lithos* **2017**, *278–281*, 43–53. [CrossRef]
159. Herzberg, C.T. Pyroxene geothermometry and geobarometry: Experimental and thermodynamic evaluation of some subsolidus phase relations involving pyroxenes in the system CaO-MgO-Al_2O_3-SiO_2. *Geochim. Cosmochim. Acta* **1978**, *42*, 945–957. [CrossRef]
160. Irvin, A.J. Geochemical and high pressure experimental studies of garnet pyroxenite and pyroxene granulite xenoliths from the Delegate basaltic pipes, Australia. *J. Petrol.* **1974**, *15*, 1–40. [CrossRef]
161. Rakotosamizanany, S.; Giuliani, G.; Ohnenstetter, D.; Rakotondrazafy, A.F.M.; Fallick, A.E. Les gisements de saphirs et de rubis associés aux basaltes alcalins de Madagascar: Caractéristiques géologiques et minéralogiques. 1ère partie: Caractéristiques géologiques des gisements. *Rev. Assoc. Fr. Gemmol. AFG* **2009**, *169*, 13–21.
162. Rakotosamizanany, S.; Giuliani, G.; Ohnenstetter, D.; Rakotondrazafy, A.F.M.; Fallick, A.E. Les gisements de saphirs et de rubis associés aux basaltes alcalins de Madagascar: Caractéristiques géologiques et minéralogiques. 2ème partie: Caractéristiques minéralogiques. *Rev. Assoc. Fr. Gemmol. AFG* **2009**, *170*, 9–18.
163. Rakotosamizanany, S. Les corindons gemmes associés au magmatisme basaltique de Madagascar. In *Rapport Final du Projet de Gouvernance des Ressources Minérales (PGRM) de Madagascar*; PGRM: Antananarivo, Madagascar, 2008.
164. Rakotondrazafy, A.F.M.; Giuliani, G.; Fallick, A.E.; Ohnenstetter, D.; Andriamamonjy, A.; Rakotosamizanany, S.; Ralantoarison, T.; Razanatseheno, M.; Offant, Y.; Garnier, V.; et al. Gem corundum deposits in Madagascar: A review. *Ore Geol. Rev.* **2008**, *34*, 134–154. [CrossRef]
165. Promprated, P.; Taylor, L.A.; Neal, C. R Petrochemistry of mafic granulite xenoliths from the Chantaburi basaltic field: Implications for the nature of the lower crust beneath Thailand. *Int. Geol. Rev.* **2003**, *45*, 383–406. [CrossRef]
166. Kornprobst, J.; Piboule, M.; Roden, M.; Tabit, A. Corundum-bearing garnet clinopyroxenites at Beni Bousera (Morocco): Original plagioclase-rich gabbros recrystallized at depth within the mantle? *J. Petrol.* **1990**, *31*, 717–745. [CrossRef]
167. Chetouani, K.; Bodinier, J.-L.; Garrido, C.J.; Marchesi, C.; Amri, I.; Targuisti, K. Spatial variability of pyroxenite layers in the Beni Bousera orogenic peridotite (Morocco) and implications for their origin. *Comptes Rendus Geoscience* **2016**, *348*, 619–629. [CrossRef]
168. Mattey, D.; Lowry, D.; Macpherson, C.G. Oxygen isotope composition of mantle peridotite. *Earth Planet. Sci. Lett.* **1994**, *128*, 231–241. [CrossRef]
169. Watt, G.; Harris, J.W.; Harte, B.; Boyd, S.R. A high-chromium corundum (ruby) inclusion in diamond from the São Luiz alluvial mine, Brazil. *Mineral. Mag.* **1994**, *58*, 490–493. [CrossRef]
170. Hutchinson, M.T.; Hursthouse, M.B.; Light, M.E. Mineral inclusions in diamonds: Associations and chemical distinctions around the 670 km discontinuity. *Contrib. Mineral. Petrol.* **2001**, *142*, 119–126. [CrossRef]
171. Hutchinson, M.T. Constitution of the Deep Transition Zone and Lower Mantle shown by Diamonds and their Inclusions. Ph.D. Thesis, University of Edinburgh, Edinburgh, UK.
172. Pivin, M.; Femenias, O.; Demaiffe, D. Metasomatic mantle origin for Mbuji-Mayi and Kundelungu garnet and clinopyroxene megacrysts (Democratic Republic of Congo). *Lithos* **2009**, *112S*, 951–960. [CrossRef]
173. Pivin, M. La Suite Complexe des Mégacristaux des Kimberlites de Mbuji-Mayi en République Démocratique du Congo: Témoins du Métasomatisme dans le Manteau Lithosphérique Sous-Continental Archéen du Craton du Congo-Kasaï. Ph.D. Thesis, Université Libre de Bruxelles, Bruxelles, Belgique, 2012.
174. Ratrimo, V. Les indices à corindons gemmes du socle précambrien de Madagascar. In *Rapport Final du Projet de Gouvernance des Ressources Minérales (PGRM) de Madagascar*; PGRM: Antananarivo, Madagascar, 2008.

175. Zhang, R.Y.; Liou, J.G.; Zheng, J.P. Ultrahigh pressure corundum rich-garnetite in garnet peridotite, Sulu terrane, China. *Contrib. Mineral. Petrol.* **2004**, *147*, 21–31. [CrossRef]
176. Padovani, E.R.; Tracey, R. A pyrope-spinel (alkremite) xenolith from Moses Rock Dyke: First known North American occurrence. *Am. Mineral.* **1981**, *66*, 741–745.
177. Shirey, S.B.; Cartigny, P.; Frost, D.J.; Keshav, S.; Nestola, F.; Nimis, P.; Pearson, D.G.; Sobolev, N.V.; Walter, M.J. Diamonds and the geology of mantle carbon. *Rev. Mineral. Geochem.* **2013**, *75*, 355–421. [CrossRef]
178. Eggler, D.H.; McCallum, M.E.; Smith, C.B. Megacryst assemblage in kimberlite from Northern Colorado and Southern Wyoming: Petrology, geothermometry-barometry and areal distribution. In *The Mantle Sample: Inclusion in Kimberlites and other Volcanics*; Boyd, F.R., Meyer, H.O.A., Eds.; Books Series Special Publication; American Geophysical Union: Washington, DC, USA, 1979; pp. 213–226. ISBN 9780875902135. Online ISBN 9781118664858. [CrossRef]
179. Stachel, T.; Harris, J.W. The origin of cratonic diamonds—Constraints from mineral inclusions. *Ore Geol. Rev.* **2008**, *34*, 5–32. [CrossRef]
180. Giuliani, G.; Fallick, A.E.; Ohnenstetter, D.; Pégère, G. Oxygen isotopes of sapphire from the French Massif Central: Implications for the origin of gem corundum in basaltic fields. *Mineral. Depos.* **2009**, *44*, 221–231. [CrossRef]
181. Dirlam, D.M.; Misiorowski, E.B.; Tozer, R.; Stark, K.B.; Bassett, A.M. Gem wealth of Tanzania. *Gems Gemol.* **1992**, *28*, 80–102. [CrossRef]
182. Le Goff, E.; Deschamps, Y.; Cocherie, A.; Guerrot, C.; Ketto, D. Structural, petrological and geochronological constraints of the Tanzanian ruby belt. In Proceedings of the 22ème RST Meeting Nancy. University of Nancy, Nancy, France, 21–24 April 2008.
183. Saul, J. Chronique de la découverte de trois mines historiques de gemmes en Afrique de l'Est. *Le Règne Minéral* **2013**, *2*, 89–92.
184. Haapala, I.; Siivola, J.; Ojanperä, P.; Yletyunen, V. Red corundum, sapphirine and kornerupine from Kittilä, Finnish Lapland. *Bull. Geol. Soc. Finland* **1971**, *43*, 221–231. [CrossRef]
185. Janardhanan, A.; Leake, B.E. Sapphirine in the Sittampundi complex, India. *Mineral. Mag.* **1974**, *39*, 901–902. [CrossRef]
186. Forestier, F.; Lasnier, B. Découverte de niveaux d'amphibolites à pargasite, anorthite, corindon et saphirine dans les schistes cristallins du Haut-Allier. *Contrib. Mineral. Petrol.* **1969**, *23*, 194–235. [CrossRef]
187. Wang, K.; Graham, I.; Martin, L.; Voudouris, P.; Giuliani, G.; Lay, A.; Harris, J.H.; Fallick, A.E. Fingerprinting Paranesti rubies through oxygen isotopes. *Minerals* **2019**, *9*, 91. [CrossRef]
188. Barot, N.; Harding, R.R. Pink corundum from Kitui, Kenya. *J. Gemmol.* **1994**, *24*, 165–172. [CrossRef]
189. Andreoli, M.A.G. Petrochemistry, tectonic evolution and metasomatic mineralizations of Mozambique belt granulites from Malawi and Tete (Mozambique). *Precambrian Res.* **1994**, *25*, 161–186.
190. Henn, U.; Bank, H.; Bank, F.H. Red and orange corundum (ruby and padparadscha) from Malawi. *J. Gemmol.* **1990**, *22*, 83–89. [CrossRef]
191. McColl, D.; Warren, R.G. The first discovery of ruby in Australia. *Aust. Mineral.* **1979**, *26*, 121–125.
192. Mortimer, G.E.; Cooper, J.A.; James, P.R. U-Pb and Rb-Sr geochronology and geological evolution of the Harts Range ruby mine area of the Arunta Inlier, central Australia. *Lithos* **1987**, *20*, 445–467. [CrossRef]
193. Aboosally, S. Update on production in Pakistan, Afghanistan. *Jew. News Asia* **1999**, *1*, 60–64.
194. Morishita, T.; Kodera, T. Finding of corundum-bearing gabbro boulder possibly derived from the Horoman peridotite complex, Hokkaido, northern Japan. *J. Mineral. Petrol.* **1998**, *93*, 52–63. [CrossRef]
195. Pratt, J.H. The occurrence and distribution of corundum in the United States. *Geol. Surv. Bull.* **1901**, *180*, 98.
196. Senoble, J.B. An expedition to Tanzania's new ruby deposit in Winza. *InColor* **2008**, 44–45. Available online: https:www.gemstone.org/incolor/Incolor07/index.html#44 (accessed on 27 June 2020).
197. Peretti, A.; Peretti, F.; Kanpraphai, A.; Bieri, W.; Hametner, C.; Günther, D. Winza rubies identified. *Contrib. Gemol.* **2008**, *7*, 1–97.
198. Pardieu, V.; Chauviré, B. Les rubis du Mozambique. *Le Règne Minéral* **2013**, *2*, 101–108.
199. Nicollet, C. Crustal evolution of the granulites of Madagascar. In *Granulites and Crustal Evolution*; Vielzeuf, D., Vidal, P., Eds.; Springer: Berlin, Germany, 1990; pp. 291–310.
200. Pardieu, V.; Sangsawong, S.; Chauviré, B.; Massi, L.; Sturman, N. Rubies from the Montepuez area (Mozambique), Bangkok, GIA Laboratory. 2013. Available online: https://www.gia.edu/doc/GIA_Ruby_Montepuez_Mozambique.pdf (accessed on 25 June 2020).

201. Vertriest, W.; Pardieu, V. Update on gemstone mining in northern Mozambique. *Gems Gemol.* **2016**, *52*, 404–409. [CrossRef]
202. *Norconsult Consortium Mineral Resources Managemant Capacity Building Project, Republic of Mozambique; Component 2: Geological Infrastructure Development Project, Geological Mapping Lot 1; Geophysical Interpretation*; Report no. B4; National Directorate of Geology: Maputo, Republic of Mozambique, 2007.
203. *Norconsult Consortium Mineral Resources Managemant Capacity Building Project, Republic of Mozambique; Component 2: Geological Infrastructure Development Project, Geological Mapping Lot 1; Sheet Explanation: 32 sheets; Scale: 1: 250000*; Report no. B6.f; National Directorate of Geology: Maputo, Republic of Mozambique, 2007.
204. Boyd, R.; Nordgulen, O.; Thomas, R.J.; Bingen, B.; Bjerkgard, T.; Grenne, T.; Henderson, I.; Melezhik, V.A.; Often, M.; Sandstad, J.S.; et al. The geology and geochemistry of the East African Orogen in Northeastern Mozambique. *S. Afr. J. Geol.* **2010**, *113*, 87–129. [CrossRef]
205. Viola, G.; Henderson, I.H.C.; Bingen, B.; Thomas, R.J.; Somethurst, M.A.; de Azavedo, S. Growth and collapse of a deeply eroded orogen: Insights from structural, geological and geochronological constraints on the Pan-African evolution of NE Mozambique. *Tectonics* **2008**, *27*, 1–31. [CrossRef]
206. Stern, R.J. Arc assembly and continental collision in the Neoproterozoic East African Orogen: Implications for the consolidation of Gondwanaland. *Annual Rev. Earth Planet. Sci.* **1994**, *22*, 319–351. [CrossRef]
207. Smith, C.P.; Gübelin, E.J.; Bassett, A.M.; Manandhar, M.N. Rubies and fancy-color sapphires from Nepal. *Gems Gemol.* **1997**, *33*, 24–41. [CrossRef]
208. Bowersox, G.W.; Foord, E.E.; Laurs, B.M.; Shigley, J.E.; Smith, C.P. Ruby and sapphire from Jegdalek, Afghanistan. *Gems Gemol.* **2000**, *36*, 110–126. [CrossRef]
209. Pêcher, A.; Giuliani, G.; Garnier, V.; Maluski, H.; Kausar, A.B.; Malik, R.M.; Muntaz, H.R. Geology and Geochemistry of the Nangimali ruby deposit area, Nanga-Parbat Himalaya (Azad Kashmir, Pakistan). *J. Asian Earth Sci.* **2002**, *21*, 265–282. [CrossRef]
210. Pham, V.L.; Pardieu, V.; Giuliani, G. Update on gemstone mining in Luc Yen, Vietnam. *Gems Gemol.* **2014**, *49*, 233–245.
211. Dunn, P.; Frondel, C. An uncommon margarite/corundum assemblage from Sterling Hill, New Jersey. *Mineral. Rec.* **1990**, *21*, 425–427.
212. Dzikowski, T.J.; Groat, L.A.; Dipple, G.M.; Marshall, D. Origin of the Revelstoke carbonate-hosted gem corundum occurrence, British Columbia, Canada. In Proceedings of the International Mineralogical Association Conference, Budapest, Hungaria, 21–27 August 2010.
213. Kyaw, T. The Igneous Rocks of the Mogok Stone Tract: Their Distributions, Petrography, Petrochemistry, Sequence, Geochronology and Economic Geology. Ph.D. Thesis, University of Yangon, Yangon, Myanmar, 2007.
214. Garnier, V.; Ohnenstetter, D.; Giuliani, G. L'aspidolite fluorée: Rôle des évaporites dans la genèse du rubis des marbres de Nangimali (Azad-Kashmir, Pakistan). *Comptes Rendus Geoscience* **2004**, *336*, 1245–1253. [CrossRef]
215. Gordon, S.G. Desilicated pegmatites. *Proc. Acad. of Nat. Sci. Phila.* **1921**, *73*, 169–192. Available online: https://www.jstor.org/stable/4063801 (accessed on 17 May 2020).
216. Lawson, A.C. Plumasite, an oligoclase corundum rock, near Spanish Peak, California University of California. *Pub. Geol. Sci.* **1903**, *3*, 219–229.
217. Seifert, A.V.; Hyrsl, J. Sapphire and garnet from Kalalani, Tanga Province, Tanzania. *Gems Gemol.* **1999**, *35*, 108–120. [CrossRef]
218. Meng, F.; Shmelev, V.R.; Kulikova, K.V.; Ren, Y. A red-corundum-bearing vein in the Rai-Iz ultramafic rocks, Polar Urals, Russia: The product of fluid activity in a subduction zone. *Lithos* **2018**, *320–321*, 302–314. [CrossRef]
219. Simonet, C. Geology of the Yellow Mine (Taita-Taveta District, Kenya) and other yellow tourmaline deposits in East Africa. *J. Gemmol.* **2000**, *27*, 11–29. [CrossRef]
220. Terekhov, E.N. REE distribution in corundum-bearing and other metasomatic rocks during the exhumation of metamorphic rocks of the belmorian belt of the Baltic Shield. *Geochem. Int.* **2007**, *45*, 364–380. [CrossRef]
221. Bindeman, I.N.; Schmitt, A.K.; Evans, D.A.D. Origin of the lowest-known $\delta^{18}O$ silicate rock on Earth in Paleoproterozoic Karelian rift. *Geology* **2010**, *38*, 631–634. [CrossRef]
222. Vysotskiy, S.V.; Ignat'ev, A.V.; Levitskii, V.I.; Nechaev, V.P.; Velivetskaya, T.A.; Yakovenko, V.V. Geochemistry of stable oxygen and hydrogen isotopes in minerals and corundum-bearing rocks in northern Karelia as an indicator of their unsual genesis. *Geochem. Int.* **2014**, *52*, 773–782. [CrossRef]

223. Spiridonov, E.M. Gemstone deposits of the former Soviet Union. *J. Gemmol.* **1998**, *26*, 111–124. [CrossRef]
224. Walker, R.J.; Prichard, H.M.; Ishiwatari, A.; Pimentel, M. The osmium isotopic composition of convecting upper mantle deduced from ophiolite chromites. *Geochim. Cosmochim. Acta* **2002**, *66*, 329–345. [CrossRef]
225. Vakhrusheva, N.V.; Ivanov, K.S. The nature and age of plagioclasites in the ultrabasic Rai-Iz massif (Polar Urals). *Dokl. Earth Sci.* **2018**, *480*, 587–590. [CrossRef]
226. Giuliani, G.; Groat, L.; Marshall, D.; Branquet, Y. Emerald deposits: A review and enhanced classification. *Minerals* **2019**, *9*, 105. [CrossRef]
227. Glodny, J.; Austrheim, H.; Molina, J.F.; Rusin, A.; Seard, D. Rb/Sr record of fluid-rock interaction in eclogites: The Marun-Keu complex, Polar Urals, Russia. *Geochim. Cosmochim. Acta* **2003**, *67*, 4353–4373. [CrossRef]
228. Bryanchaninova, N.I.; Makeev, A.B.; Zubkova, N.V.; Filipov, N.V. Sodium-strontium mica $Na_{0.50}Sr_{0.25}Al_2(Na_{0.25}\square_{0.75})Al_{1.25}Si_{2.75}O_{10}(OH)_2$ from Rubinovyi Log. *Dokl. Earth Sci.* **2004**, *395*, 260–265.
229. Korzhinskii, D.S. *Theory of Metasomatic Zoning*; Clarendon Press: Oxford, UK, 1970.
230. Lacroix, A. *Les Gisements de Phlogopites de Madagascar et les Pyroxénites qui les Renferment: Ann. Géol. Serv. Min.*; Imprimerie officielle: Tananarive, Madagascar, 1941.
231. Lacroix, A. *Minéralogie de Madagascar, tome 1*; Challamel A.: Paris, France, 1922.
232. Raith, M.M.; Rakotondrazafy, R.; Sengupta, P. Petrology of corundum-spinel-sapphirine-anorthite rocks (sakenites) from the type locality in Southern Madagascar. *J. Metam. Geol.* **2008**, *26*, 647–667. [CrossRef]
233. Devouard, D.; Raith, M.M.; Rakotondrazafy, R.; El-Ghozzi, M.; Nicollet, C. Occurrence of musgravite in anorthite-corundum-spinel-sapphirine rocks (sakenites) from south Madagascar: Evidence for a high-grade metasomatic event. In Proceedings of the 18th General Meeting of the International Mineralogical Association, Edinburgh, Scotland, 1–6 September 2002.
234. Keulen, N.; Kokfelt, T.F. The Homogenization Team. A Seamless, Digital, Internet-Based Geological Map of South-West and Southern West Geenland, 1:100.000, 61. Geological Survey of Danemark and Greenland. 2011, pp. 300–364. Available online: http://geuskort.geus.dk/gisfarm/svgrl.jsp (accessed on 5 September 2018).
235. Keulen, N.; Schumacher, J.C.; Næraa, T.; Kokfelt, T.F.; Scherstén, A.; Szilas, K.; van Hinsberg, V.J.; Schlatter, D.M.; Windley, B.F. Meso- and Neoarchaean geological history of the Bjørnesund and Ravns Storø Supracrustal Belts, southern West Greenland: Settings for gold enrichment and corundum formation. *Precambrian Res.* **2014**, *254*, 36–58. [CrossRef]
236. Polat, A.; Fryer, B.J.; Appel, P.W.U.; Kalvig, P.; Kerrich, R.; Dilek, Y.; Yang, Z. Geochemistry of anorthositic differentiated sills in the Archean (~2970 Ma) Fiskenæsset Complex, SW Greenland: Implications for parental magma compositions, geodynamic setting, and secular heat flow in arcs. *Lithos* **2011**, *123*, 50–72. [CrossRef]
237. Yakymchuk, C.; Szilas, K. Corundum formation by metasomatic reactions in Archean metapelite, southern West Greenland—Identification of exploration vectors for ruby deposits within high-grade greenstone belts. *Geosci. Front.* **2018**, *9*, 727–749. [CrossRef]
238. Appel, C.C.; Appel, P.W.U.; Rollinson, H.R. Complex chromite textures reveal the history of an early Archaean layered ultramafic body in West Greenland. *Mineral. Mag.* **2002**, *66*, 1029–1041. [CrossRef]
239. Poulsen, M.D.; Paulick, H.; Rosa, D.; van Hinsberg, V.J.; Petersen, J.; Thomsen, L.L. Follow-up on Ujarassiorit mineral hunt finds and outreach activities, South-East Greenland. In *Review of Survey Activities 2014*; Bennike, O., Garde, A.A., Watt, W.S., Eds.; Geological Survey of Denmark and Greenland Bulletin: Greenland, Danemark, 2015; Volume 33, pp. 53–56. Available online: www.geus.dk/publications/bull (accessed on 25 June 2020).
240. Ralantoarison, L.T. Les Corindons Associés aux Roches Métamorphiques du Sud de Madagascar: Le Gisement de Saphirs de Sahambano (Sud-Est d'Ihosy). Master's Thesis/Mémoire de DEA, Université d'Antananarivo, Antananarivo, Madagascar, 2006.
241. Ralantoarison, T.; Offant, Y.; Andriamamonjy, A.; Giuliani, G.; Rakotondrazafy, A.F.M.; Ohnenstetter, D.; Schwarz, D.; Fallick, A.E.; Razanatseheno, M.; Rakotosamizanany, S.; et al. Les saphirs multicolores de Sahambano et Zazafotsy, région granulitique d'Ihosy, Madagascar. *Rev. Assoc. Fr. Gemmol. AFG* **2006**, *158*, 4–13.
242. Andriamamonjy, S.A. Importance des fluides sur la métallogénie des gisements de saphirs et de rubis dans le domaine granulitique de haute température du Sud de Madagascar: Cas de Zazafotsy et d'Ambatomena. Ph.D. Thesis, Université d'Antananarivo, Antananarivo, Madagascar, 2010.

243. Andriamamonjy, A.; Giuliani, G.; Ohnenstetter, D.; Rakotondrazafy, A.F.M.; Ralantoarison, T.; Ranatsenho, M.M.; Rakotosamizanany, S.; Fallick, A.E. Les corindons d'âge néoprotérozoïque de Madagascar. *Le Règne Minéral* **2013**, *2*, 109–117.
244. Feneyrol, J.; Giuliani, G.; Ohnenstetter, D.; Fallick, A.E.; Martelat, J.-M.; Monié, P.; Dubessy, C.; Rollion-Bard, C.; Le Goff, E.; Malisa, E.; et al. Worldwide tsavorite deposits: New aspects and perspectives. *Ore Geol. Rev.* **2013**, *53*, 1–25. [CrossRef]
245. Malik, R.H. *Geology and Ressource Potential of Kashmir Ruby Deposits*; Unpublished Report; Azad-Kashmir Mineral and Industrial Development Corporation: Muzaffarabad (Azad-Kashmir), Pakistan, 1994.
246. Ray, H.N.; MacRae, G.P.; Cain, L.J.; Malloch, K.R. *New South Wales Industrial Minerals Data Base*, 2nd ed.; Geological Survey of New South Wales: Sydney, Australia, 2013.
247. Oakes, G.; Barron, L.M.; Lishmund, S.R. Alkali basalts and associated volcanoclastic rocks as a source of sapphire in Eastern Australia. *Aust. J. Earth Sci.* **1996**, *43*, 289–298. [CrossRef]
248. Sutherland, F.L.; Coenraads, R.R. An unusual ruby-sapphire-sapphirine-spinel assemblage from the Tertiary Barrington volcanic province, New South Wales. *Austr. Mineral. Mag.* **1996**, *60*, 623–638. [CrossRef]
249. Webb, G.; Sutherland, F.L. Gemstones of New England. *Austr. J. Mineral.* **1998**, *4*, 115–121.
250. Abduryiyim, A.; Sutherland, F.L.; Coldham, T. Past, present and future of Australian gem corundum. *Austr. Gemmol.* **2012**, *24*, 234–242.
251. Vichit, P.; Vudhichativanich, S.; Hansawek, R. The distribution and some characteristics of corundum-bearing basalts in Thaïland. *J. Geol. Soc. Thailand* **1978**, *M4*, 1–27.
252. Liu, Y.; Lu, R. Ruby and sapphire from Muling, China. *Gems Gemol.* **2016**, *52*, 98–100.
253. Sutherland, F.L.; Graham, I. *Geology of the Barrington Tops Plateau; It s Rocks, Minerals and Gemstones, New South Wales, Australia*; The Australian Museum Society: Sydney, Australia, 2003; p. 56.
254. Birch, W.D. Gem corundum from the St Arnaud district, western Victoria, Australia. *Austr. J. Mineral.* **2008**, *14*, 73–78.
255. Schwarz, D.; Schmetzer, K. Rubies from the Vatomandry area, eastern Madagascar. *J. Gemmol.* **2001**, *27*, 409–416. [CrossRef]
256. Rasimanana, G. Caractérisations Géochimiques et Géochronologiques de Trois Épisodes Magmatiques (Crétacé, Miocène Terminal et Quaternaire) à Madagascar Liés aux Phénomènes de Rifting. Ph.D. Thesis, Université d'Orsay, Paris, France, 1996.
257. Chapin, M.; Pardieu, V.; Lucas, A. Mozambique: A ruby discovery for the 21st century. *Gems Gemol.* **2015**, *51*, 44–54. [CrossRef]
258. Hsu, T.; Lucas, A.; Pardieu, V. Mozambique: A Ruby Discovery for the 21st Century. GIA Research News. 2014. Available online: https://www.gia.edu/gia-news-research-mozambique-expedition-ruby-discovery-new-millennium (accessed on 25 June 2020).
259. Kane, R.E.; Kammerling, R.C. Status of ruby and sapphire mining in the Mogok Stone Tract. *Gems Gemol.* **1992**, *28*, 152–174. [CrossRef]
260. Kammerling, R.C.; Scarratt, K.; Bosshart, G.; Jobbins, E.A.; Kane, R.E.; Gübelin, E.J.; Levinson, A.A. Myanmar and its gems-an update. *J. Gemmol.* **1994**, *24*, 3–40. [CrossRef]
261. Garnier, V.; Giuliani, G.; Ohnenstetter, D.; Schwarz, D.; Kausar, A.B. Les gisements de rubis associés aux marbres de l'Asie centrale et du Sud-est. *Le Règne Minéral* **2006**, *67*, 17–48.
262. Kane, R.E.; McClure, S.F.; Kammerling, R.C.; Khoa, N.D.; Mora, C.; Repetto, S.; Khai, N.D.; Koivula, J.I. Rubies and fancy sapphires from Vietnam. *Gems Gemol.* **1991**, *27*, 136–155. [CrossRef]
263. Keller, P. *Gemstones of East Africa*; Geoscience Press: Phoenix, AZ, USA, 1992.
264. Henn, U.; Milisenda, C.C. Neue Edelsteinvorkommen in Tansania: Die region Tunduru-Zongea. *Z Deut. Gemmol. Gesell.* **1997**, *46*, 29–43.
265. Pardieu, V.; Vertriest, W.; Weeramonkhonlert, V.; Raynaud, V.; Atikarnsakul, U.; Perkins, R. Sapphires from the Gem Rush Bemainty Area, Ambatondrazaka (Madagascar). GIA Research News. 2017. Available online: https://www.gia.edu/gia-news-research/sapphires-gem-rushbemainty-ambatondrazaka-Madagascar (accessed on 20 June 2020).
266. Pardieu, V.; Sangsawong, S.; Detroyat, S. Gem News International: Rubies from a new deposit in Zahamena National Park, Madagascar. *Gems Gemol.* **2015**, *51*, 454–456.

267. Thirangoon, K. Ruby and Pink Sapphire from Aappaluttoq, Greenland: Status of Ongoing Research, GIA, Bangkok, Thailand. 2009. Available online: https://www.gia.edu/gia-news-research-nr32309 (accessed on 20 June 2020).
268. Palke, A.C. Coexisting rubies and blue sapphires from major world deposits: A brief review of their mineralogical properties. *Minerals* **2020**, *10*, 472. [CrossRef]
269. Palke, A.C.; Breeding, C.M. The origin of needle-like inclusions in natural gem corundum: A combined EPMA, LA-ICP-MS, and nanoSIMS investigation. *Am. Mineral.* **2017**, *102*, 1451–1461. [CrossRef]
270. Franz, G.; Vyshnevskyi, O.; Taran, M.; Khomenko, V.; Wiedenbeck, M.; Schiperski, F.; Nissen, J. A new emerald occurrence from Kruta Balka, Western Peri-Azovian region, Ukraine: Implications for understanding the crystal chemistry of emerald. *Am. Mineral.* **2020**, *105*, 162–181. [CrossRef]

Permissions

All chapters in this book were first published in MDPI; hereby published with permission under the Creative Commons Attribution License or equivalent. Every chapter published in this book has been scrutinized by our experts. Their significance has been extensively debated. The topics covered herein carry significant findings which will fuel the growth of the discipline. They may even be implemented as practical applications or may be referred to as a beginning point for another development.

The contributors of this book come from diverse backgrounds, making this book a truly international effort. This book will bring forth new frontiers with its revolutionizing research information and detailed analysis of the nascent developments around the world.

We would like to thank all the contributing authors for lending their expertise to make the book truly unique. They have played a crucial role in the development of this book. Without their invaluable contributions this book wouldn't have been possible. They have made vital efforts to compile up to date information on the varied aspects of this subject to make this book a valuable addition to the collection of many professionals and students.

This book was conceptualized with the vision of imparting up-to-date information and advanced data in this field. To ensure the same, a matchless editorial board was set up. Every individual on the board went through rigorous rounds of assessment to prove their worth. After which they invested a large part of their time researching and compiling the most relevant data for our readers.

The editorial board has been involved in producing this book since its inception. They have spent rigorous hours researching and exploring the diverse topics which have resulted in the successful publishing of this book. They have passed on their knowledge of decades through this book. To expedite this challenging task, the publisher supported the team at every step. A small team of assistant editors was also appointed to further simplify the editing procedure and attain best results for the readers.

Apart from the editorial board, the designing team has also invested a significant amount of their time in understanding the subject and creating the most relevant covers. They scrutinized every image to scout for the most suitable representation of the subject and create an appropriate cover for the book.

The publishing team has been an ardent support to the editorial, designing and production team. Their endless efforts to recruit the best for this project, has resulted in the accomplishment of this book. They are a veteran in the field of academics and their pool of knowledge is as vast as their experience in printing. Their expertise and guidance has proved useful at every step. Their uncompromising quality standards have made this book an exceptional effort. Their encouragement from time to time has been an inspiration for everyone.

The publisher and the editorial board hope that this book will prove to be a valuable piece of knowledge for researchers, students, practitioners and scholars across the globe.

List of Contributors

Ying Jiang and Guanghai Shi
State Key Laboratory of Geological Processes and Mineral Resources, China University of Geosciences, Beijing 100083, China

Liguo Xu
Geological Museum of China, Beijing 100034, China

Xinling Li
Xinjiang Uygur Autonomous Region Product Quality Supervision and Inspection Institute, Xinjiang 830004, China

Kehan Zhao and Feng Bai
School of Gemmology, China University of Geosciences (Beijing), Beijing 100083, China

Matthew F. Hardman, David G. Pearson, Yan Luo and Chiranjeeb Sarkar
Earth and Atmospheric Sciences Department, University of Alberta, 1-26 Earth Sciences Building, Edmonton, AB T6G 2E3, Canada

Mandy Y. Krebs
Earth and Atmospheric Sciences Department, University of Alberta, 1-26 Earth Sciences Building, Edmonton, AB T6G 2E3, Canada
Gemological Institute of America, 5355 Armada Drive, Carlsbad, CA 92008, USA

Andrew J. Fagan
Hummingbird Geological Services, Vancouver, BC V5Z 1R9, Canada

Andrey K. Litvinenko
Department of Mineralogy and Gemology, Russian State Geological Prospecting University (MGRI), Miklukho-Maklai str. 23, 117485 Moscow, Russia

Tobias Häger, Roman E. Botcharnikov and Wolfgang Hofmeister
Institut für Geowissenschaften, Johannes Gutenberg-Universität Mainz (JGU), J.-J.-Becher-Weg 21, 55128 Mainz, Germany

Yuri A. Kostitsyn, Alina V. Somsikova and Tatiana V. Romashova
Vernadsky Institute of Geochemistry and Analytical Chemistry, Russian Academy of Sciences (GEOKHI RAS), Kosygin str. 19, 119991 Moscow, Russia

Elena S. Sorokina
Institut für Geowissenschaften, Johannes Gutenberg-Universität Mainz (JGU), J.-J.-Becher-Weg 21, 55128 Mainz, Germany
Vernadsky Institute of Geochemistry and Analytical Chemistry, Russian Academy of Sciences (GEOKHI RAS), Kosygin str. 19, 119991 Moscow, Russia

Thomas Ludwig
Institut für Geowissenschaften, Universität Heidelberg, Im Neuenheimer Feld 234-236, 69120 Heidelberg, Germany

Valery Murzin and Daria Kiseleva
A.N. Zavaritsky Institute of Geology and Geochemistry, Ural Branch of Russian Academy of Sciences, Vonsovskogo str., 15, 620016 Ekaterinburg, Russia

Sergei Sustavov
Department of Mineralogy, Petrology and Geochemistry, Ural State Mining University, Kuybysheva str., 30, 620144 Ekaterinburg, Russia

Aleksander Kissin and Irina Gottman
A.N. Zavaritsky Institute of Geology and Geochemistry, Ural Branch of Russian Academy of Sciences, Vonsovskogo str., 15, 620016 Ekaterinburg, Russia
Department of Mineralogy, Petrology and Geochemistry, Ural State Mining University, Kuybysheva str., 30, 620144 Ekaterinburg, Russia

Wim Vertriest and Aaron Palke
Gemological Institute of America, 5355 Armada Drive, Carlsbad, CA 92008, USA

Leander Franz
Department of Environmental Sciences, Mineralogy and Petrology, University of Basel, Bernoullistrasse 30, 4056 Basel, Switzerland

Hao A. O. Wang, Walter A. Balmer and Michael S. Krzemnicki
Swiss Gemmological Institute SSEF, Aeschengraben 26, 4051 Basel, Switzerland

Myint Myat Phyo
Department of Environmental Sciences, Mineralogy and Petrology, University of Basel, Bernoullistrasse 30, 4056 Basel, Switzerland
Swiss Gemmological Institute SSEF, Aeschengraben 26, 4051 Basel, Switzerland

Marcel Guillong
Institute of Geochemistry and Petrology, ETH Zurich, Clausiusstrasse 25, 8092 Zürich, Switzerland

Alfons Berger
Institute of Geological Sciences, University of Bern, Baltzerstrasse 1+3, 3012 Bern, Switzerland

Gaston Giuliani
Géosciences Environnement, Université Paul Sabatier, GET/ IRD et Université de Lorraine, C.R.P.G./C.N.R.S., 15 rue Notre-Dame-des-Pauvres, BP 20, 54501 Vandoeuvre, France

Lee A. Groat
Earth, Ocean and Atmospheric Sciences, University of British Columbia, Vancouver, BC V6T 1Z4, Canada

Anthony E. Fallick
Isotope Geosciences Unit, Scottish Universities Environmental Research Centre (S.U.E.R.C.), Rankine Avenue, East Kilbride, Glasgow G75 0QF, UK

Isabella Pignatelli
Georessources, Université de Lorraine, UMR 7539 CNRS-UL, BP 70239 Vandoeuvre, France

Vincent Pardieu
Field gemmologist & Consultant at VP Consulting, Manama, Bahrain

Index

www.ingramcontent.com/pod-product-compliance
Lightning Source LLC
LaVergne TN
LVHW082203200926
848484LV00008B/2208